DATA MINING FOR GENOMICS AND PROTEOMICS

WILEY SERIES ON METHODS AND APPLICATIONS IN DATA MINING

Series Editor: **Daniel T. Larose**

Discovering Knowledge in Data: An Introduction to Data Mining • Daniel T. Larose

Data-Mining on the Web: Uncovering Patterns in Web Content, Structure, and Usage • Zdravko Markov and Daniel Larose

Data Mining Methods and Models • Daniel Larose

Practical Text Mining with Perl • Roger Bilisoly

Knowledge Discovery with Support Vector Machines • Lutz Hamel

Data Mining for Genomics and Proteomics: Analysis of Gene and Protein Expression Data • Darius M. Dziuda

DATA MINING FOR GENOMICS AND PROTEOMICS

Analysis of Gene and Protein Expression Data

DARIUS M. DZIUDA

WILEY

A JOHN WILEY & SONS, INC., PUBLICATION

Published by John Wiley & Sons, Inc., Hoboken, New Jersey
Published simultaneously in Canada

Library of Congress Cataloging-in-Publication Data:

Dziuda, Darius M.
Data mining for genomics and proteomics : analysis of gene and protein expression data / Darius M. Dziuda.
p. cm.
Includes bibliographical references and index.
ISBN 978-0-470-16373-3 (cloth)
1. Genomics—Data processing. 2. Proteomics—Data processing. 3. Data mining. I. Title.
QH441.2.D98 2010
572.8′6—dc22

2009052129

Printed in the United States of America

10 9 8 7 6 5 4 3 2 1

To Dorota and Mateusz

CONTENTS

PREFACE

By combining data mining, statistics, and computer science with computational biology, bioinformatics, genomics, proteomics and personalized medicine, and then adding other familiar and emerging fields we will portray an inclusive interdisciplinary environment in which a growing number of us enjoy working. This book is written for students, practitioners, and researchers involved in biomedical or data mining projects that use gene expression or protein expression data. This group includes students of bioinformatics, data mining, computational biology, biomedical sciences, and other related areas. In addition, this book is directed to anyone interested in efficient methods of knowledge discovery based on data sets consisting of thousands or millions of variables and often only dozens or hundreds of observations, whether the data is related to biomedical research or to such other fields as market, financial, or physical modeling.

Even experienced data miners may be surprised to learn that some well established methods and procedures fail in such high-dimensional $p \gg N$ situations. New methods are constantly being developed, making it even more difficult for biomedical researchers and bioinformaticians to select appropriate methods to use for a particular project. No single method works optimally in all situations. Some approaches are efficacious for some study goals but inappropriate for others. For example, we should not drive biomarker discovery with unsupervised learning methods. However, methods like clustering are so popular that they are used indiscriminately, not only for studies seeking new taxonomic information, for which they are a perfect choice, but also often as primary methods in situations in which supervised approaches should be used. The confusion among biomedical researchers is visible in the large body of papers describing experiments based on methods that are inappropriate for particular study goals or for the data at hand. Such experiments provide either misleading conclusions or only anecdotal evidence.

One of the primary goals of this book is to provide clear guidance on when to use which methods and why. Such subjects as why some of the popular approaches can lead to inferior solutions, and sometimes even to the worst possible results will be examined. Among the topics discussed are: univariate versus multivariate approaches, supervised versus unsupervised methods, selecting proper methods for feature selection, regularization of biomarker discovery leading to more stable and generalizable classifiers, ensemble-based estimation of the misclassification error rate, identification of the *Informative Set of Genes* containing all information significant for class differentiation. The book covers all aspects of gene and protein expression analysis—technology, data preprocessing and quality assessment, basic exploratory analysis, unsupervised and supervised learning algorithms. Special attention is given to multivariate biomarker discovery leading to parsimonious and generalizable classifiers.

Students of my course on *Data Mining for Genomics and Proteomics*, a part of the Central Connecticut State University (CCSU) data mining program, have diverse backgrounds, from molecular biology PhDs to those with no biology background. This book is written for an equally broad audience. The only assumption is that the reader is interested in the subject. As a result, for different readers different parts of this book may be seen as introductory and different parts as advanced. All efforts have been made to start each subject from its basics.

The book comprises five main chapters with the sixth chapter containing examples of hands-on analysis of real-world data sets. Chapter 1 provides an introduction to basic, mostly biological, terms. Chapter 2 focuses on microarray technology and basic analysis of gene expression data, including low-level preprocessing methods, probe set level preprocessing and filtering, basic exploratory analysis, and such unsupervised learning methods as cluster analysis, principal component analysis, and self-organizing maps.

Chapters 3 and 4 are the core of the book, covering multivariate methods for feature selection, biomarker discovery, and identification of generalizable and interpretable classifiers. Such biomarkers and classifiers can be designed and used for early medical diagnosis, for tailoring therapy selection to prediction of individual response to available treatment modalities, for assessing treatment progression, for drug discovery, and in any other areas that can benefit from parsimonious multivariate markers identified from a very large number of variables.

Chapter 3 starts with an overview of biomarker discovery and classification, followed by the coverage of feature selection methods, and then descriptions of selected supervised learning algorithms. Important differences between univariate and multivariate approaches as well as between supervised and unsupervised methods are discussed. Three learning algorithms are described in detail. *Linear discriminant analysis* represents classical and still powerful parametric methods that should be in the portfolio of any data miner and bioinformatician. *Support vector machines* are newer but already well-established methods for designing both linear and nonlinear classifiers. *Random forests* represent recent ensemble-based approaches. In addition, two other learning algorithms are described—*k-nearest neighbors* and *artificial neural networks*, useful in some situations, even if they are usually not the first choice in biomarker discovery. A discussion of the bootstrap and ensemble-based approaches culminates with defining the *modified bagging* schema that allows for building ensembles of classifiers based on randomized training sets generated by stratified sampling without replacement.

Chapter 4 defines the *Informative Set of Genes* as a set containing all information significant for the differentiation of classes represented in training data, which can be used in two complementary capacities. First, it facilitates biological interpretation of class differences. Second, in combination with ensembles of classifiers generated with the use of the *modified bagging* schema, it allows the identification of robust biomarkers. The method introduced in Chapter 4 allows for the optimization of feature selection that leads to identification of parsimonious and generalizable multivariate biomarkers with plausible biological interpretation.

Chapter 5 addresses the analysis of protein expression data. Data mining methods for feature selection, biomarker discovery and classification are the same in proteomics as those described, in Chapters 3 and 4, for genomics. For that reason, Chapter 5 focuses on proteomic technologies and on the analytical steps that are intrinsic to proteomics. The covered technologies include protein microarrays, two-dimensional gel electrophoresis and MALDI-TOF and SELDI-TOF mass spectrometry. Among the discussed analytical methods are: preprocessing of mass spectrometry data, and associating biomarker peaks with proteins.

The exercises included in Chapters 2–4 give readers an opportunity to put the methods they have learned to practical use. To make it easier and more enjoyable, many of these exercises are covered by studies presented in Chapter 6.

Finally, Chapter 6 provides examples of the analysis of real-world gene expression data sets that illustrate the practical application of the methods described in this book. The examples illustrate such tasks as designing a multi-stage classification schema, combining two gene expression data sets into one training set, identification of the *Informative Set of Genes* and its primary expression patterns, and using ensembles of classifiers built with the *modified bagging* schema to identify parsimonious and generalizable biomarkers.

DARIUS M. DZIUDA

Bethany, Connecticut
June 2009

ACKNOWLEDGMENTS

Thanks to Dr. Daniel Larose, Director of the Data Mining Program at Central Connecticut State University (CCSU) for encouraging me to develop a course on *Data Mining for Genomics and Proteomics*, and then suggesting extension of my lectures into a book. Thanks to Dr. Timothy Craine, Chair of the Department of Mathematical Sciences of CCSU for all of his support. Thanks to all of my colleagues in the department for a great and inspiring working environment.

Thanks to Dr. Daniel S. Miller for reading the manuscript, and for his much appreciated comments and suggestions and the numerous discussions that greatly improved this book. Thanks to David LaPierre for proofreading the manuscript and for helpful suggestions. Thanks to Dr. Yi Zhou for being always ready to share his biomedical expertise and for our productive brain storming sessions. Thanks to Dr. Preethi Gunaratne for providing biological interpretation for one of the biomarker discovery exercises included in this book.

Thanks to Paul Petralia, senior editor at Wiley, Michael Christian, and all folks at Wiley for their professional support.

Finally, special thanks to my wife Dorota for our quarter-century journey together and to my son Mateusz—thank you both, from the bottom of my heart, for your patience and love that make my efforts worthwhile.

D. M. D.

CHAPTER 1

INTRODUCTION

Data mining for genomics and proteomics belongs to an interdisciplinary and relatively new field of bioinformatics, which evolves so rapidly that it is difficult to predict the extent and pace of the changes. Biology, or more generally life sciences, can now be considered information sciences. They are changing from disciplines that deal with relatively small data sets to research fields overwhelmed by a large number of huge data sets. Two main triggers are the source of these changes. The first was *the Human Genome Project*. As the result of research sparked by this project, we now have a large and growing library of organisms with already sequenced genomes. The second was a new technology—genomic microarrays—that allows for the quick and inexpensive measurement of gene expression level for thousands of genes simultaneously.

These and other changes have occurred during the last ten years or so. Before then, biologists and biomedical researchers were dealing with data sets typically consisting of dozens or perhaps hundreds of biological samples (patients, for example) and dozens of variables. The number of samples was typically greater than the number of variables. "Traditional" statistical methods were used and researchers did not have to think about *heuristic approaches* to overcome *the curse of dimensionality* as we do today.

Typical data sets generated with the use of current microarray technologies include many thousands of variables and only dozens or hundreds of biological samples. When exon arrays are more widely used or when protein chip technologies allow for direct quantification of the protein expression level on the whole-human-proteome scale, we may routinely analyze data sets with more than a million variables. The traditional univariate approach—*one-gene or one-protein-at-a-time*—is no longer sufficient. Different approaches are necessary and multivariate analysis has to become a standard one. There is nothing wrong with using the univariate analysis, but if research stops at that point, as is the case in some studies, a huge amount of generated data may be heavily underused, and potentially important biomedical knowledge not extracted. Here is where data miners should be involved.

Today, hardly any study involving high throughput gene or protein expression data is performed exclusively by biologists or biomedical scientists. Although few research groups realize the importance of including data miners in their studies, the role of a relatively new breed of scientists called *bioinformaticians* is indisputable. The bioinformaticians are not necessarily data miners, although data mining should

Data Mining for Genomics and Proteomics. By Darius M. Dziuda

be one of their required skills. There is a small but growing population of scientists who were majoring in bioinformatics. However, most of the experienced bioinformaticians are still either biologists who learned computer science and statistics, or computer scientists familiar with biology. The interdisciplinary nature of this field is expansive and involves many other disciplines, like physiology, clinical sciences, mathematics, physics, and chemistry.

Since this book is written for students and practitioners of data mining and bioinformatics as well as biomedical sciences, there may be some terms that are well known to some readers but new for others. We will start with a short explanation of some basic, mostly biological, terms that are relevant for our data mining focus.

1.1 BASIC TERMINOLOGY

1.1.1 The Central Dogma of Molecular Biology

The central dogma of molecular biology was originally introduced by Francis Crick in 1957 at a symposium of the Society for Experimental Biology in London and then published in 1958 (Crick 1958). The dogma[1] states that once the detailed sequence information has passed into protein it cannot be transferred to nucleid acid or protein. Crick's original description of the dogma (Crick 1958) was:

> ***The Central Dogma***
>
> *This states that once 'information' has passed into protein* ***it cannot get out again****. In more detail, the transfer of information from nucleic acid to nucleic acid, or from nucleic acid to protein may be possible, but transfer from protein to protein, or from protein to nucleic acid is impossible. Information means here the* ***precise*** *determination of sequence, either of bases in the nucleic acid or of amino acid residues in the protein.*

Please note the qualification that 'information' here means the *sequential* information. Understood in its original meaning, the central dogma is still one of the fundamental ideas of molecular biology. Although introduced as a speculative idea, the central dogma holds true as well as there are plausible arguments that it is rather unlikely for it to be reversed (Crick 1970; Crick 1988).

The central dogma is quite often confused with the standard pathway of information flow from DNA to RNA to protein. To address misunderstandings about the dogma, Crick explained it in relation to three classes of transfers of sequential information: general transfers (ones that commonly occur), special transfers (may occur in special situations), and unknown transfers. The central dogma is about the *unknown transfers*—protein to protein, protein to DNA, and protein to RNA—and it postulates that these transfers never occur (Crick 1970).

[1]Francis Crick admits in his autobiography (Crick 1988) that he was criticized for using the word *dogma* in the situation where the word *hypothesis* would be more appropriate. Crick argues, however, that he already used the term hypothesis in the *sequence hypothesis* introduced at the same time (Crick 1958) and wanted to emphasize the more powerful and central position of the "dogma."

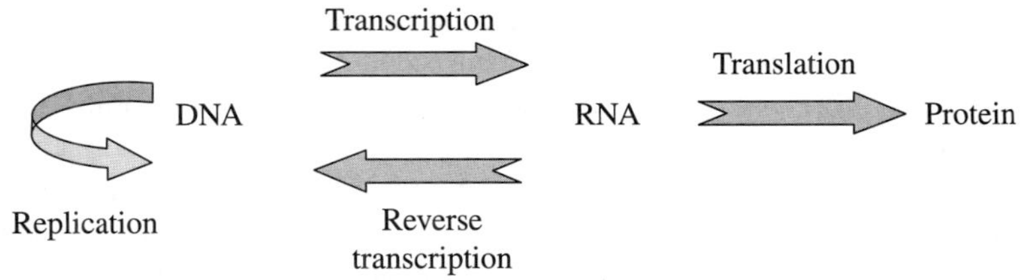

Figure 1.1: The basic flow of sequential information: DNA to DNA (replication), DNA to RNA (transcription), RNA to protein (translation), and RNA to DNA (reverse transcription). The central dogma of molecular biology states that "*once (sequential) information has passed into protein it cannot get out again*" (Crick 1970).

The current knowledge of information transfer is consistent with the central dogma. The standard pathway of information flow describes the process, in which proteins are synthesized based on DNA information (Fig. 1.1): DNA is ***transcribed*** into RNA, and then RNA—or more precisely, mRNA (messenger RNA)[2]—is ***translated*** into protein. These are two of the Crick's general transfers. The third one is *replication* (DNA to DNA). Special transfers are represented by *reverse transcription* (RNA to DNA). There is no evidence for the *unknown transfers*.

In humans (and other eukaryotic[3] organisms), transcription takes place in the cell nucleus, and translation in the cytoplasm—outside the nucleus. Most human genes contain noncoding sequences (*introns*), which have to be removed from mRNA before it is translated into protein. The process that eliminates introns is called *splicing*. During translation, which takes place at ribosomes, the mRNA sequential information is translated into a string of amino acids that are used to synthesize the protein. First, the mRNA sequence is divided into three-letter codons representing amino acids. Subsequently, the amino acids are linked together to create the protein. Translation ends when one of the stop codons is encountered, and the newly created protein leaves the ribosome.

1.1.2 Genome

The term *genome* can be understood either as the complete set of genetic information or as the entire set of genetic material contained in an organism's cell. When applied to the human genome, this definition includes both the *nuclear genome* and the *mitochondrial genome.*[4] Nevertheless, in the area of gene expression analysis, we often use the term genome as referring to the nuclear genome only, that is, understood as the complete DNA sequence of one set of the organism's chromosomes.

[2]Crick referred to RNA since mRNA had yet to be discovered when he formulated the dogma.

[3]Eukaryotic cells are cells that have a nucleus. Eukaryotes, that is, organisms with eukaryotic cells, may be unicellular or multicellular (e.g., fungi, plants, and animals). Prokaryotes are unicellular organisms (such as bacteria) that have no nuclei. *Pro* in the term means "prior to" and *karyot* means "nucleus". The prefix *eu* means "true" (Garrett and Grisham 2007).

[4]The mitochondrial genome represents organellar genomes carried by cells of most eukaryotic organisms. Another example of organellar genomes is the *chloroplast genome* in plants.

In humans, every nucleated cell (circulating mature red blood cells have no nucleus) contains the nuclear genome organized within the nucleus into the DNA molecules called chromosomes—22 pairs of autosomes (nonsex chromosomes) and two heterosomes (sex chromosomes). The length of the human genome sequence is about 3×10^9 nucleotides (base pairs).[5] For two randomly selected humans, the order of nucleotides in their genomes is about 99.9% identical (it is more than that if the two are related). However, 0.1 percent of the three billion bases amounts to three million places where two such genomes differ. The space for different DNA sequences is huge, beyond huge actually (up to $4^{3000000}$).[6] We are different; no two humans (with the exception of identical twins) have identical DNA.

The Human Genome Project was an international research program aimed at obtaining a high-quality sequence of the euchromatic (i.e., gene-rich) portion of the human genome. The project was initiated in 1990 and was officially completed in 2003. In February 2001, two draft versions of the human genome sequence were simultaneously announced and published, one from the International Human Genome Sequencing Consortium (International Human Genome Sequencing Consortium 2001) and the other from Celera Genomics (Venter et al. 2001). Each of the drafts contained over 100,000 gaps and the draft sequences were missing about 10% of the euchromatic portion of the genome (Stein 2004). In 2003, the 'finished' human genome sequence was announced. It covered 99% of the euchromatic portion of the human genome (and about 94% of the total genome), had only 341 gaps,[7] and contained only about one error per 100,000 bases (International Human Genome Sequencing Consortium 2004).

1.1.3 Proteome

A simplified definition could state that the *proteome* is the complete set of protein products expressed by the genome (see the central dogma). However, unlike the genome that can be considered a rather stable entity, the proteome constantly changes (mainly due to protein–protein interactions and changes in a cell's environmental conditions). Furthermore, the set of proteins expressed in a cell depends on the type of cell. Thus, a proteome can also be interpreted as a snapshot of all the proteins expressed in a cell or a tissue at a particular point in time. This means that depending on the context, we may refer to the single proteome of an organism (i.e., the *complete proteome* understood as

[5]This is the length of the *haploid* human genome, that is, the length of the DNA sequence of the 24 distinct chromosomes, one from each pair of the 22 autosomes and the 2 heterosomes. The *diploid* human genome, including the DNA sequence from all 46 chromosomes, would have about 6×10^9 nucleotides, or six gigabases (6 Gb). The latest technological advances allowed for sequencing the diploid genome of James Watson in two months (Wheeler et al. 2008).

[6]This number is the upper limit of potentially significant differences since not all changes in the DNA sequence are necessarily associated with differences in functions.

[7]Many of these gaps were associated with segmental duplications that could not be sequenced with available technologies. This was also one of the reasons why The Human Genome Project did not target the heterochromatic (gene-poor) regions of the human genome, which contain highly repetitive sequences. The estimated size of the gaps was: about 2.8×10^7 bases in the euchromatic portion and about 2.0×10^8 bases in the heterochromatic portion of the genome.

the set of all protein products that can be expressed in the organism) or to many *cellular proteomes* that are qualified by location and time.

The number of proteins in the human proteome is much larger than the number of the underlying protein-coding genes. Currently, the number of protein-coding genes in the human genome is estimated to be under 21,000 (Clamp et al. 2007). The current estimate for the size of the complete human proteome is about one million proteins. Why are there more proteins than genes if each protein is synthesized by reading the sequence of a gene? This is due to such events as *alternative splicing* of genes and *post-translational modifications* of proteins.[8]

The Human Proteome Organisation (HUPO) plans to identify and characterize all proteins in the complete human proteome. However, due to the scale and complexity of this task, the goal of the first phase of the Human Proteome Project (HPP) is limited to the identification of one representative protein for each protein-coding human gene. After this first stage, the catalogue of human proteins will be extended to eventually include all protein isoforms (Uhlen 2007; Pearson 2008; Service 2008).

1.1.4 DNA (Deoxyribonucleic Acid)

DNA is a nucleic acid that encodes genetic information. DNA is capable of self-replication and synthesis of RNA. In 1944, Oswald Avery and colleagues identified DNA as the genetic material of inheritance (Avery et al. 1944). In 1953, James Watson and Francis Crick proposed and described a model of DNA as the double helical structure[9] (Watson and Crick 1953a, 1953b).

DNA consists of two long chains[10] (strands) of nucleotides twisted into a *double helix* (Fig. 1.2) and joined by hydrogen bonds between the complementary nucleotide bases. Each of the strands has a sugar phosphate backbone, to which the bases are covalently attached.

There are four types of nucleotides in DNA and each of them contains a different base: *adenine* (A), *guanine* (G), *cytosine* (C), or *thymine* (T).[11] The four bases are letters of the alphabet in which genetic information is encoded. Strings of these letters are read only in one direction. Each of the two DNA strands has its polarity—the 5′ end and the 3′ end[12]—and the two complementary strands are bound with the opposite

[8] Alternative splicing is described in Section 1.1.8. Post-translational modifications are various chemical alterations that proteins may undergo after their synthesis. Examples of these modifications include the enzymatic cleavage (cutting a protein into smaller functional proteins), covalent attachment of additional biochemical groups (e.g., phosphorylation, methylation, oxidation, acetylation), and forming covalent bonds between proteins.

[9] Watson and Crick admitted that their work was stimulated by unpublished experimental results (X-ray crystallographic data) and ideas of Rosalind Franklin and Maurice Wilkins; a part of these results was published in the same issue of *Nature* as the Watson and Crick main paper (Franklin and Gosling 1953; Wilkins et al. 1953).

[10] The entire DNA in one human cell (in all 46 chromosomes) is about two meters long, yet fits into a cell nucleus, which is 2–3 micrometers wide.

[11] Due to their chemical structure, A and G nucleotides are purines, whereas C and T are pyrimidines.

[12] These ends correspond to 5′ and 3′ carbon atoms in a ribose residue, which are not linked to another nucleotide (Singleton 2008).

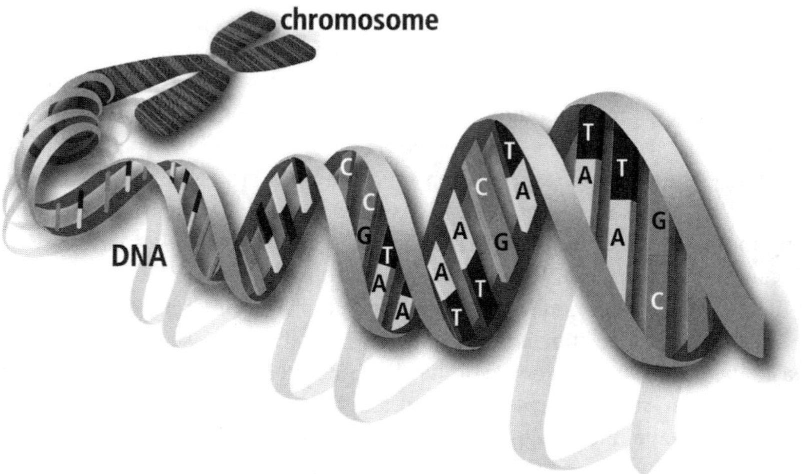

Figure 1.2: The DNA double helix (courtesy: The U.S. Department of Energy Genome Programs, http://genomics.energy.gov). The DNA structure includes two antiparallel deoxyribose-phosphate helical chains, which are connected via hydrogen bonds between complementary bases on the chains (base pairs). The base pairs "rungs of the ladder" are spaced 0.34 nm apart. This double helix structure repeats itself every 10 base pairs, or 3.4 nm (Watson and Crick 1953a; Garrett and Grisham 2007). (See color insert.)

directionality (i.e., they are *antiparallel*). The sequence of nucleotides determines the genetic information. In nature, base pairs form only between A and T and between G and C. Therefore, the sequence of bases in one strand of DNA can be deduced from the sequence of bases in its complementary strand (which enables DNA replication). During transcription, the sequential information contained in DNA is transferred into the single-stranded RNA.[13]

The latest discoveries indicate that the majority of human DNA is transcribed into various RNA transcripts (The ENCODE Project Consortium 2007). The messenger RNA (mRNA) is the only type of these RNA transcripts that are subsequently translated into proteins. Consequently, these discoveries indicate that there are few unused sequences in the human genome.

1.1.5 RNA (Ribonucleic Acid)

RNA is a nucleic acid molecule similar to DNA but containing the sugar component *ribose* rather than deoxyribose[14] and the *uracil* (U) base instead of thymine (T) in DNA. RNA is formed upon a DNA template, but is almost always single-stranded

[13]The DNA strand that has the same sequence as the synthesized transcript RNA (except for the replacement of Ts by Us) is called the *sense* strand. The other DNA strand is called the *antisense* strand (Campbell and Farrell 2006).

[14]The prefix *deoxy* means that an oxygen atom is missing (H instead of OH) from one of the ribose carbon atoms.

and has a much shorter sequence of nucleotides. There are many kinds of RNA molecules, with the three main classes of cellular RNA being: mRNA—messenger RNA; tRNA—transfer RNA; and rRNA—ribosomal RNA. Other classes of RNAs include microRNA (miRNA) and small interfering RNA (siRNA), both of which can regulate gene expression during and after transcription. Since only mRNAs are translated into proteins, some genes do not encode proteins but various RNA functional products.

Uracil is very similar to thymine and they both carry the same information. One may ask why there is thymine in DNA, but uracil in RNA. One of the common DNA mutations is chemical degradation of cytosine that forms uracil. If uracil was a valid base in DNA, cellular repair mechanisms could not efficiently correct such mutations. Thymine is then more appropriate for the stability of long-lived DNA. For short-lived RNA molecules, for which quantity is more important than long-term stability, uracil is more appropriate since it is energetically less expensive to produce than thymine.

1.1.6 mRNA (messenger RNA)

Messenger RNA (mRNA) is the type of RNA that contains coding information for protein synthesis. Messenger RNA is transcribed from a DNA template (a sequence of bases on one strand of DNA), and subsequently undergoes maturation that includes splicing. The mature mRNA is then transported from the cell nucleus to the cytoplasm where proteins are made during the process called translation. During translation, which takes place at the cytoplasm's ribosomes, the mRNA information contained in the sequence of its nucleotides is translated into amino acids. Amino acids are the building blocks of proteins. First, the mRNA sequence is divided into three-letter codons representing amino acids. Subsequently, the amino acids are linked together to produce a polypeptide. Translation ends when one of the stop codons is encountered. A functional protein is the result of properly folded polypeptide or polypeptides.

1.1.7 Genetic Code

The four bases of mRNA—adenine (A), guanine (G), cytosine (C), and uracil (U)—are the "letters" of the *genetic code*. Amino acids, which are building blocks of proteins, are coded by three-letter words of this code, called *codons*. The genetic code (Table 1.1) defines the relation between mRNA codons and amino acids of proteins.

With the four letters of the genetic code, there are $4^3 = 64$ possible triplets (codons). All 64 codons are meaningful: 61 of them code for amino acids and three (UAA, UAG, and UGA) serve as the *stop codons* signaling the end of the protein. The stop codons are called *nonsense codons* as they do not correspond to any amino acid. AUG codes for methionine, but is also the *start codon*—a part of the initiation signal to start protein synthesis.

Since there are 61 *sense* codons and only 20 amino acids in proteins (see Table 1.2), the genetic code is *degenerate*, which means that almost all amino acids

TABLE 1.1: Genetic Code

First Base in Codon	Second Base in Codon: U	C	G	A	Third Base in Codon
U	Phe	Ser	Cys	Tyr	U
U	Phe	Ser	Cys	Tyr	C
U	Leu	Ser	Trp	*Stop*	G
U	Leu	Ser	*Stop*	*Stop*	A
C	Leu	Pro	Arg	His	U
C	Leu	Pro	Arg	His	C
C	Leu	Pro	Arg	Gln	G
C	Leu	Pro	Arg	Gln	A
G	Val	Ala	Gly	Asp	U
G	Val	Ala	Gly	Asp	C
G	Val	Ala	Gly	Glu	G
G	Val	Ala	Gly	Glu	A
A	Ile	Thr	Ser	Asn	U
A	Ile	Thr	Ser	Asn	C
A	*Met*	Thr	Arg	Lys	G
A	Ile	Thr	Arg	Lys	A

The gray background denotes *four-fold degeneracies*—codons with irrelevant third base. Sixty one out of 64 codons code for 20 amino acids. Three codons are the stop codons (UAA, UAG, and UGA). The AUG is the start codon, but it also codes for methionine. Codons that code for the same amino acid (e.g., the six codons coding for Leu, or leucine) are called *synonymous codons*.

(18 out of 20) are associated with more than one codon. Methionine (Met) and tryptophan (Trp) are exceptions—each of them is coded by only one codon.

When the third base of a codon is irrelevant, that is, four codons with the same first and second base code for the same amino acid, we have *four-fold degeneracy* (sometimes called *third-base degeneracy*). The four-fold degenerate families of codons (such as UC* or CU*, where the symbol * denotes an irrelevant third base) are marked in Table 1.1 with the gray background. Another main family of degeneracy is *two-fold degeneracy*, when two codons with a different base in the third position are associated with the same amino acid (for instance, UG[U|C], where the [U|C] notation means the third base is either U or C).

TABLE 1.2: Amino Acids

<table>
<tr><th>Amino acid</th><th>3-letter code</th><th>1-letter code</th><th>Number of codons</th><th>Codons</th></tr>
<tr><td>Alanine</td><td>Ala</td><td>A</td><td>4</td><td>GC*</td></tr>
<tr><td>Arginine</td><td>Arg</td><td>R</td><td>6</td><td>CG*, AG[A|G]</td></tr>
<tr><td>Asparagine</td><td>Asn</td><td>N</td><td>2</td><td>AA[U|C]</td></tr>
<tr><td>Aspartic acid</td><td>Asp</td><td>D</td><td>2</td><td>GA[U|C]</td></tr>
<tr><td>Cysteine</td><td>Cys</td><td>C</td><td>2</td><td>UG[U|C]</td></tr>
<tr><td>Glutamic acid</td><td>Glu</td><td>E</td><td>2</td><td>GA[A|G]</td></tr>
<tr><td>Glutamine</td><td>Gln</td><td>Q</td><td>2</td><td>CA[A|G]</td></tr>
<tr><td>Glycine</td><td>Gly</td><td>G</td><td>4</td><td>GG*</td></tr>
<tr><td>Histidine</td><td>His</td><td>H</td><td>2</td><td>CA[U|C]</td></tr>
<tr><td>Isoleucine</td><td>Ile</td><td>I</td><td>3</td><td>AU[U|C|A]</td></tr>
<tr><td>Leucine</td><td>Leu</td><td>L</td><td>6</td><td>CU*, UU[A|G]</td></tr>
<tr><td>Lysine</td><td>Lys</td><td>K</td><td>2</td><td>AA[A|G]</td></tr>
<tr><td>Methionine</td><td>Met</td><td>M</td><td>1</td><td>AUG</td></tr>
<tr><td>Phenylalanine</td><td>Phe</td><td>F</td><td>2</td><td>UU[U|C]</td></tr>
<tr><td>Proline</td><td>Pro</td><td>P</td><td>4</td><td>CC*</td></tr>
<tr><td>Serine</td><td>Ser</td><td>S</td><td>6</td><td>UC*, AG[U|C]</td></tr>
<tr><td>Threonine</td><td>Thr</td><td>T</td><td>4</td><td>AC*</td></tr>
<tr><td>Tryptophan</td><td>Trp</td><td>W</td><td>1</td><td>UGG</td></tr>
<tr><td>Tyrosine</td><td>Tyr</td><td>Y</td><td>2</td><td>UA[U|C]</td></tr>
<tr><td>Valine</td><td>Val</td><td>V</td><td>4</td><td>GU*</td></tr>
</table>

Twenty amino acids that are building blocks of proteins. Amino acids are coded by one to six codons. For example, arginine is coded by six codons, CG* and AG[A|G]. The CG* notation means four codons starting with CG, that is, [CGU, CGC, CGG, CGA], and AG[A|G] means that the first and second bases are AG, and the third base is either A or G, [AGA, AGG].

1.1.8 Gene

For the purpose of mining gene expression data, we could use the following description of a gene:

> *A gene is the segment of DNA, which is transcribed into RNA, which may then be translated into a protein. Human (and other eukaryotic) genes often include non-coding sequences (introns) between coding regions (exons). Such genes encode one or more functional products, which are proteins or RNA transcripts.[15] Each gene is associated with a promoter sequence, which initiates gene transcription and regulates its expression.*

This description is satisfactory for the analysis of gene expression microarray data. Nevertheless, as a definition of the gene this description may be incomplete. The same is true for other definitions of the gene that were proposed during the last hundred years. Although it may be true that none of those definitions has been

[15] A transcript is the RNA product of the gene transcription process.

generally accepted (Falk 1986; Rheinberger and Müller-Wille 2008), some of them were popular enough to prevail in their time. Here are a few examples of the concept of a gene as it was evolving in time (Gerstein et al. 2007; Rheinberger and Müller-Wille 2008).

- *A gene as a discrete unit of heredity that determines a characteristic of an organism as well as the heritability of this characteristic.*
- *A gene as a stretch of DNA that codes for one protein.*
- *A gene as "a DNA segment that contributes to phenotype/function. In the absence of demonstrated function a gene may be characterized by sequence, transcription or homology"* (Wain et al. 2002; HUGO Gene Nomenclature Committee 2008).

New discoveries have been forcing changes to the concept of a gene. The following are just a few among such discoveries:

- *Splicing*—the process of removing introns from the primary RNA transcript and using only exons to form the mature mRNA (see Fig. 1.3). Splicing redefined the gene as including a series of exons (coding sequences) separated by introns (noncoding sequences).
- *Alternative splicing*—generally, combining different subsets of gene exons to form (code for) different mRNA transcripts (see Fig. 1.4). Alternative splicing invalidated the "one gene—one protein" paradigm.
- Nonprotein-coding genes—genes that "*encode functional RNA molecules*[16] *that are not translated into proteins*" (HUGO Gene Nomenclature Committee 2008).
- Overlapping protein-coding genes—genes sharing the same DNA sequence. New exons have been discovered far away from the previously known gene locations, some of them within the sequence of another gene (Pennisi 2007). It is possible for a gene to be completely contained within an intron of another gene; it is also possible for two genes to share the same stretch of DNA without sharing any exons (Gerstein et al. 2007).
- Nonprotein-coding genes sharing the same DNA sequence with protein-coding genes (Gingeras 2007).
- Some RNA transcripts are composed from exons belonging to two genes.
- Transcribed *pseudogenes*. A pseudogene is a sequence of DNA that is almost identical to a sequence of an existing and functional gene but is or appears to be inactive, for instance, due to mutations that make it nonfunctional (Ganten and Ruckpaul 2006; Singleton 2008). However, the Encyclopedia of DNA Elements project (The ENCODE Project Consortium 2007) revealed recently that some pseudogenes—although by definition lacking protein coding potential—can produce RNA transcripts that could potentially have some regulatory functions (Gerstein et al. 2007; Gingeras 2007; Zheng et al. 2007).

[16]Some of these nonprotein-coding RNA transcripts are well known functional RNA molecules, for example ribosomal RNAs (rRNA) or transfer RNAs (tRNA). Others include recently discovered microRNAs (miRNA) and small interfering RNAs (siRNA); both of them can regulate expression of specific genes (Gingeras 2007).

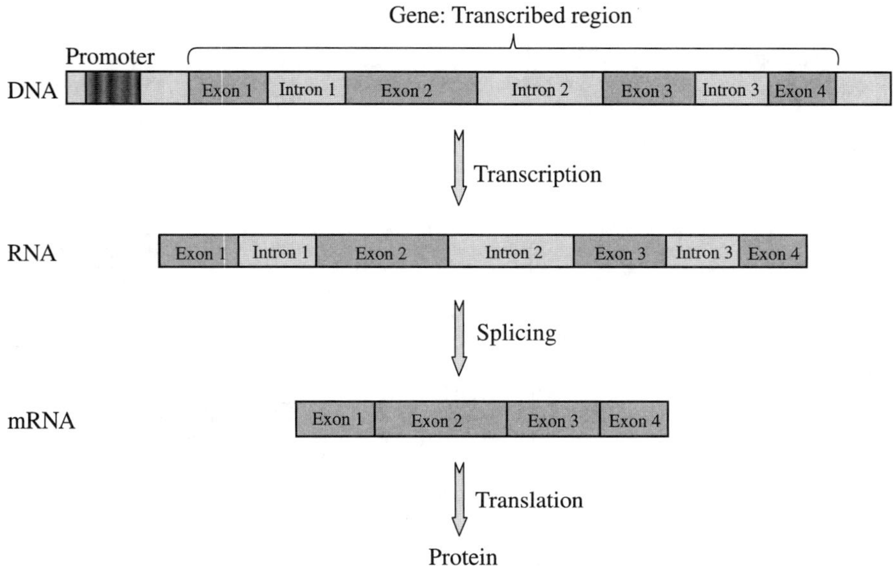

Figure 1.3: A human protein-coding gene: the structure, transcription, splicing, and translation. The top row shows a schematic structure of the gene. The transcribed region consists of exons that are interrupted by introns (on average, introns are about 20 times longer than exons). The promoter associated with the gene is a regulatory sequence that facilitates initiation of gene transcription and controls gene expression. Enhancers and silencers are other gene-associated regulatory sequences that may activate or repress transcription of the gene (they are not shown here as they are often located distantly from the transcribed region). The gene's exons and introns are first transcribed into a complementary RNA (called nuclear RNA, primary RNA transcript or pre-mRNA). Then, the splicing process removes introns, joins exons, and creates mRNA (called also mature mRNA). The mRNA travels from the nucleus to the cytoplasm where, in ribosomes, it is translated into a protein. (See color insert.)

Since the concept of a gene has been changed over and over again, we may consider one of two options: (i) stop using this concept at all, or (ii) formulate a gene definition in a way that is either quite general (and perhaps vague) or deliberately verbose in an attempt to cover all possibilities. Here are two recently proposed gene definitions:

- "*A gene is a union of genomic sequences encoding a coherent set of potentially overlapping functional products.*" (Gerstein et al. 2007)
- "*A gene is a discrete genomic region whose transcription is regulated by one or more promoters and distal regulatory elements and which contains the information for the synthesis of functional proteins or non-coding RNAs, related by the sharing of a portion of genetic information at the level of the ultimate products (proteins or RNAs).*" (Pesole 2008)

These two definitions provide different answers to the question "*What is a gene?*" Consequently, if we were able to enumerate all the genes in the human

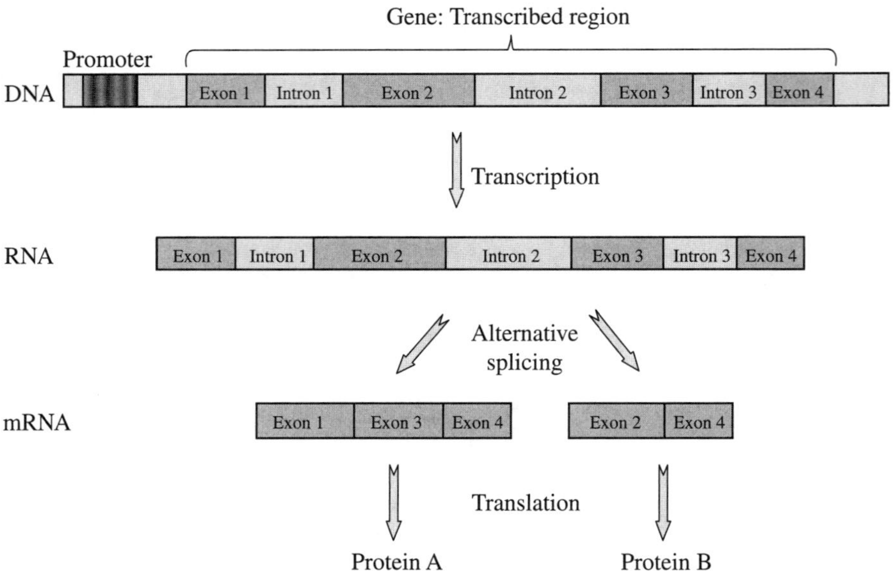

Figure 1.4: A human protein-coding gene: an example of alternative splicing. During alternative splicing of the primary RNA transcript, different subsets of exons are joined to create two (or more) different mRNA isoforms. These mRNA isoforms are then translated into usually distinct proteins. The majority of human protein-coding genes undergo alternative splicing (Stamm 2006). (See color insert.)

genome, these definitions would also lead to different answers to the question "*How many genes are there in the human genome?*" To simplify enumeration of human genes, we may start with counting only protein-coding genes. A recently performed analysis of the three most popular human gene databases[17] suggests that there may be fewer than 21,000 protein-coding genes in the human genome (Clamp et al. 2007).

Although cells of different tissues have the same genome, only a limited and tissue-specific subset of genes is expressed in each tissue. While tissues can be identified by general patterns of their gene expression, these patterns are not constant. Gene expression levels in a cell change in time and in response to changes in a cell's environmental conditions. Changes in the expression profile (turning on or off some genes or changing expression levels of some of the expressed genes) alter protein production and may result in significant changes in tissue functioning; for instance, in transforming a normal tissue into a cancerous one.

1.1.9 Gene Expression and the Gene Expression Level

Gene expression is the process that converts information encoded in a gene into functional products of the cell (see the central dogma). In the context of genomic projects

[17]The three databases are: Ensembl (Flicek et al. 2008), RefSeq (Pruitt et al. 2007), and Vega (Wilming et al. 2008).

involving gene expression microarray data, we are interested in the *gene expression level* that refers to the number of copies of RNA transcripts created by transcription of a particular gene (for the protein-coding genes, this is the number of mRNA transcripts generated from the gene) at a given time. The genes that are expressed include the protein-coding genes as well as the genes coding for RNA functional products. Most often, the gene expression analysis focuses, however, on the expression level of the protein-coding genes.

Although the same DNA is contained in every cell of an organism, cells of different tissues are different. The differences result from different levels of gene expressions, that is, different patterns of gene activations. If a gene is active (expressed), the protein (or proteins) encoded by the gene is (are) synthesized in the cell. The higher the gene expression level, the more of the protein is produced. There may be various reasons for a gene to be over-expressed, under-expressed or not expressed at all—hereditary factors, environmental factors, or their combinations. The hereditary factors may mean a mutation in a single gene, simultaneous mutations in a set of genes, chromosomal abnormalities, etc. If this prevents or significantly influences the production of one or more important proteins, we have hereditary diseases (such as cystic fibrosis for a single gene mutation, or Down syndrome when there are three copies of chromosome 21). On the other hand, many diseases are not necessarily related to genetic abnormalities, but are caused by the environmental factors that are changing the expression level of otherwise "normal" genes. This may be more complicated when such changes in the gene expression level are still moderated by some gene mutations, which by themselves are not causing any diseases. And here is the promise of genomics—identification of gene expression patterns associated with a disease and linking the patterns to underlying biological and environmental factors may lead to cure or prevention.

1.1.10 Protein

Protein is a biological macromolecule composed of one or more chains of amino acids synthesized in the order determined by the DNA sequence of the gene coding for the protein. Short chains of amino acids are called *peptides* and longer ones *polypeptides*. Since there is no precise distinction between them, we can say that the term *polypeptide* is used for chains longer than several dozen amino acids. We can define protein as a molecule composed of one or more polypeptide chains (Garrett and Grisham 2007).

A protein folds into a three-dimensional structure, which determines the function of the protein. The way the protein folds into its three-dimensional form is determined by its sequence of amino acids. Proteins are essential components of all living cells and each of them has a unique function. Examples of proteins include enzymes, hormones, and antibodies.

The number of proteins in the human proteome is much larger than the number of protein-coding genes in the human genome. This is mostly due to *alternative splicing* (when more than one protein is produced by different combinations of exons of the same gene) and *post-translational modifications* of proteins.

1.2 OVERLAPPING AREAS OF RESEARCH

Biomedical research based on the analysis of gene or protein expression data is an interdisciplinary field including molecular biology, computational biology, bioinformatics, data mining, computer science, statistics, mathematics, . . . and the list of overlapping disciplines is far from being exhaustive. New terms are constantly coined. Furthermore, the same term may have different meanings in some of the overlapping areas. Here are a few terms that are important for the subject of this book.

1.2.1 Genomics

Genomics can be understood as the systematic analysis of genome-scale data in order to expand biomedical knowledge. Genomic studies investigate structure and function of genes and do this simultaneously for all the genes in a genome. Investigating patterns of gene expression—one of the main themes of this book—is an important part of *functional genomics* that focuses on analysis of genes and their products at the functional level.

Although there are opinions that genomics and genetics should be combined and considered together, it is not currently the case and genetics is still understood as the study of genes in the context of inheritance.

1.2.2 Proteomics

Proteomics is the study of functions and structures of proteins. Proteomics is often seen as the next stage of research after genomics, and as the area that should give us more direct insight into biological processes (since proteins are direct players in the cell physiology whereas genes are mostly intermediate entities). Proteomics seems to be much more complicated than genomics, the main reason being that proteomes are constantly changing and that different cells of the same organism may have different proteomes.

1.2.3 Bioinformatics

One may find many definitions of *bioinformatics* and it is not unusual for them to limit this field to the areas researched by the definition authors. Defining it more generally, we may say that bioinformatics is the science of managing, analyzing, mining, and interpreting biological data. Bioinformatics, computational biology, and data mining are overlapping in their use of mathematical, statistical, and computer science tools and methods to analyze large sets of biological data.

1.2.4 Transcriptomics and Other -omics . . .

Though the term *genomics* is already well established in biomedical sciences, the term itself is an "*unusual scientific term because its definition varies from person to person*" (Campbell and Heyer 2007). Once genomics became popular, many new related *-omics* terms were created. Whether their areas belong to genomics or not is pretty much an open question. For example, *transcriptomics* is most often defined[18] as the area covering analysis of gene expression data (expressed genes are called *transcripts*). One could then say that the focus of this book is transcriptomics. Nevertheless, we will use the widely recognized term *genomics*, which has been used for this kind of investigations since the beginning of high-throughput gene expression data analysis.

1.2.5 Data Mining

We define *data mining* as efficient ways of extracting new information or new knowledge from large data sets or databases. Some more classical definitions limit data mining to extraction of 'useful information.' However, in biomedical research any new information reflecting underlying biological processes can be potentially useful. Furthermore, translating the extracted new information into new biomedical knowledge is often a nontrivial task and the data mining definition (and methods) should be extended as well to include the information-to-knowledge stage of investigations.

[18] There are other definitions of transcriptomics, one—for example—defines it as the study of transcription process itself.

CHAPTER 2

BASIC ANALYSIS OF GENE EXPRESSION MICROARRAY DATA

Microarray Technology

Analysis of gene expression microarray data:

- Preprocessing
- Exploratory data analysis
- Unsupervised learning (taxonomy-related analysis)

"The bottleneck is no longer in the generation of data, but in our ability to make sense of it."

—(Seidel 2008)

2.1 INTRODUCTION

Since one of the main topics of this book is data mining of gene expression data, we will start with a short overview of the microarray technology and the steps leading from raw microarray data to the gene expression matrix. One could say that higher-level preprocessing, exploratory data analysis, and mining of gene expression data start *after* we have the gene expression levels available in the form of a single number representing the abundance of a gene transcript in a biological sample. However, knowing where data come from and what methods were applied to convert the raw data into this "starting point" expression data is very important for assessing the quality of the data, validity of experimental design and selection of particular approaches.

Once the gene expression matrix is created as the result of low-level preprocessing of the raw microarray data, we are ready for the higher-level analysis. Although data mining methods have to be selected depending upon the goals of a

Data Mining for Genomics and Proteomics. By Darius M. Dziuda

study, there are common preliminary tasks performed for most studies utilizing gene expression data. Among these tasks are: higher-level preprocessing (such as scaling or transforming the data), quality assessment of gene expression data, and filtering out noise and the variables with unreliable measurements. Basic exploratory analysis is commonly the first step in analyzing preprocessed and filtered gene expression data. As this analysis is limited to the univariate approach, its main goal is not to answer any specific questions, but to give us some initial feeling about, and understanding of, the data at hand. Though selection of the analytical approaches (for instance, whether to use unsupervised or supervised learning) is dictated by the data and the goals of a study, the exploratory data analysis may provide information helpful in deciding which of the appropriate methods to use (for example, whether to use a parametric or a nonparametric supervised learning algorithm).

We want to stress the importance of selecting the right methods for the task. This should be obvious, but it is not necessarily the case for some biomedical studies reported in the literature. All too often, the data at hand is analyzed with a tool that happens to be available and popular, whether or not it is the proper tool to achieve the goal of a study. This, unfortunately, usually leads to under-usage of the data and the reporting of inferior results in situations where the information or knowledge sought could have been extracted from the data had appropriate methods and tools been used.

We will cover this subject in Chapter 3, here is just one example to think about. Consider clustering methods—they are popular and are implemented in many software packages. They are so popular that they are used in many studies, whether the clustering approach is appropriate for the study goals or not. Imagine a study, whose goal is to identify a multivariate biomarker differentiating states of a disease. For this goal clustering (or any other unsupervised method) is inappropriate—supervised data mining methods should be used. Biomarker discovery will be one of the main subjects of Chapter 3. The current chapter will focus on the preprocessing of gene expression data, the basic exploratory analysis (resulting in a univariate list of significant genes), and unsupervised learning (the taxonomy-related analysis, for which clustering is among the primary methods to use).

2.2 MICROARRAY TECHNOLOGY

The term *microarray* (Schena et al. 1995) is widely used for chip-based high-throughput technologies utilized to study biomedical samples. The most commonly used chips are DNA microarrays. Currently, they allow for the simultaneous measurement of the gene expression level on a whole-genome scale. Simultaneous quantification of expression for a large number of genes was very important for biological studies to evolve beyond the *one-gene-at-a-time* paradigm. Greatly simplified, a microarray is a small glass slide[1] with a large number of *spots* containing single-stranded DNA sequences. The spots are organized in a rectangular grid, and each

[1]Instead of glass, it could be a silicon chip or a nylon membrane.

cell of the grid contains DNA material (DNA, cDNA,[2] or oligonucleotide[3]) representing one gene. Generally, these spotted gene-specific sequences are called probes.[4]

Once the microarray is available, the target mRNA (extracted from a biological sample we want to investigate) is labeled with a fluorescent dye. The microarray is then washed with a solution of the target mRNA. The mRNA molecules, which find a spot with the complementary DNA sequence, will hybridize[5] and stick to the array. The remaining nonhybridized solution will be washed away. Since the target is labeled with a dye, a laser scanner can be used to measure the fluorescent signal emitted by each spot. The signal intensity of a spot is related to the abundance of mRNA corresponding to the represented gene.

A single DNA microarray can simultaneously provide information about the expression level of thousands of genes. By analyzing the signal intensity across multiple microarrays, multigene expression patterns characteristic of the states, diseases, or phenotypes represented by groups of samples may be discovered.

The Affymetrix GeneChip® arrays (Affymetrix, Santa Clara, CA) and spotted arrays (also referred to as *cDNA arrays* or *two-color arrays*[6]) are among the most popular microarray platforms. The spotted arrays are more flexible since their design may be customized for a particular study. However, this flexibility makes it difficult to compare the results of such customized studies. In biomarker discovery and gene expression based classification, we strive for large data sets. The standardization offered by such platforms like the Affymetrix GeneChip allows for combining data from different studies, or at least for using another study's data as an independent test set. For this reason, we focus mainly on the Affymetrix gene expression microarrays. In addition, we will mention a relatively new microarray platform, *bead-based* technology, which is gaining popularity.

2.2.1 Spotted Microarrays

There are two types of spotted microarrays: the *in-house* arrays and the *pre-spotted* arrays (Elvidge 2006). The spotted arrays of the first type were usually manufactured "in house" by research groups. Originally, relatively long cDNA sequences were spotted onto glass slides. Later, however, cDNAs were replaced by much shorter oligonucleotide probes (25–70 nucleotides in length), which are easier to produce. The in-house production of spotted arrays is very flexible as the arrays can be custom designed for each study. The experiments require, however, rather expensive

[2]Complementary DNA, cDNA, is DNA synthesized from a mature (fully spliced) mRNA template.

[3]Oligonucleotides are short sequences of nucleotides (typically 20–50 base pairs). Since they are sequences of single-stranded DNA or RNA, they can be used as probes for detecting complementary DNA or RNA (Ganten and Ruckpaul 2006). In the slang of the science, oligonucleotides are often referred to as *oligos*.

[4]A probe is a tethered nucleic acid with a known sequence (Phimister 1999). In GeneChip arrays, probes are pieces of single-stranded DNA bound to small sections of the chip known as features.

[5]Hybridization is the process in which two complementary single-stranded nucleic acids are combined into a single molecule.

[6]Spotted arrays can be used with one to four colors (one to four different samples per array), but most often two colors are used to investigate the relative expression level between two samples labeled with different colors.

spotting tools. Such equipment is not necessary for the pre-spotted arrays offered by a number of manufacturers. Though the pre-spotted arrays are less flexible, they require only a scanner to perform experiments. However, neither of these two types of spotted arrays can match the density and the number of features[7] available on such arrays as the Affymetrix GeneChip microarrays.

The spotted arrays are commonly used for simultaneous processing of multiple samples on a single array. Usually, one to four samples can be processed on an array in order to evaluate the relative levels of gene expression in the samples. In practice, most often two samples are processed on a single array, and these multichannel arrays are often referred to as *two-color* or *two-channel* arrays. The multichannel design allows for decreasing the experimental variation in the data since both the case and control samples are processed simultaneously on the same slide. Each of them would be labeled with a different fluorescent dye [such as Cy3 (green) and Cy5 (red)] and then each signal intensity would be read by appropriate scanners.

2.2.2 Affymetrix GeneChip® Microarrays

In the Affymetrix GeneChip microarrays, the probe sequences are synthesized *in situ* on quartz surfaces (wafers). Using mask-based photolithographic technology[8] (similar to that used in manufacturing computer chips), the probe sequence is built base by base.[9] Since this technology allows for millions of features to be synthesized on a single microarray, more than one feature can be used to detect each target transcript (gene or exon). This redundancy allows a single gene to be represented by several different oligo sequences (corresponding, for example, to various parts, or exons, of the gene).

Until recently, most studies based on microarray data were focusing on the overall expression of a gene, which means they analyzed the expressions computed at the probe set level (see Table 2.1). A probe on an array is for one exon, so different probes representing the same gene may represent different splice variants; it is estimated that some 40–60% of human genes may have alternative splice variants (Modrek and Lee 2003). Very few studies utilizing gene expression arrays such as the Affymetrix GeneChip HG-U95 or HG-U133 microarrays looked into changes in the expression of different splice variants. Studies interested in splice variants had to use the probe-level expression and additional information on mapping probes to exons. The situation changed with the introduction of exon arrays. According to Affymetrix (Affymetrix 2005a), their goal for the first generation of the exon arrays was "*to interrogate each potential exon with one probe set over the entire genome on a single array.*" The exon arrays allow for the exon-level expression

[7]Features—in the context of microarray technology, features are single segments or grid cells that make up the microarray. Some microarrays have millions of features, which may be as small as 5 × 5 micrometers. A feature may contain millions of copies of the same DNA probe and is designed specifically for only one gene or exon target in the sample.

[8]Other technologies can also be used to create *in situ* arrays. Among such commercial technologies are: Agilent's ink-jet method of printing 60-mer oligos on glass surfaces and Roche NimbleGen's maskless photolithography using small rotating mirrors (Seidel 2008).

[9]Affymetrix probes are usually 25-mers (25 nucleotides in length).

TABLE 2.1: Basic Terms Related to Technology of Affymetrix GeneChip Arrays

Feature	A square section of a microarray, as small as 5 micrometers by 5 micrometers (as of 2008), that holds a unique type of single-stranded DNA (*probe*). A feature contains millions of copies of the same probe.
probe	A 25 base long oligonucleotide sequence (25-mer) representing a single target (a gene or an exon) on an Affymetrix GeneChip microarray. The probes are synthesized directly on a glass array, one layer at a time. The probe should be unique for its target—the probe sequence is complementary to a section of the targeted gene and should not match any other part of the genome. When the array is scanned, the probe intensity (also called the spot intensity) corresponds to the average signal from all copies of the probe in one feature.
probe pair	A set of two probes, one *perfect match* probe and one corresponding *mismatch probe*. The mismatch probe is based on the same 25-mer as its perfect match probe but with the middle base (the 13th base) changed to prevent the target gene to match it perfectly. In the most recent Affymetrix arrays, the mismatch probes are not used, thus these arrays do not have probe pairs.
probe set	A set of probes designed for the same target. For example, on *HG-U133* microarrays, a gene is represented by eleven probe pairs—11 perfect match probes (with sequences complementary to sequences from eleven different parts of the gene), and eleven mismatch probes (used to assess the non-specific binding). On the *Human Exon 1.0 ST Array*, the majority of probe sets include four perfect match probes and no mismatch probes.

analysis (thus a more straightforward analysis of different splice variants, or isoforms, of a gene) as well as for the gene-level expression analysis (when multiple exon-level signals are used to compute the gene-level expression).

Although Affymetrix microarrays cover genomes of many organisms, we will focus on the arrays for the human genome. Below are descriptions of a few recent human genome arrays. Table 2.2 shows how the genome coverage and technological characteristics of Affymetrix microarrays have been evolving over time.

Human Genome U133 Plus 2.0 microarray

On this microarray (Affymetrix 2003), 22 different oligo sequences[10] are used to detect a single transcript. The set of 22 oligos is called a *probe set*. Each probe set includes 11 sequences designed as the *perfect match* (PM) to the target transcript (the target gene sequence), with the remaining 11 oligos being 'mismatched' by changing a single base in the middle of the corresponding perfect match sequence to its complementary base. Each perfect match probe is paired with its corresponding *mismatch probe* (MM), and the pair is referred to as a *probe pair*. Eleven probe pairs designed for a single transcript constitute a probe set. Hybridization specific to the targeted gene is represented by the intensity readings from the PM probes, whereas

[10]Older generations of Affymetrix GeneChip arrays used different numbers of probe pairs dedicated to one transcript (from 11 to 20 probe pairs, i.e., from 22 to 40 different oligos).

TABLE 2.2: Comparison of Main Characteristics of Selected Affymetrix Gene Expression Microarrays

Microarray	GeneChip Hu6800 (4 arrays)	Human Genome U95 Set (5 arrays)	Human Genome U133 Set (2 arrays)	Human Genome U133 2.0 Plus Array	Human Exon 1.0 ST Array	Human Gene 1.0 ST Array
Number of probes (features)				1.3 million	5.5 million	0.76 million
Number of probe sets	1700–1800 per array	12,000 per array	44,000	54,000	1.4 million	Over 28,000 (genes)
PM probes per probe set	20	16–20	11	11	4	26 probes per gene (median)
MM probes per probe set	20	16–20	11	11	0	0
Background estimation		by using MM probes	by using MM probes	by using MM probes	by using median intensity of dedicated background probes	by using median intensity of dedicated background probes
Feature size	50 microns	20 microns	18 microns	11 microns	5 microns	5 microns
Probe		25-mer oligo	25-mer oligo	25-mer oligo	25-mer oligo	25-mer oligo
Year released	1998	2000	2001	2003	2005	2007

nonspecific hybridization can be calculated from the MM probes. The GeneChip *Human Genome U133 Plus 2.0* microarray (released in 2003) contains about 1.3 million features (each feature of the size 11 × 11 micrometers) and 54,000 probe sets[11] (each probe set includes 11 PM/MM probe pairs).

HT HG-U133+ PM Array Plate

The recently introduced Affymetrix Array Plate technology enables parallel processing of multiple samples on a single plate of arrays. The GeneChip *HT HG-U133+ PM Array Plate* (Affymetrix 2008a) uses the same genomic content as the *Human Genome U133 Plus 2.0* microarray, but allows for simultaneous processing of 24 or 96 samples. There are, however, some design differences between the plate arrays and the U133 Plus 2.0 chip. The plate arrays include only the perfect match (PM) probes, and most of the probe sets have only nine (rather than eleven) PM probes.

Human Exon 1.0 ST microarray

Advances in the microarray technology allowed for a decrease in feature size, thus increasing the number of features per array. Introduced in October 2005, the GeneChip *Human Exon 1.0 ST Array* (Affymetrix 2005a) allows for the exon-level analysis on a whole genome scale. This microarray contains over 5 million features (with the feature size of 5 micrometers[12]) and about 1.4 million probe sets. Each probe set includes four PM probes[13] and no MM probes. With the Affymetrix's departure from the mismatched probes, the background signal may be evaluated by using the median signal intensity of up to 1,000 specially designed *background probes* with the same GC content. The probe sets correspond to exons. The array covers about 1.4 million exons and about one million exon clusters, which are sets of overlapping exon variants. Therefore, probe set annotations associate each probe set with an exon and exon cluster. Furthermore, each probe set is associated with a transcript cluster, which roughly corresponds to a gene. The 1.4 million probe sets of the array are grouped into about 300,000 transcript clusters (which means that many transcript clusters may represent one gene).

Instead of the mismatch probes specific for the perfect match probes, the *Human Exon 1.0 ST Array* contains two collections of background probes—the *antigenomic* collection (probes with sequences not represented in the human genome) and the *genomic* collection (also called surrogated mismatch probes; they are mismatch probes of genomic sequences that are less likely to be expressed). Background correction of each PM probe is

[11] Since probe sets do not always correspond to genes, it is preferable—at least for this and older arrays—to use the *probe set* term.

[12] As the semiconductor industry uses submicrometer lithography, we can expect further reduction in the microarray feature size.

[13] According to Affymetrix (Affymetrix 2008b), about 90 percent of probe sets on the *Human Exon 1.0 ST Array* have four PM probes, with the remaining about 10 percent having less than four PM probes.

based on comparing its signal to the median signal of all the background probes that have the same GC content as the PM probe. Since the 25-mer probes may contain 0–25 Gs or Cs, each of the two collections of the background probes consists of 26 sets, each of the sets having about 1000 probes with the same GC content.

The probes on the Affymetrix ST arrays are designed in the *antisense* orientation. The *ST* in the array name means *sense target* since targets in the *sense* orientation can hybridize to the *antisense* probes (Affymetrix 2008b). This is different than for earlier Affymetrix arrays (such as the *Human Genome U133 Plus 2.0* microarray), which were designed for targets in the antisense orientation. Another important novelty in the design of the ST arrays is whole-transcript coverage, which means that the probes are distributed across the entire length of the gene (its mRNA transcript), whereas previous designs (now called 3′ based expression arrays) were biased toward the 3′ end of the gene.

Human Gene 1.0 ST microarray

At the time of this writing, the GeneChip *Human Gene 1.0 ST* microarray (Affymetrix 2007) is the newest whole-transcript Affymetrix array (introduced in April 2007) focusing on gene-level expressions for the human genome. Over 28,000 "well-annotated genes" are represented. At the probe level, there is a significant overlap between this microarray and the *Human Exon 1.0 ST* microarray (about 80 percent of this array's probes are present on the exon array), although this gene-oriented array does not include probes that do not correspond to well-annotated sequences. As there are fewer probes per exon, this chip is smaller and less expensive than the exon microarray. The *Human Gene 1.0 ST* microarray consists of over 760,000 *perfect match* probes, which are distributed across transcribed regions of the represented genes. The number of probes per gene can vary significantly depending on the number of exons in a gene and the gene sequence composition. The median number of probes per gene is 26. The *antigenomic* collection of the background probes (the same 26 sets of up to 1000 probes as on the exon array) is used to estimate the background signal for each probe.

2.2.3 Bead-Based Microarrays

In bead-based microarrays, oligonucleotides are attached to microbeads, which are then deposited into wells on an array. Illumina's BeadArray Technology (Illumina, Inc., San Diego, California), represented by such chips as the *HumanWG-6* and *HumanHT-12* Expression BeadChips, is an example of this approach. The BeadArray technology uses 50-base long oligonucleotides as gene-specific probes (capture sequences). Each probe is coupled with a short address sequence,[14] uniquely identifying the capture sequence of the probe. Hundreds of thousands of identical synthetic oligonucleotides (each composed of the capture sequence and the address sequence) are covalently

[14]The address sequences have low similarity to human genomic sequences (Gunderson et al. 2004).

attached to a silica microbead of three micrometers in diameter. Thousands of beads are then deposited into wells on microarray substrates, which are either bundles of optical fibers or silicon wafers.[15] However—unlike in the Affymetrix GeneChip arrays, where the identity of each synthesized probe is predefined by its location on the array—the beads are randomly distributed over the wells. This approach simplifies the microarray manufacturing process and facilitates flexible and high density designs. While the target mRNA hybridizes to the 50-base capture sequence of the oligonucleotide, the address part of the oligonucleotide (unique for each bead type) is used to decode the identity of the probe (Gunderson et al. 2004; Barnes et al. 2005; Illumina 2008a). The latest BeadArray whole-genome expression chip, the *HumanHT-12* Expression BeadChip, provides coverage of over 25,000 sequences representing genes, gene candidates and splice variants. There are more than 48,000 probes per sample, and a single *HumanHT-12* BeadChip consists of 12 arrays allowing for simultaneous processing of twelve samples (Illumina 2008b).

2.3 LOW-LEVEL PREPROCESSING OF AFFYMETRIX MICROARRAYS

When microarrays are scanned,[16] the image of each array is created. For Affymetrix microarrays, the image of a chip is stored in a DAT[17] file. Probe intensities resulting from the image analysis are stored in a CEL file. Additional information, such as the association of probe set IDs with probes or probe pairs, is stored in a CDF library file (specific for a particular microarray type rather than for an individual experiment).

The goal of *low-level preprocessing* (also called the *low-level analysis* or the *probe-level analysis*) is to deliver the measurement of gene expression level and to assess the reliability of this measurement. A single value representing the abundance of the target transcript (a gene or an exon) needs to be derived from the intensities of the probes designed specifically for this transcript. For example, for GeneChip arrays which include the mismatch probes, the gene expression level needs to be calculated from the intensities of the 11–20 probe pairs of the probe set designed for the gene. The resulting expression level of the gene may be associated with the qualitative *detection call* (P – present, A – absent, M – marginally present).

Low-level preprocessing of Affymetrix arrays consists of three main steps:

- background adjustment;
- normalization;
- summarization at the probe set level.

[15]Illumina uses fabric optic bundles in its 96-sample Array Matrix format. The BeadChip format uses planar silica slides (Illumina 2008a).

[16]To compare two or more populations of samples we need multiple arrays. For the multichannel (multi-color) microarrays, more than one sample is processed on a single array. For the synthesized oligonucleotide arrays (such as Affymetrix GeneChip arrays), one array is used for one sample.

[17]More precisely, in a *.DAT file where * denotes the file name and DAT is the file name extension.

TABLE 2.3: Main Characteristics of Selected Methods for Low-Level Preprocessing of Affymetrix Arrays

Method	Background correction	Summarization	Normalization	Notes
MAS5	Subtracts spatially adjusted background PM–IM	Robust average (one-step Tukey's biweight)	Scaling arrays to the same trimmed mean value	Single array analysis Calculates detection calls Limited ability to detect small changes between arrays
RMA	Global correction based on a convolution model MM probes ignored	Robust multiarray summarization fitting a linear model	Quantile normalization	Multiple array analysis Global background adjustment does not account for probe affinity
GCRMA	Accounts for probe affinity using probe sequence information Uses MM probes	Robust multiarray summarization fitting a linear model	Quantile normalization	Multiple array analysis
PLIER	Accounts for probe affinity Fits multiplicative model to PM–MM, but also assumes that MM error is the reciprocal of PM error		Quantile normalization	Multiple array analysis

Furthermore, quality assessment is a very important task at all levels of data preprocessing (this will be covered in the next section).

The main goals of the *background adjustment* step are: correction for background noise and adjustment for nonspecific binding. There are indications that the background adjustment step has the largest effect on the preprocessing results (Irizarry et al. 2006). *Normalization* aims at reducing the nonbiological variation (both within and between arrays) and is usually performed under the assumptions that only a relatively small number of genes are differentially expressed and that they are equally likely to be under- or over-expressed. Once the probe-level signal intensities are determined, the *summarization* step combines them into a single expression level for a probe set.

There are many methods for the low-level preprocessing of Affymetrix arrays and new methods are constantly reported. However, we have yet to see a single preprocessing method that would be widely accepted as the best. "*Conflicting reports have been published comparing the more popular methods*" (Irizarry et al. 2006). A likely reason for this situation is that the advantages and disadvantages of particular approaches are not necessarily invariable; they may depend on idiosyncrasies of the processed data set. It is possible that the "best" method will have a collection of approaches and will dynamically select those that are most appropriate for a particular data set. (Not to add to the user's confusion, the assessment of the data and the selection of the approach should be done automatically.) Benchmarks have been developed specifically for the comparison of algorithms for low-level preprocessing of Affymetrix chips (Cope et al. 2004; Irizarry et al. 2006). Although they are very valuable, we should remember that comparisons performed for specific data sets and specific arrays are not necessarily generalizable (Irizarry et al. 2006). Nevertheless, it seems that achieving a balance between precision and accuracy is important in the overall performance of an algorithm (Seo and Hoffman 2006). There are indications that algorithms implementing a probe-specific background correction (such as GCRMA or PLIER) provide a good balance of accuracy and precision (Irizarry et al. 2006).

The three currently most popular preprocessing methods (ordered by the time of their introduction) are: *MAS5* (Affymetrix 2001; Affymetrix 2002), *RMA* (Irizarry, Hobbs et al. 2003) and *GCRMA* (Wu et al. 2004). The MAS5 method is the oldest among these and seems to have gradually being phased out by *PLIER* (Affymetrix 2005b)—a newer method from Affymetrix. Table 2.3 summarizes the main characteristics of these methods which are discussed in the following sections.

Some data miners (or bioinformaticians) have a tendency to treat initial preprocessing as a "black box," and assume that the real analysis starts when the gene expression level for each gene has already been calculated. Although one could understand this approach, in real research projects it is important to know how the gene expressions were calculated. Improper preprocessing methods (or their assumptions or parameters) may have a significant impact on the results of an entire study.

2.3.1 MAS5

The *MAS5* algorithm (Affymetrix 2001; Affymetrix 2002) performs the background adjustment and summarization steps independently for each array. Then all the arrays are scaled to have the same trimmed mean expression level. The MAS5 algorithm is included in the *Affymetrix Expression Console* software (Affymetrix 2006).

MAS5 Background Adjustment

Adjustment for Background

To estimate the background intensity, MAS5 divides each array into K equally sized rectangular zones, with $K = 16$ as the default number of zones. For each zone k, $k = 1, \ldots, K$, the mean and the standard deviation of the lowest two percent of the probe intensities in the zone are used to estimate the zone's mean background b_k and noise n_k (the background variability). Subsequently, for each probe on the array the weighted averages of all b_k and n_k values are calculated,

$$b(x, y) = \frac{\sum_{k=1}^{K} w_k(x, y) b_k}{\sum_{k=1}^{K} w_k(x, y)}, \tag{2.1}$$

$$n(x, y) = \frac{\sum_{k=1}^{K} w_k(x, y) n_k}{\sum_{k=1}^{K} w_k(x, y)}, \tag{2.2}$$

where

- (x, y) defines the location of the probe on the array,
- $w_k(x, y)$, $k = 1, \ldots, K$, are the weights based on the squared Euclidean distance $d_k^2(x, y)$ between the probe location (x, y) and the center of each zone k,

$$w_k(x, y) = \frac{1}{d_k^2(x, y) + \varsigma}, \tag{2.3}$$

with a smooth factor ς (the default value is $\varsigma = 100$).

The intensity $I(x, y)$ of each probe on the array is adjusted by subtracting the local background value $b(x, y)$ calculated for the probe. To avoid negative intensities, the adjusted intensity cannot be less than a fraction φ of the noise value $n(x, y)$ calculated for the probe location (with a default value of $\varphi = 0.5$),

$$I_{adjusted}(x, y) = \max\,[I(x, y) - b(x, y), \varphi n(x, y)]. \tag{2.4}$$

MAS5 Summarization and Detection Calls

Adjustment for Nonspecific Binding

The previous version of this tool (MAS4) operated under the assumption that most of the signal coming from the mRNA hybridized to the PM probes will correspond to the gene targeted by the probes, and that the amount of mRNA only partially bound to the sequence of the PM probes (the nonspecific binding) will be represented by the amount of mRNA bound to the MM probes. The MAS4 algorithm (not used anymore) calculated the signal simply as PM−MM. There was a problem with this approach—some twenty to thirty percent of the MM signals were greater than their corresponding PM signals resulting in data sets with negative intensities. To avoid negative intensities at the probe set level, the MAS5 algorithm introduces the *Ideal Mismatch* IM, which is never greater than PM. For each probe pair l of probe set k, the value of the ideal mismatch signal IM_{kl} is determined in the following way:

$$IM_{kl} = \left\{ \begin{array}{ll} MM_{kl} & \text{when } MM_{kl} < PM_{kl} \\ \dfrac{PM_{kl}}{2^{SB_k}} & \text{when } MM_{kl} \geq PM_{kl} \text{ and } SB_k > \tau_c \\ \dfrac{PM_{kl}}{2^{\left(\frac{\tau_c}{1+(\tau_c - SB_k)/\tau_s}\right)}} & \text{when } MM_{kl} \geq PM_{kl} \text{ and } SB_k \leq \tau_c \end{array} \right\} \tag{2.5}$$

where

- SB_k is the *specific background ratio* representing the average ratio of the PM signal to the MM signal for probe set k consisting of n_k probe pairs l, $l = 1, \ldots, n_k$. One-step Tukey's biweight algorithm T_{BW} is used to calculate the robust average unaffected by outliers (Affymetrix 2002),

$$SB_k = T_{BW}(\log_2 PM_{kl} - \log_2 MM_{kl}, \quad l = 1, \ldots, n_k), \tag{2.6}$$

- τ_c is the contrast parameter used to decide whether the SB_k ratio is large enough to use it for the IM_{kl} calculation (the default value is $\tau_c = 0.03$),
- τ_s is the scale parameter used to assign to IM_{kl} a value slightly less than PM_{kl} (the default value is $\tau_s = 10$).

The first case in (2.5) represents the best situation when the ideal mismatch IM is probe-pair-specific (and equal to MM). In the second case, the IM is not specific to the probe pair, but is based on probe set-specific information. The goal of the third case is to set IM to a value slightly less than the PM.

Once the ideal mismatch IM_{kl} is calculated, it is subtracted from the corresponding PM_{kl} intensity. The result represents the adjusted intensity for probe pair l of probe set k.

Summarization

To summarize the probe pair intensities into a single expression value for probe set k, one-step Tukey's biweight estimator of the robust mean[18] is calculated on the $\log_2$ scale,

$$\log_2 signal(k) = T_{BW}(\log_2(PM_{kl} - IM_{kl}), \quad l = 1, \ldots, n_k). \tag{2.7}$$

However, the expression values reported by MAS5 are in the original scale rather than in the $\log_2$ scale.

Detection Calls

The MAS5 probe set signals are coupled with detection calls (Absent, Marginal, and Present) based on the probability that a gene is expressed (or, more precisely, on the probability that the target gene is expressed and its expression level can be reliably determined). For probe set k, the Wilcoxon signed rank test is used to calculate p-value for the following one-tailed hypothesis test (Affymetrix 2002):

$$\begin{aligned} H_0&: \ median(R_l - \tau) \leq 0 \\ H_a&: \ median(R_l - \tau) > 0 \end{aligned} \tag{2.8}$$

where

- R_l is the discrimination score for probe pair l of the probe set,[19] calculated on the raw intensity values as

$$R_l = \frac{PM_l - MM_l}{PM_l + MM_l}, \tag{2.9}$$

[18]This method first calculates the median of the n_k adjusted probe intensities $\log_2(PM_{kl} - IM_{kl})$. The distance of each probe intensity from the median is used to determine the probe contribution to the average. Data points that are far from the median (potential outliers) contribute less to the average.

[19]Saturated probe pairs are excluded from the detection call analysis. If all probe pairs of a probe set are saturated, the probe set is assigned the *Present* call and the p-value is set to zero.

- τ is the threshold parameter (the default value is $\tau = 0.015$) used to prevent false calls when PM is only slightly greater than the MM.[20]

To make the detection call for the probe set, the p-value calculated for the probe set is compared to two significance levels, α_1 and α_2,

$$DetectionCall = \left\{ \begin{array}{ll} \text{Present} & \text{if } p\text{-value} < \alpha_1 \\ \text{Marginal} & \text{if } \alpha_1 \leq p\text{-value} < \alpha_2 \\ \text{Absent} & \text{if } p\text{-value} \geq \alpha_2 \end{array} \right\}. \tag{2.10}$$

The significance levels α_1 and α_2 are user-adjustable parameters with default values reported in (Affymetrix 2002) of $\alpha_1 = 0.04$ and $\alpha_2 = 0.06$. In the Affymetrix Expression Console software (Affymetrix 2006) these default values are set to $\alpha_1 = 0.05$ and $\alpha_2 = 0.065$.

A present call (P) means that the gene is expressed. An absent call (A) means that the gene is either not expressed or that the amount of target could not be reliably determined.

There is no direct relation between the signal level and the detection call (if there were, we would not need detection calls). Therefore, detection calls can be interpreted as the probability that a gene is expressed (at any level) and also as the reliability of the signal measurement.

MAS5 Normalization

The MAS5 normalization procedure is limited to scaling each array to the same trimmed mean. The scaling is performed after summarization. First, the trimmed mean $\bar{x}_{trim}(i)$ is calculated for each array i by excluding the lowest two percent and highest two percent of the probe set expression values and averaging all the remaining intensities. Then each array is scaled to the target signal T (the default value is $T = 500$) by multiplying each expression on array i by the scaling factor $s(i)$ calculated for the array as

$$s(i) = \frac{T}{\bar{x}_{trim}(i)}. \tag{2.11}$$

By analyzing each array independently, the MAS5 algorithm cannot take into consideration probe-specific affinities across arrays. This reduces the algorithm's ability to detect small changes between arrays (and between the differentiated phenotypes) when compared to multichip approaches. The Affymetrix *Expression Console* User Guide (Affymetrix 2006) suggests that "*The primary use of the MAS 5.0 algorithm is to obtain a quick report regarding the performance of the arrays and to identify any obvious problems before submitting the final set of arrays to one of the multichip analysis methods (RMA, PLIER).*" A short description of PLIER is in Section 2.3.4. Detection calls are, however, to the advantage of the MAS5 algorithm since they are very useful in filtering out probe sets with unreliable measurements.

[20]Probe pairs with $|PM - MM| < \tau$ are not used for the detection call analysis.

2.3.2 RMA

The *Robust Multichip Analysis* (RMA) method (Bolstad et al. 2003; Irizarry, Bolstad et al. 2003; Irizarry, Hobbs et al. 2003; Bolstad 2007, 2008), performs a background adjustment using only the PM probe intensities, and then performs quantile normalization and a robust multichip summarization.

RMA Background Adjustment

RMA background adjustment is performed under the following assumptions:

- each array has a common mean background,
- the MM mismatch probes are ignored,
- the observed perfect match intensity Y is modeled as a convolution of the exponentially distributed signal S and the normally distributed background B,

$$Y = S + B, \tag{2.12}$$

  where

  - S and B are independent random variables,
  - the S distribution is $\exp(\alpha)$,
  - the B distribution is $N(\mu, \sigma^2)$, truncated at 0 to avoid negative background noise values.

The background-adjusted estimate of the true signal is given by the expected value of S given the observed value of Y,

$$E(S|Y = y) = a + b\frac{\phi\left(\frac{a}{b}\right) - \phi\left(\frac{y-a}{b}\right)}{\Phi\left(\frac{a}{b}\right) + \Phi\left(\frac{y-a}{b}\right) - 1}, \tag{2.13}$$

where

- $a = y - \mu - \sigma^2\alpha$
- $b = \sigma$
- $\Phi(\ )$ denotes the standard normal distribution function
- $\phi(\ )$ denotes the standard normal density function.

According to Bolstad (Bolstad 2004, 2007), for most Affymetrix data sets, we may assume $\phi(\frac{y-a}{b}) \approx 0$ and $\Phi(\frac{y-a}{b}) \approx 1$, and so Equation 2.13 can be simplified to

$$E(S|Y = y) = a + b\frac{\phi\left(\frac{a}{b}\right)}{\Phi\left(\frac{a}{b}\right)}. \tag{2.14}$$

The RMA background adjustment procedure estimates the model parameters α, μ, and σ, and replaces the probe PM intensities Y with their background-corrected values S.

RMA Quantile Normalization

The RMA algorithm assumes that the probe-level signal intensities should be similarly distributed over all biological samples of an experiment.[21] Quantile normalization is performed to adjust the intensities in a way that they are identically distributed for all the arrays. Assume that we have probe-level intensity data for N microarrays and L probes on each array. The data can be represented by an $N \times L$ matrix with N columns representing biological samples and L rows representing probes. The quantile normalization algorithm may be described in the following way:

- Independently sort each column in ascending order.
- Calculate the mean value for each row of such sorted matrix (note that after sorting, the rows represent quantiles of the intensity distribution rather than intensities for one probe).
- Replace all values in a row by the mean value calculated for the row.
- Re-sort each column to its original order.

This kind of quantile normalization is referred to as *complete* data normalization because it uses all the experiment arrays and does not need a reference array.

RMA Robust Multiarray Summarization

Once the probe intensities are corrected for background and then normalized, they need to be summarized to a single expression value for each probe set of each array. The RMA summarization procedure first $\log_2$ transforms the normalized probe intensities, and then, for each probe set, fits the following model (Irizarry, Hobbs et al. 2003; Bolstad 2007):

$$y_{li} = \mu_i + \alpha_l + e_{li}, \tag{2.15}$$

where

- y_{li} is the background-adjusted, normalized and $\log_2$ transformed intensity value for probe l on array i,
- μ_i represents the array effect (the $\log_2$ scale expression level of the probe set for array i),
- α_l represents the probe affinity effect for probe l in the probe set,
- e_{li} is the error term (the residual for the probe l on array i).

The RMA summarization procedure uses the median polish algorithm to fit the model (2.15) subject to the constraint imposed for identifiability,

$$\sum_l \alpha_l = 0. \tag{2.16}$$

For each probe set, the starting point is the matrix of y_{li}'s with row l representing probe l of the probe set and column i representing array i. The results are the estimates

[21]This assumption is not necessarily true for genes expressed at higher levels. Furthermore, forcing the distributions to be equal introduces additional noise.

of the μ_i values, with each μ_i representing the probe set expression for array i. The algorithm can be described in the following way:

- Repeat
    - Calculate the median for each row (probe) and subtract it from the row values.
    - Calculate the median for each column (array) and subtract it from the column values.

  until all row and column medians are zero or the changes are small. The resulting matrix is the matrix of residuals.[22]
- Subtract the residual matrix from the original matrix of y_{li}'s. The result is the matrix of fitted values.
- Calculate the means of the column values in the fitted matrix. The means are the probe set expression values for the arrays represented by the columns.
- Repeat all the above steps for each probe set.

Since RMA performs the global background adjustment, it does not account for the fact that different probes may have different susceptibility for nonspecific hybridization. This may result in underestimating the fold changes for the low expressed probe sets. The GCRMA algorithm is a modification of RMA, which addresses this problem.

2.3.3 GCRMA

The *GCRMA* low-level preprocessing method (Wu et al. 2004; Wu and Irizarry 2005) is a version of RMA that makes use of probe sequence information (stronger bonding of GC pairs) at the background adjustment step. The normalization and summarization steps of GCRMA are the same as for RMA.

GCRMA Background Adjustment

The GCRMA algorithm uses both the PM and MM signals and assumes that—for any probe pair on an array—the observed signals may be described by the following model:

$$\begin{aligned} PM &= O + N_{PM} + S \\ MM &= O + N_{MM} \end{aligned} \tag{2.17}$$

where

- O represents the optical noise, which is assumed to follow the normal distribution $N(\mu_O, \sigma_O^2)$,
- N_{PM} and N_{MM} represent the nonspecific binding; it is assumed that $\log(N_{PM})$ and $\log(N_{MM})$ follow a bivariate normal distribution with means μ_{PM} and μ_{MM}, and equal variances σ_N^2,
- S represents the target expression signal.

[22]The residuals can be used for quality control.

To perform background adjustment, GCRMA estimates the model parameters under the following assumptions:

- The optical noise and nonspecific binding are independent.
- $\sigma_O^2 \ll \sigma_N^2$, thus the optical noise is approximately constant and can be estimated by the minimum observed intensity on the array.
- $\log(N_{PM})$ and $\log(N_{MM})$ have a correlation of 0.7 across probe pairs.

The mean values μ_{PM} and μ_{MM} of the non-specific binding depend on the probe sequence and are modeled as a smooth function of the probe affinity α, with α calculated as the sum of the position-related base effects,

$$\alpha = \sum_{k=1}^{25} \sum_{l \in \{A,C,G,T\}} \mu_l(k)\, \mathbf{1}_{b_k=l}, \tag{2.18}$$

where

- k denotes the position along the 25-base long sequence of the probe,
- l denotes the base letter,
- b_k is the base at position k,
- $\mathbf{1}_{b_k=l} = \begin{Bmatrix} 1 & \text{when } b_k = l \\ 0 & \text{otherwise} \end{Bmatrix}$,
- $\mu_l(k)$ is the contribution of base l in position k to the probe affinity.

In practice, the $\mu_l(k)$ can be estimated from all the data of an experiment or may be hard-coded in a particular implementation of the GCRMA algorithm (Bolstad 2008). Once the parameters of the model are estimated, the probe intensity can be calculated as the expected value of S given the observed signals PM and MM.

2.3.4 PLIER

According to Affymetrix (Affymetrix 2005b), the *Probe Logarithmic Intensity Error* (PLIER) algorithm builds upon many recently published concepts. Similar to GCRMA, it utilizes probe specific affinity across all arrays and quantile normalization. Like MAS5, it summarizes the weighted probe intensities. Although PLIER makes a counter-intuitive assumption that the error term associated with the MM probe is the reciprocal of the PM error term, it performs quite well when compared to other algorithms (Therneau and Ballman 2008).

2.4 PUBLIC REPOSITORIES OF MICROARRAY DATA

2.4.1 Microarray Gene Expression Data Society (MGED) Standards

The main obstacles for early studies based on gene expression microarray data were related to the availability of large and good quality training and test data sets. As we are dealing with many thousands of variables (probe sets, genes, exons), a proper *sample size* (in the statistical meaning), or *the number of independent biological samples* in the experiment (or in each of the differentiated classes), is very important for the statistical and scientific validity of the results. Other problems were related to the reproducibility of experiments and reusability of their data. Different data formats and often insufficient annotations and descriptions of experiments made it rather difficult to verify and compare the published studies. Furthermore, finding independent (also large enough and of good quality) test data sets was practically impossible without direct collaboration with other groups generating such data for the same area of biomedical research. Please note that the gene expression levels resulting from microarray experiments are not measured in any objective units.[23] Therefore, sufficient information about the data and their processing is crucial for the interpretation and comparison of results and for the integration of data from different experiments (Brazma et al. 2001).

The situation started to change when the Microarray Gene Expression Data Society (MGED, www.mged.org) initiated the development and promotion of standards for storing and sharing microarray based gene expression data and study results. Among such standards are MIAME (*Minimum Information About a Microarray Experiment*), MAGE-ML (*Microarray Gene Expression Markup Language*), and the recently preferable MAGE-TAB—a spreadsheet format for MIAME-compliant microarray experiment information. MIAME is a conceptual standard describing the minimum information content required for proper interpretation and verification of microarray experiments whereas MAGE-ML and MAGE-TAB are standards defining formats of MIAME-compliant data and experiment descriptions.

MIAME

The *Minimum Information About a Microarray Experiment* (MIAME) standard (Brazma et al. 2001; MGED_Society 2008b) requires that the following information be provided for publications based on microarray experiments.

1. The raw data resulting from each microarray image analysis (such as CEL files for Affymetrix arrays).
2. The final data after preprocessing, for instance the results of MAS5 or RMA preprocessing. This should be the gene expression matrix that is analyzed in the study.
3. The essential information about sample annotation and experimental factors. All information necessary for proper interpretation of the experimental results and for eventual replication of the experiment.
4. The experimental design including relationships between samples, microarrays, and data files.
5. A description of the microarray design (such as probe sequence information and its database accession numbers). This is especially important

[23]In the future, we may have technologies measuring the number of copies of each transcript in a cell, but we are not there yet.

for experiments using customized arrays. For standard commercial arrays, such as those by Affymetrix, this information is provided by the manufacturer.

6. The experimental and data processing protocols. This is basically the information usually provided in the Methods section of a paper. Any non-standard protocols should be described in a way that allows an understanding of how the experiment has been performed and how the data have been analyzed.

Although the MIAME standard does not *require* that the data be in any specific format, it does *recommend* the use of either the MAGE-TAB or MAGE-ML formats.

MAGE-ML

The *Microarray Gene Expression Markup Language* (MAGE-ML) is an XML-based data format for sharing MIAME-compliant data and information about microarray experiments (Spellman et al. 2002). MAGE-ML is the XML representation of the microarray gene expression object model (MAGE-OM) developed as a part of the MGED initiative. Although the MAGE-ML format has been used by many tools and databases, it has not been universally accepted, mainly due to its complexity (Rayner et al. 2006). The MAGE-ML format is still used (at the time of this writing), but the newer, spreadsheet-based MAGE-TAB format is recommended as a replacement for the MAGE-ML format.

MAGE-TAB

MAGE-TAB (*Microarray Gene Expression Tabular*) is a simple spreadsheet-based (or, more generally, tab-delimited) format for sharing MIAME-compliant data. It does not require an understanding of XML and can be used instead of the more complex MAGE-ML format. The MAGE-TAB standard defines four types of files necessary to describe a microarray experiment (Rayner et al. 2006):

- Investigation Description Format (IDF). An IDF tab-delimited text file provides general information about the experiment (such as a brief description of the study and its protocols, contact information, and bibliography).
- Array Design Format (ADF). Using a tab-delimited format, an ADF text file describes the design of an array type used in the experiment (for example, information about the probe sequences, their locations on the array and annotations). For standard commercial array types, it may simply provide a reference to array information in a public repository.
- Sample and Data Relationship Format (SDRF). An SDRF file provides all information required by the MIAME standard, which is not included in the other MAGE-TAB files. In particular, it describes experimental design and relationships between samples, arrays, and data.
- Raw and preprocessed data files. The raw data should be provided as binary or ASCII files in their native formats (for example, CEL files for

Affymetrix arrays). The preprocessed data files may be provided either in their native formats or as tab-delimited text files in the *data matrix* format. This data matrix format is similar to our gene expression matrix described in the next section since its rows correspond to genes and columns to biological samples. However, the columns of a MAGE-TAB data matrix have references to objects defined in SDRF files (for example, references to CEL files associated with the column samples).

The MAGE-TAB format is currently recommended by the MGED Society as the best practice approach (MGED_Society 2008a).

2.4.2 Public Databases

Thanks to the MGED Society initiatives including open letters to the scientific journals (Ball et al. 2002, 2004), most scientific journals accepting papers based on gene expression microarray data now require—as a part of the publication process—the submission of the microarray data to one of the public repositories adhering to the MIAME standards. The MGED Society recommends three such repositories: ArrayExpress, Gene Expression Omnibus (GEO) and CiBEX[24] (Center for Information Biology Gene Expression Database at DNA Data Bank of Japan). The GEO and ArrayExpress databases are rapidly growing and already include thousands of microarray experiments.

The availability of well-annotated microarray data in standard formats allows for the easy access, querying and sharing of data. It makes possible integration of data from different studies into larger training sets or finding quality test sets to validate the outcomes of new experiments.

2.4.2.1 Gene Expression Omnibus (GEO)

Gene Expression Omnibus[25] (Barrett et al. 2005, 2007) is a public repository of high-throughput genomic and proteomic data, primarily MIAME-compliant gene expression microarray data. It was established in 2000 at the National Center for Biotechnology Information (NCBI). Experimental data can be submitted to GEO using either interactive web forms or batch deposit of files in such formats as spreadsheets, text SOFT (Simple Omnibus Format in Text) files or MINiML (MIAME Notation in Markup Language) XML files. The submission may be kept confidential until the paper is published. The data are stored in the form of three basic record types:

- *Platform*—description of the array.
- *Sample*—description of a biological sample and results of its hybridization (such as probe set signal levels and detection calls).
- *Series*—description of the experiment performed on a group of samples.

[24]http://cibex.nig.ac.jp

[25]www.ncbi.nlm.nih.gov/geo/

Based on the submitted experiments (series), GEO curators organize data into higher level objects represented by the DataSet and Profile record types:

- *DataSet*—a collection of biologically comparable samples that were processed on the same platform and whose measurements are the results of processing and calculations consistent across the DataSet.
- *Profile*—a gene-centered view of the DataSet; expression level of a single gene across all DataSet samples.

2.4.2.2 ArrayExpress

ArrayExpress[26] (Parkinson et al. 2007, 2009) is a public database of microarray experiments and gene expression profiles that was established in 2002 at European Bioinformatics Institute (EBI). ArrayExpress has currently three components:

- *ArrayExpress Repository*, which is a MIAME-compliant database of microarray data, such as the original data supporting publications. Pre-publication data can be submitted as private and then made publicly available as soon as the paper is published. New experiments can be submitted to the repository either via the online *MIAMExpress* tool or uploaded as spreadsheets (MAGE-TAB is the preferred spreadsheet format). The user interface allows for efficient browsing and querying of the repository data. The results may be filtered and sorted. Selected raw or preprocessed data may be downloaded. For example, to retrieve data in a form similar to our gene expression matrix, but with both signal and detection call information (if available), we would select the *Quantitation Types* options corresponding to the signal and detection calls and then the sequence identifier under the *Design Element Properties.* If we save the displayed matrix as a text file, it can be opened in Excel.
- *ArrayExpress Warehouse*, which is a database of gene-indexed expression profiles selected from the ArrayExpress repository. The selection is based on MIAME-compliance and the quality of data annotations. The selected experiments are re-annotated and curated for consistency with the warehouse environment. Users can query the warehouse profiles by gene name, accession numbers or other annotations.
- *ArrayExpress Atlas*, which is a new summary database allowing for querying the curated and ranked gene expression data across multiple experiments and conditions. Search results are linked to more detailed gene and experiment information in the Warehouse and to raw data in the Repository. Since the Atlas database is an extension of the Warehouse of gene expression profiles, we may expect that the two tools will be combined into a single ArrayExpress component.

2.5 GENE EXPRESSION MATRIX

After low-level preprocessing of microarray data, gene expression data can be presented as a matrix with N columns representing samples and p rows representing

[26]www.ebi.ac.uk/microarray-as/ae/
www.ebi.ac.uk/microarray-as/aer/entry

probe sets (see Table 2.4). If necessary, meta level data—such as assignment of samples to classes—may be incorporated within the matrix or supplied as a separate file.[27]

Samples

Columns of the gene expression matrix represent *biological samples*. The data for each biological sample (one column) come from one microarray. In statistics, the term *sample* has a well-defined meaning—a *statistical sample* would refer to the entire group of *biological samples* selected from one of the investigated populations. However, when we are using statistics in such an interdisciplinary area as genomics, the term *sample* has a different—though also well-defined—meaning for biomedical researchers. To communicate efficiently, we will use the term *sample* as corresponding to a *biological sample* rather than in its statistical meaning, hence ***sample* = *biological sample.***

Variables

Rows of the gene expression matrix represent variables—probe sets or genes.[28] Names like *1053_at*, *1255_g_at* or *91952_at* are Affymetrix probe set IDs, which correspond to genes (annotations for probe sets can be downloaded from www.affymetrix.com). Table 2.5 shows an example of probe set annotations.

Classes

Classes represent different phenotypes, diseases, or—more generally—states we want to differentiate. In statistics, they would be called *populations*. Take the following examples:

- Two classes: Class 1—patients with breast cancer, Class 2—healthy women.
- Six classes: Six populations of patients with six subtypes of acute lymphoblastic leukemia.

Class information is crucial for supervised learning, but not necessary or not available for unsupervised learning.

Each data value (a cell in the gene expression matrix) represents the expression level of a gene (row) in a sample (column). This single number for a probe set/sample combination represents the abundance of the mRNA target[29]—in the sample—corresponding to the gene represented by the probe set.[30] The number by

[27]The gene expression matrix itself does not have to have this meta-data information about assignment of samples to classes. It is common practice to include the class name (disease, state, cell line, etc.) in the sample name. Another option is to use a separate text file with the meta-data.

[28]Here we are focusing on the gene-level analysis of gene expression data. When exon arrays are used to investigate the expression of different splice variants, probe sets correspond to exons, which in turn correspond to genes.

[29]Technically, the target of oligonucleotide arrays is cRNA, but it represents mRNA. The gene is represented *after* introns are removed (you may also look up *ORF* definition). The oligos printed on the array are selected to be specific for the targeted gene (since the oligos are only 25-base pair long, a gene may be represented by several different oligos for greater specificity).

[30]After low-level preprocessing, one expression level is calculated for all probes of a probe set. Since the probe set represents a gene, this expression level can be called the *gene's signal.*

TABLE 2.4: Gene Expression Matrix: *N* samples (columns), *p* variables (rows), and *J* classes

	Class 1							Class 2	***		Class *J*	
	Sample 1	Sample 2	Sample 3	Sample 4	Sample 5	***	***	***	***	***	Sample $N-1$	Sample N
1053_at	92.1	52.4	74.4	63.9	75.8						114.9	122.8
117_at	14.9	26.6	45.1	40.8	40.9						27.0	40.9
121_at	313.9	230.8	359.0	414.1	525.0						243.8	351.4
1255_g_at	23.7	32.0	40.4	29.4	10.0						14.7	4.5
1294_at	93.9	187.2	257.4	185.3	164.2						328.9	444.0
1316_at	58.1	42.3	94.1	66.9	46.9						45.1	36.7
***												
***												
***												
91684_g_at	73.5	80.3	96.5	39.8	63.3						68.1	30.6
91703_at	113.3	9.9	177.0	257.7	7.1						65.8	9.5
91816_f_at	36.8	42.3	10.8	56.9	21.6						27.9	38.1
91826_at	73.3	65.7	32.2	15.1	17.9						5.0	18.1
91920_at	105.2	137.7	195.1	153.2	154.3						142.3	128.6
91952_at	152.4	403.6	144.8	150.4	76.1						92.0	100.4

The first column identifies the variables (probe sets representing genes). Generally, only one row at the top of the table is required to identify samples. For supervised learning, it is convenient to add one more row defining the assignment of samples to classes (which means that the class name, for example "Class 1" here, will be repeated in the first row of the columns corresponding to samples belonging to this class). Alternatively, this meta-data information may be supplied in a separate file.

TABLE 2.5: Example of Probe Set Annotations

Probe Set ID	Representative Public ID	UniGene ID	Gene Title	Gene Symbol	Chromosomal Location	Entrez Gene	SwissProt	RefSeq Transcript ID	Gene Ontology Biological Process	Gene Ontology Cellular Component	Gene Ontology Molecular Function
1053_at	M87338	Hs.647062	replication fac	RFC2	chr7q11.23	5982	B5BU07 /// E	NM_002914	0006260 // D	0005634 // nu	0000166 //
1316_at	X55005	Hs.724	thyroid hormo	THRA	chr17q11.2	7067	A8K206 /// A	NM_003250	0001502 // c	0005634 // nu	0003677 //
1494_f_at	M33318	Hs.250615	cytochrome P	CYP2A6	chr19q13.2	1548	A6NE97 /// A	NM_000762	0055114 // o	0005783 // er	0004497 //
1729_at	L41690	Hs.460996	TNFRSF1A-a	TRADD	chr16q22	8717	B2RDS3 ///	NM_003789	0006915 // a	0005737 // cy	0004871 //
200595_s_at	NM_003750	Hs.523299	eukaryotic tra	EIF3A	chr10q26	8661	B1AMV5 ///	NM_003750	0006412 // tr	0005634 // nu	0003743 //
200600_at	NM_002444	Hs.87752	moesin	MSN	chrXq11.2-q12	4478	A8MZ70 /// F	NM_002444	0006928 // c	0001931 // ur	0005102 //
81811_at	AI744451	Hs.675200	—	—	—	—	—	—	—	—	—
89948_at	AI743331	Hs.716563	PDX1 C-termi	PCIF1	chr20q13.12	63935	Q9H4Z3	NM_022104	—	0005634 // nu	0005515 //
91617_at	AI028241	Hs.643452	DiGeorge syn	DGCR8	chr22q11.2	54487	B2R8G1 /// (	NM_022720	0031053 // p	0005622 // int	0003723 //
91682_at	AI571298	—	—	—	—	—	—	—	—	—	—
91684_g_at	AI571298	Hs.632041	exosome con	EXOSC4	chr8q24.3	54512	Q9NPD3	NM_019037	0006364 // rR	0000178 // ex	0000175 //
91952_at	AI363375	Hs.443636	hypothetical p	LOC90379	chr19p13.12	90379	Q66K64	NM_138353	0006511 // u	—	—

itself is not that important, what is important is the relative expression level of a gene in different samples.

2.5.1 Elements of Gene Expression Microarray Data Analysis

Experienced data miners are aware of the fact that there is no automatic data mining. Properly performed data mining projects have to include input from experts. Data

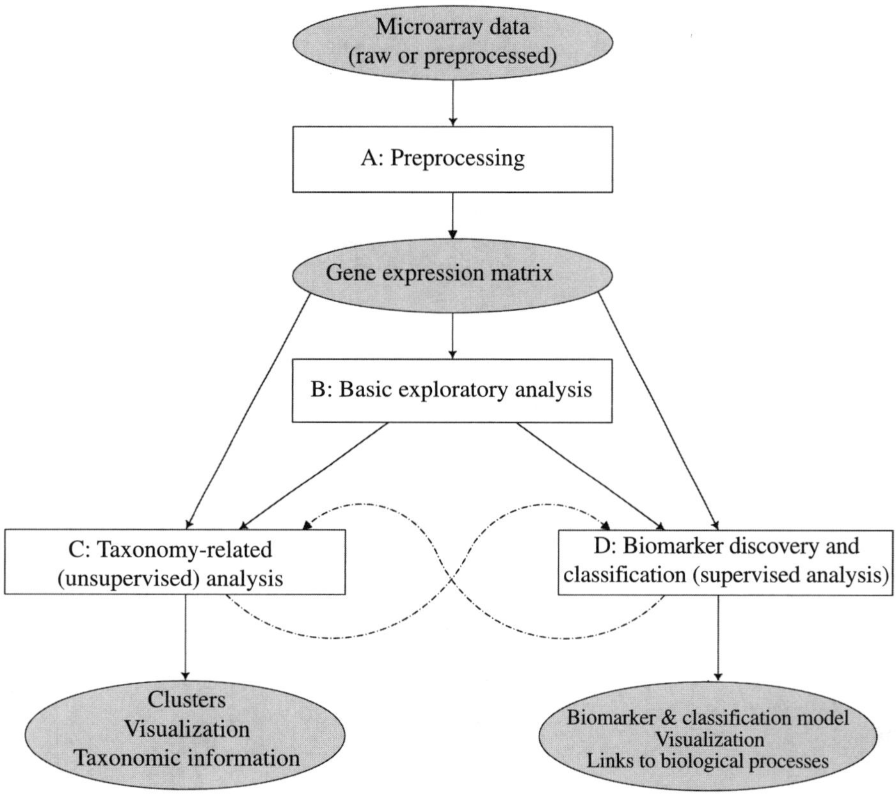

Figure 2.1: Elements of microarray gene expression data analysis—an example. Input data can be in a raw form (e.g., the CEL files resulting from microarray image analysis) or in a preprocessed form (e.g., probe set level expression data downloaded from a public repository). Additional preprocessing, quality control, and filtering is usually necessary depending on the requirements of the study. The gene expression matrix is the basic data form used for either supervised or unsupervised analysis. Basic exploratory analysis is commonly performed to gain some information and feeling about the data. Unsupervised learning is the main approach for studies seeking new taxonomic information. For supervised studies, unsupervised analysis is sometimes performed as a part of the exploratory data analysis (the dashed line from element C to D). However, unsupervised analysis should *not* be used for variable selection or dimensionality reduction preprocessing for supervised analysis because it will most likely result in the discarding of important discriminatory information. The dashed line from element D to C represents the use of unsupervised analysis to find associations between genes identified during the supervised analysis (for instance, to cluster genes selected into the *Informative Set of Genes*—see Chapter 4).

mining enables the finding of patterns in data, but which patterns are found and what research value they possess depends on the way we formulate the research problem and whether we use appropriate data mining methods. Consequently, there is no single workflow common for all studies based on gene expression data. The selection of methods used in a study depends on its scientific goals as well as on the properties of the available data. Some initial steps (like *low-level* preprocessing and *gene expression level* preprocessing described below) are, however, always performed. Figure 2.1 shows the most common elements of microarray data analysis. After preprocessing (the element A), any of the elements (or their sequences) may be performed.

2.6 ADDITIONAL PREPROCESSING, QUALITY ASSESSMENT, AND FILTERING

> *"Data pre-processing does not mean that the data should be tortured until they confess. As a general rule of thumb, the analyst should pre-process the data as little as possible and as much as necessary."*
>
> —(Berrar et al. 2007)

Depending on the source of our data, as well as on the methods and scope of their low-level preprocessing, additional preprocessing at the gene (probe set) expression level may be necessary. If the data come from our own microarray experiment, it is likely that all of the necessary preprocessing and at least some quality control have already been done at the low-level preprocessing step. However, if we start our analysis at the gene expression level (e.g., after downloading the gene expression matrix from one of the public repositories), it is a good idea to check whether the data conform to the quality requirements of our particular study goals and methods as well as whether any additional preprocessing is necessary. It is recommended that data miners do not treat the gene expression matrix data as the result of some "black box" process, but be familiar with the advantages and disadvantages of different preprocessing approaches. Sometimes, if we use publicly available data and the goals or approaches of our analysis are different from those originally published, it may be better to start with the raw data (the CEL files) and perform our own low-level preprocessing. In any case, additional preprocessing may include normalization, cross-chip scaling, or transformation, and it should include gene expression level quality control.

Normalization

If performed at this level, the goals of normalization are the same as for low-level preprocessing—to reduce effects of systematic technical and experimental variation.

Cross-Chip Scaling

Scaling means global linear normalization. If we use arrays from different experiments (although with the same microarray type and the same or similar preprocessing), we would like to scale their expressions to the same level, for instance to the same value of a trimmed mean. The scaling should be done using the original expression levels rather than their log values. Often this is done when we include an independent test data set in the experiment.

Transformation

This usually means logarithmic transformation. *To transform or not to transform?* There have been long discussions about blanket logarithmic transformation of microarray data. Currently, the common practice is to log-transform the microarray data, and recently the $\log_2$ transformation is very popular. We need to note, however, that it is unlikely that any single approach would be optimal in all situations. Depending on the data and the assumptions of a particular statistical model, different transformations (or maybe even no transformation at all) may be preferable.

The primary reason for the transformation of microarray data is to stabilize the variation across the range of signal intensities. The general tendency of microarray data is that higher intensities are associated with higher variation. This *heteroskedasticity* of the data violates the assumption of a constant variance, which is made by some popular statistical models (such as analysis of variance). A nonlinear transformation of microarray data may eliminate, or at least reduce, the heteroskedasticity. It is, however, not clear which transformation, if any, could be universally the most appropriate. The often recommended, and widely used, logarithmic transformation stabilizes (or approximately stabilizes) the variance at higher intensities, but it inflates the variance at very low intensities (Durbin et al. 2002; Huang et al. 2004). Theoretically, for each study we could determine the relationship between the measured intensity and its variance, and use this information to identify the most appropriate transformation. A more practical approach would be to restrict our choice to a few transformations and select the one that minimizes some measure of the variance heterogeneity. The generalized logarithmic transformation (Durbin et al. 2002; Huber et al. 2002; Lin et al. 2008) and Box-Cox power transformation (Box and Cox 1964) are among the plausible candidates.

Stabilizing the variance means that multiplicative errors are converted into additive errors.

- A *multiplicative error* means that the standard deviation increases with the intensity and the coefficient of variation (CV) is similar at different intensity levels.
- An *additive error* means that the standard deviation is similar for different intensity levels, thus the CV decreases as the intensity increases.
- The *coefficient of variation* is defined as

$$CV = \left(\frac{\text{standard deviation}}{\text{mean}} * 100\right)\%. \tag{2.19}$$

Many popular low-level preprocessing methods perform logarithmic transformation and return the log-transformed probe set level intensities.

Some methods return the intensity values in their original scale; if we want to work with the logged ones, we would need to perform the transformation as an additional preprocessing step.

Let us also note that the logarithmic transformation is convenient for the interpretation of data generated by two-channel microarrays. The data points are usually represented by the ratio of the intensities measured for the two samples processed simultaneously on a single array. The same convenience applies to situations where we are interested in the ratio of intensities measured on different single-channel arrays. Logarithmic transformation treats the ratios symmetrically, whether their values are greater or less than one. In effect, the ratios are converted into differences. The base 2 logarithms are most convenient when we are particularly interested in the fold changes[31] that are powers of 2. For example, a fourfold change in either direction will be represented by the distance of 2 ($\log_2(4) = 2$ and $\log_2(0.25) = -2$) from the no-change state ($\log_2(1) = 0$). Another reason for the popularity of the base 2 logarithmic transformation could be related to the fact that the intensity values are typically measured on a scale from zero to 64K $-$ 1 (i.e., $2^{16} - 1$).

2.6.1 Quality Assessment

Quality assessment should span all stages of microarray preprocessing—image analysis, probe-level and gene expression-level preprocessing. Its main goals are to identify (and eventually eliminate) low-quality arrays and outliers, and evaluate the quality of the preprocessing steps. Most procedures for quality checking are based on a graphical presentation of the data and their characteristics.

Image Analysis

Visual inspection of the scanned array images can reveal problems like increased or decreased intensities in some regions, nonuniform background, other spatial patterns, defective spots, dust particles, etc. Nonvisual methods also exist for the identification of problems with spatial intensity distribution. For instance, if the low and high intensity spots can be separated solely by their spatial coordinates, the array should be eliminated (Amaratunga and Cabrera 2004). Software procedures for the automatic assessment of individual spot quality can flag problematic spots by evaluating their geometric properties such as shape regularity, uniformity, the ratio of signal area to spot area, or the displacement from the expected grid location (Drăghici 2003).

Box Plots

Box plots (graphical presentations of quartiles, the interquartile range, the range of values, and outliers; see Figs 2.2 and 2.3) are popular for comparing array intensity distributions before and after a preprocessing step. They are especially useful for the evaluation of normalization results. Scatterplots may also be useful if we need

[31]Using fold change as the only criterion for filtering genes is strongly discouraged. For more information on filtering refer to Section 2.6.2.

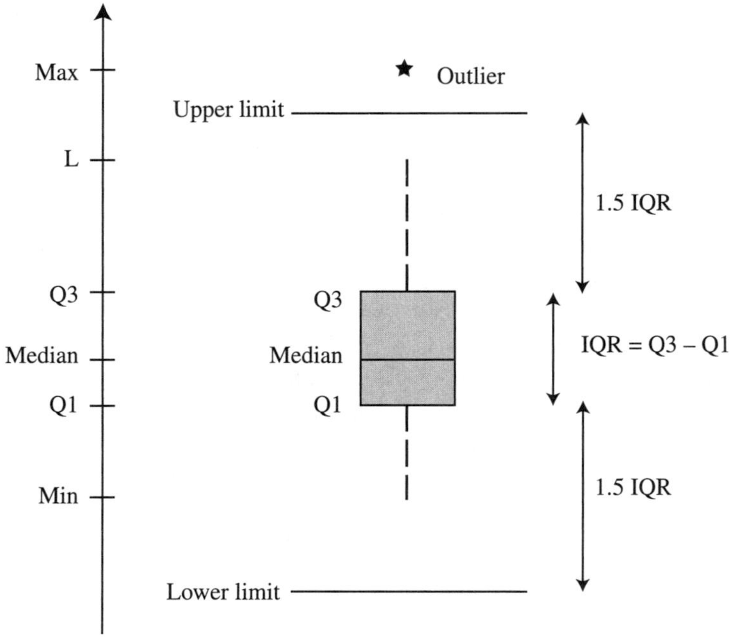

Figure 2.2: Anatomy of a box plot. The box plot is based on the *five-number summary* of a data set: Smallest value (Min), First quartile (Q1), Second quartile (Median), Third quartile (Q3), and Largest value (Max). Since the box is drawn from the first to the third quartile, it contains the middle 50 percent of the data. The *interquartile range* IQR is defined as the difference between the third and first quartiles. *The upper and lower limits* are located 1.5 IQR above Q3 and below Q1 respectively. The dashed lines (*whiskers*) connect the box with the largest and smallest data values that are within the limits. The data values above the upper limit or below the lower limit are considered *outliers*.

to focus on the intensities of a specific array before and after a preprocessing step or to compare two arrays at the same step.

Box plots are also a convenient way to visually compare the distributions of various parameters used for the array quality assessment (such as RLE or NUSE described later in this section). Besides the evaluation of individual arrays, box plots may be used to detect systematic technical problems with the experiment. By coloring boxes (representing arrays) according to experiment or sample conditions, we may be able to identify trends or quality problems associated with such properties like the hybridization date and time, specific groupings of samples, different sample handling, different lab sites, etc. (Brettschneider et al. 2008).

Histograms

Histograms (with bars) or smoothed histograms (with bars and a density trace approximating the empirical density function) can be used to visualize the distribution of gene intensities (across all samples or across the samples of a class) or the distribution of the array intensities (across all genes). Histograms enable the evaluation of assumptions

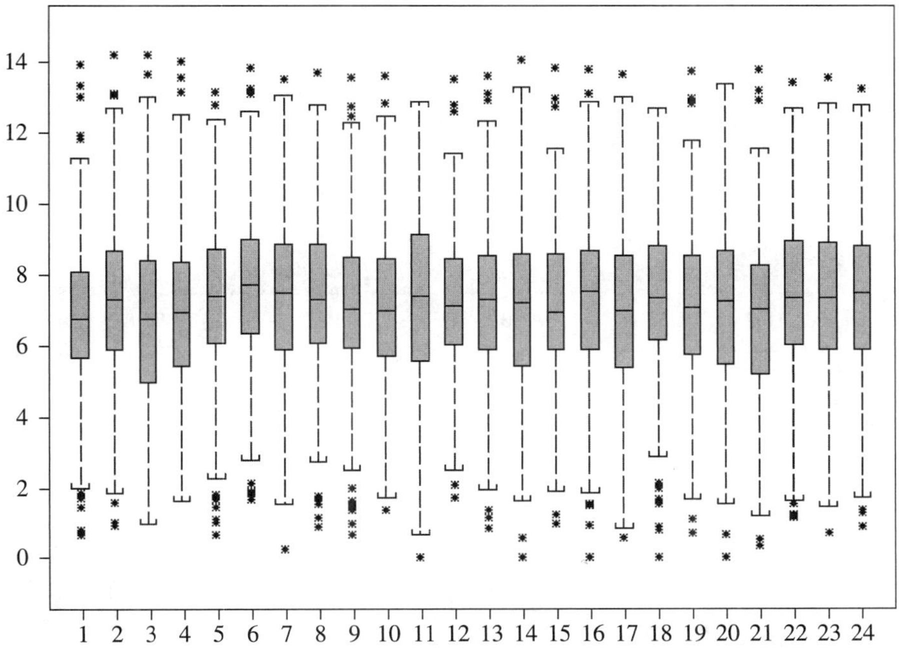

Figure 2.3: An example of box plot. The box plot of expression values of 176 genes of the *Informative Set of Genes* for 24 MLL samples of the training set combining two data sets (for more information about this experiment, see Section 6.6 of Chapter 6).

about a particular intensity distribution or for evaluation of changes in the distribution shape after normalization (see Fig. 2.4).

MA Plots

MA plots can visualize the relationship between the relative $\log_2$ intensity and the mean $\log_2$ intensity (Dudoit et al. 2002). The plots may be used at various stages of preprocessing to detect nonbiological variability, to assess the necessity for normalization or to evaluate the results of normalization. Assume that we want to visualize the relationship between the probe set level intensities of two arrays a_i, $i = 1, 2$ with the p probe sets on each of them. Let the intensities of each array be represented by a p-dimensional vector $\mathbf{x}_i$ with elements x_{ki}, $k = 1, \ldots, p$,

$$\mathbf{x}_i = \begin{bmatrix} x_{1i} \\ x_{2i} \\ \vdots \\ x_{pi} \end{bmatrix}. \tag{2.20}$$

The MA plot is a scatterplot of the p probe set-associated points. The vertical axis M represents the difference between the $\log_2$ intensities (or $\log_2$ of the ratio of the original scale intensities), and the horizontal axis A represents the mean value of the two $\log_2$ intensities. Thus, probe set k, $k = 1, \ldots, p$, is represented by the point with

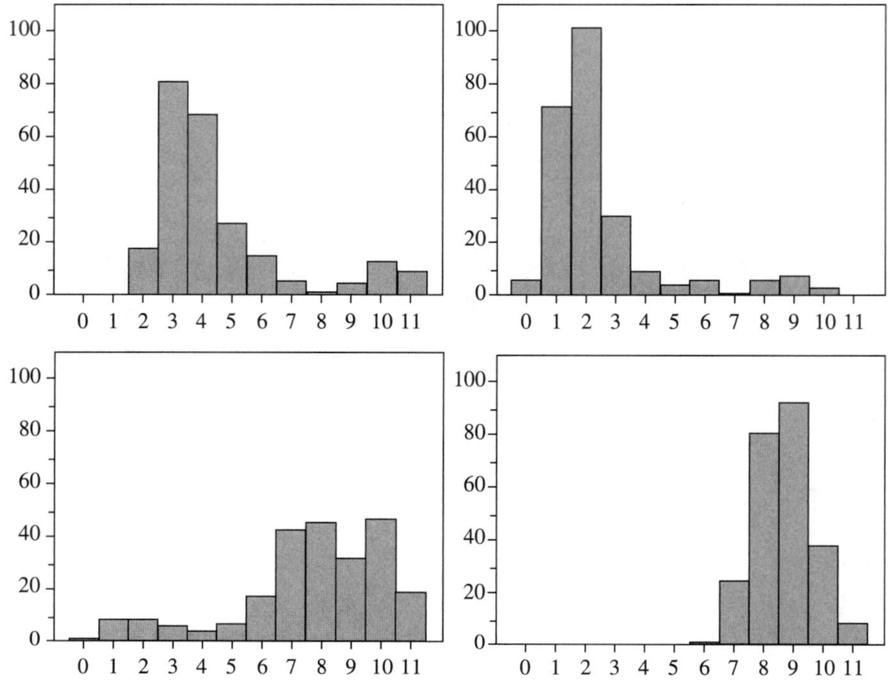

Figure 2.4: An example of histograms showing the distribution of $\log_2$ transformed intensities of four genes across the samples of a training data set. The horizontal axis represents the $\log_2$ transformed expression level of a gene. The vertical axis, the height of each bar, shows the frequency—the number of training samples with the gene expression level falling into a particular bin (the interval of expression values represented by the bin).

coordinates (M_k, A_k), where

$$\begin{aligned} M_k &= \log_2(x_{k1}) - \log_2(x_{k2}) \\ &= \log_2\left(\frac{x_{k1}}{x_{k2}}\right), \\ A_k &= \frac{1}{2}[\log_2(x_{k1}) + \log_2(x_{k2})] \\ &= \log_2 \sqrt{x_{k1} \cdot x_{k2}}\ . \end{aligned} \tag{2.21}$$

For experiments with many arrays, $i = 1, \ldots, N$, such pairwise comparisons of the arrays may be impractical. Instead, we may compare each vector $\mathbf{x}_i$ (each array) to the vector $\mathbf{x}_*$ of the median values calculated over all N arrays (a virtual reference array). The coordinates of the point representing probe set k on such an MA plot for array i are:

$$\begin{aligned} M_k &= \log_2(x_{ki}) - \log_2(x_{k*}), \\ A_k &= \frac{1}{2}[\log_2(x_{ki}) + \log_2(x_{k*})]. \end{aligned} \tag{2.22}$$

Assuming that only a small number of probe sets have significantly different expressions and that a similar number of them are over- and under-expressed, the "cloud" on the MA plot should be centered around the $M = 0$ horizontal line (see Fig. 2.5). A common practice is to show smooth lines on the MA plots representing the quartiles of the M values (Q1, the median, and Q3) over the range of the A values. The lines allow for the quick visual evaluation of the distribution of the $\log_2$ intensities as a function of the A values.

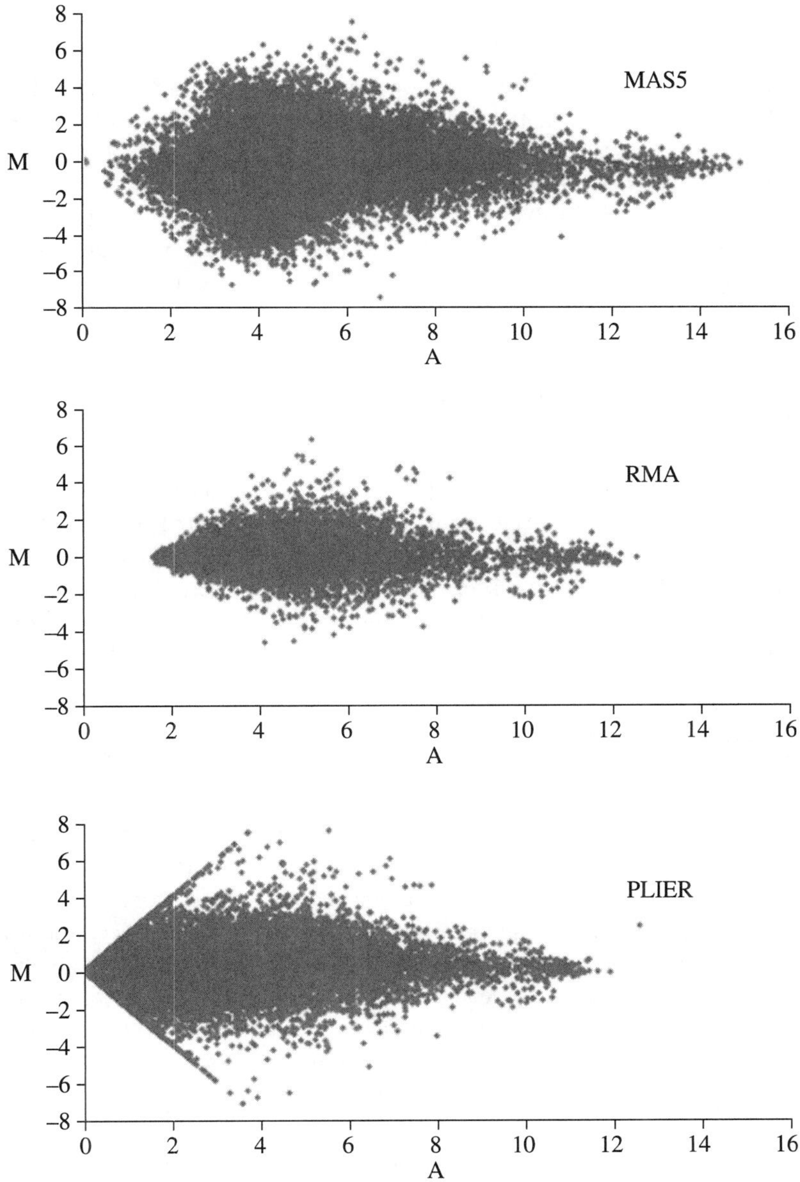

Figure 2.5: An example of MA plots for the same two microarrays selected from the data set preprocessed with three low-level preprocessing methods: MAS5, RMA, and PLIER.

Relative Log Expression (RLE)

The *relative log expression* (RLE) compares the expression level of a probe set on a particular array to the median expression level for this probe set over all arrays. The RLE values correspond to the M values on the MA plot. The box plots representing the distribution of the RLE values on an array can be used to identify low quality arrays—those whose RLE values have a large interquartile range or are not centered around zero (Bolstad 2007).

Normalized Unscaled Standard Error (NUSE)

For preprocessing methods that produce a matrix of residuals (such as the RMA algorithm described in Section 2.3.2), we may use the residuals for quality assessment. The *normalized unscaled standard error* (NUSE) can be calculated for each probe set by normalizing the standard error of the probe set intensities across arrays to a median value equal to one. The distribution of NUSE values on an array can be used for array quality assessment. The box plots of NUSE values that have a higher median (and a larger interquartile range) indicate lesser quality arrays (Brettschneider et al. 2008).

Clustering Samples (or PCA Visualization)

Quality assessment based on clustering samples may be used to check whether the samples group together by their biological characteristics. As with colored box plots, the points representing biological samples may be assigned different colors depending upon conditions of sample processing. Serious quality problems may be indicated when the samples are clustered predominantly by factors other than their biological characteristics (for instance, when they cluster by the time of their hybridization). Similar results may be achieved by visualizing the samples in 2D or 3D space using the first two or three principal components. For more information on clustering and principal component analysis refer to Section 2.8 of this chapter.

2.6.2 Filtering[32]

Once we are satisfied with the quality and the representation of the preprocessed expression data, we usually eliminate some probe sets. The gene expression matrix may include tens of thousands of probe sets. Not all of these data are useful. Although criteria for filtering out some probe sets may differ from study to study, the goals of such filtering are similar—to eliminate the probe sets (i.e., the rows of the gene expression matrix) whose expression measurements are not reliable or represent experimental noise. Many genes in a tissue are not expressed at biologically significant levels. The rows of the gene expression matrix associated with such genes can be treated as experimental noise and a potential source of false positives. They should be identified and removed before further analysis.

[32]Note that this kind of *filtering* is not related to univariate filtering by the significance level of differential expression. It is also not related to filter methods for feature selection. The filtering here aims at removing unreliable variables from the data set.

Some preprocessing methods associate the expression values with indicators of the measurement reliability (for instance, the detection calls of the MAS5 algorithm). Such indicators can be used directly to filter out unreliable probe sets. In the lack of such indicators, filtering by the expression level may be performed. Examples of filtering criteria may include the following:

- Filtering by the *fraction of Present calls*[33] in a class.

  The genes that have at least the predetermined fraction of Present calls in at least one of the classes are retained. This is equivalent to removing the genes for which the percent of Present calls in each class is less than the threshold fraction. The threshold should not exceed 50 percent of the class size; otherwise important discriminatory information could be removed. This criterion requires that in order to keep the gene in the analysis, it should be reliably expressed in at least one of the differentiated phenotypic classes (i.e., it should have at least the threshold fraction of Present calls in at least one class). Hence, potentially important discriminatory genes that are expressed in only one class will be preserved. For supervised analysis, this criterion is preferable over criteria based on the fraction or the number of Present calls across all classes because such criteria are not necessarily associated with the biologically defined groups of samples. Obviously, this criterion cannot be applied to unsupervised analysis for which we do not have class information.

- Filtering by the *number of Present calls* across all the samples (and across the classes in the case of supervised analysis).

  The genes with fewer than the predetermined threshold number of Present calls across all the samples are removed. Simple and common criteria for this filtering approach are: to eliminate all probe sets that have no Present calls, or to eliminate probe sets with all Absent calls. If we have the class information, then thresholds that use a nonzero number of Present calls should be determined in relation to the size of the smallest class. The threshold number should not be greater than half of the smallest class size. Otherwise genes important for discrimination could be removed.

- Filtering by the *average expression level* in a class.

  Usually, some more or less arbitrary level of experimental noise is used to define the filtering threshold. Genes that do not have an average expression level greater than the threshold in any of the classes are eliminated.

- Filtering by the *maximal expression level.*

  Genes with all expression values (across all the samples) below the experimental noise threshold are eliminated.

- Filtering by the *range of expression values.*

  Genes with an amplitude of expression (max – min) less than the specified experimental noise threshold will be removed. At the same threshold level, this criterion

[33]Details on MAS5 detection calls (Present, Absent, and Marginal) can be found in Section 2.3.1.

will eliminate all the low-expressed genes that would be filtered out by the *maximal expression level* criterion. However, it will also eliminate genes expressed at any level for which the range of their expression values is below the noise threshold (the very low variance genes).

- Filtering by the *fold change*.

  Using this criterion as the only filtering criterion is not recommended. The probe sets with unreliable expression measurements at low expression levels can generate large fold changes that have no biological meaning. If we want to use the fold change as one of the filtering criteria, it should be done after unreliable variables are removed, preferably with the *fraction of Present calls* criterion.

As could be expected, filtering by detection calls provides better results than filtering by the expression value. McClintick and Edenberg performed a comprehensive comparison of these two filtering approaches when applied to MAS5 and RMA preprocessed data (McClintick and Edenberg 2006). Their results indicate that filtering by the fraction of Present calls in a class is efficient in removing unreliable variables when applied to MAS5 preprocessed data as well as when applied to data preprocessed by other algorithms. It is superior to filtering by the average expression level (whether the expressions are calculated by the MAS5 or RMA algorithms). Though the Absent detection calls are more likely to be assigned to low level expressions than to high level ones, filtering by the expression level removes the low expressed genes whether their measurements are reliable or not. On the other hand, filtering by the fraction of Present calls targets unreliable variables and does not affect the genes with relatively low but reliable expression measurements. McClintick and Edenberg suggest that the threshold fraction of Present calls be associated with the number of samples per class. For smaller experiments, the filtering should be more aggressive, with the threshold as large as 25 percent or even 50 percent. For large experiments, it may be sufficient to remove only those probe sets that have no Present calls.

Since probe sets with all (or almost all) Absent calls as well as probe sets expressed at very low levels are very likely to represent experimental noise, it may be preferable to filter expression data using more than one criterion—for instance, filtering by the fraction of Present calls in a class and filtering by the maximal expression level[34] (which specifically targets genes with consistently low expression). If our preferred preprocessing method does not provide detection calls, it may be advantageous to also preprocess the data with the MAS5 algorithm and use its detection calls for filtering our otherwise preprocessed data.

2.7 BASIC EXPLORATORY DATA ANALYSIS

The preprocessing and filtering steps result in the gene expression data ready for analysis (see Fig. 2.6). Although not always necessary (for instance, for biomarker discovery, we could jump directly to multivariate analysis), it is common practice to start the analysis of gene expression data with univariate *exploratory*

[34] In practice, there is often a large overlap in the subsets of probe sets removed by these two criteria.

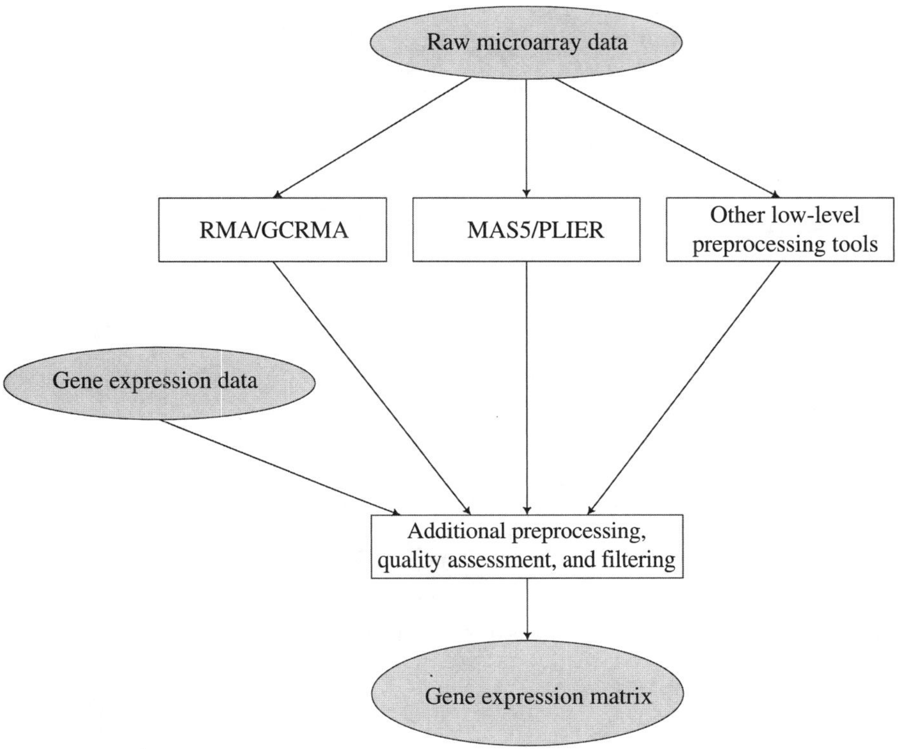

Figure 2.6: The preprocessing, quality control, and filtering of gene expression data—an example. The methods named in the top boxes represent selected examples rather than an exhaustive list of methods.

data analysis. Since univariate analysis is performed under the assumption that variables are independent (which is definitely not true for gene expression data), it could rarely answer any serious scientific questions. Usually then, this analysis does not have any specific research goals; it is used for researchers to get familiar with the data, maybe to learn about some structures in the data, and to check whether the data is consistent with the study expectations. Different univariate statistical tests may be utilized, but the most common are the t test and the ANOVA F test, or their derivatives such as the ones used in *SAM* (Significance Analysis of Microarrays—basically, a two-sample t test with the multiplicity adjustment using the false discovery rate) or in *Limma* (a moderated t test or ANOVA F test using empirical Bayes methods to "borrow information between genes" for experiments with only a few samples) (see Fig. 2.7).

Univariate analysis results in an ordered list of probe sets with an assigned significance of differential expression between the compared classes. The statistical significance of each gene (probe set) should be adjusted for multiple testing. Additional calculations for each probe set may be performed at this step (such as the mean expression for each class or the N-fold). All these results may give us some *feeling* about the data.

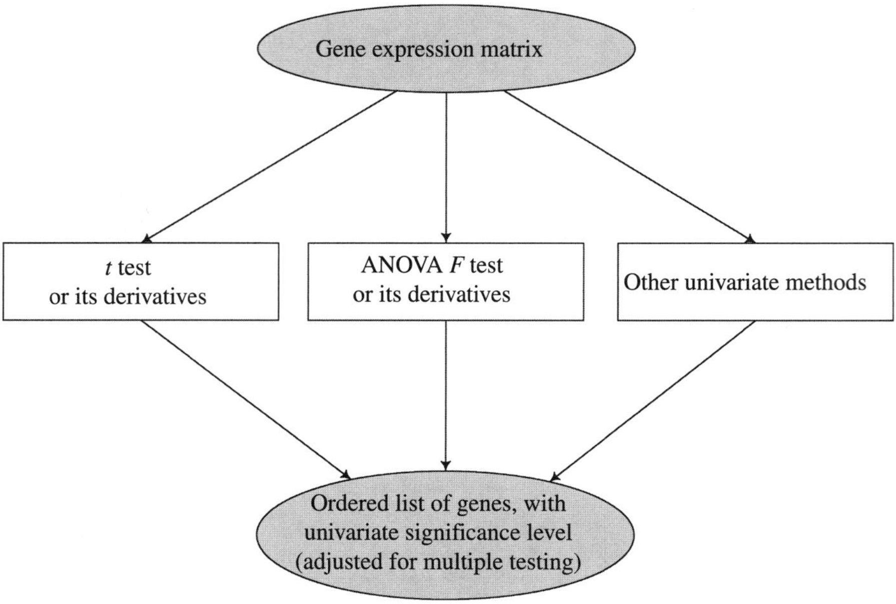

Figure 2.7: Basic exploratory analysis of gene expression data—an example.

A few basic terms:

p-value—the probability of getting a particular or more extreme value of the test statistic under the assumption that the null hypothesis is true. A smaller p-value gives us stronger evidence against the null hypothesis. We reject the null hypothesis if the p-value is less than the predetermined significance level α of the test. Sometimes, we may prefer to report only the p-value (rather than the decision about rejecting or not rejecting the null hypothesis) since it does not require determination of the significance level in advance.

Significance level $\boldsymbol{\alpha}$ (also known as the significance of the test)—the probability of rejecting the null hypothesis when in fact this hypothesis is true. This is the same as the probability of making a Type I error. By selecting α we are stating our tolerance for making a Type I error. If the sample-based probability of making a Type I error (p-value) is less than our tolerance for making a Type I error (α), we reject the null hypothesis. We do not know if our decision (of rejecting the null hypothesis) is correct, but we do know the a priori probability that it is incorrect is not greater than α.

Type I error—rejecting the true null hypothesis.

Type II error—failing to reject the false null hypothesis.

2.7.1 *t* Test

One of the classical t tests may be used to identify differentially expressed genes in two-class experiments. Depending on whether we can or cannot assume equal variances of gene expression in both differentiated populations, we will use either the t test for equal variances or the t test for unequal variances.

2.7.1.1 *t* Test for Equal Variances

This t test makes the normality and homoscedasticity assumptions, that is, that the data in both populations (represented by the biological samples in the two classes of the training data set) is normally and independently distributed and that the population variances are equal (or similar). It evaluates the difference in the mean expression level of a particular gene between the two populations,[35]

$$\begin{aligned} H_0&: \quad \mu_1 = \mu_2 \\ H_a&: \quad \mu_1 \neq \mu_2 \end{aligned} \tag{2.23}$$

using the following test statistic

$$t = \frac{\overline{x}_1 - \overline{x}_2}{s_p\sqrt{\dfrac{1}{n_1} + \dfrac{1}{n_2}}}, \tag{2.24}$$

where

- $\overline{x}_1$ and $\overline{x}_2$ are the mean expression values of the gene in each class of the training data,
- s_p is the pooled estimate of the standard deviation of the gene expression,

$$s_p = \sqrt{\frac{(n_1 - 1)s_1^2 + (n_2 - 1)s_2^2}{n_1 + n_2 - 2}}, \tag{2.25}$$

- n_1 and n_2 represent the number of biological samples in each class,
- s_1^2 and s_2^2 are the observed variances of gene expression in each class.

The test statistic (2.24) follows a t distribution with $n_1 + n_2 - 2$ degrees of freedom.

2.7.1.2 *t* Test for Unequal Variances

Often we cannot make the homoscedasticity assumption. For data with unequal population variances we will use the following test statistic

$$t = \frac{\overline{x}_1 - \overline{x}_2}{\sqrt{\dfrac{s_1^2}{n_1} + \dfrac{s_2^2}{n_2}}}. \tag{2.26}$$

[35] A two-tail test is appropriate since we do not make any *a priori* assumptions about the nature of a difference between the two means.

This test statistic follows the t distribution with df degrees of freedom, where

$$df = \frac{\left(\frac{s_1^2}{n_1} + \frac{s_2^2}{n_2}\right)^2}{\frac{1}{n_1 - 1}\left(\frac{s_1^2}{n_1}\right)^2 + \frac{1}{n_2 - 1}\left(\frac{s_2^2}{n_2}\right)^2}. \tag{2.27}$$

If the study included only a single gene, the p-value of either of the t tests would represent the observation based or posterior probability of making a Type I error (i.e., the observation based probability of rejecting the null hypothesis when in fact this hypothesis is true). The gene would be declared differentially expressed, if the p-value is less than α, the significance level of the test. However, in gene expression studies we perform as many tests as the number of genes in the training data set. Therefore, this "raw" p-value has to be adjusted for multiple testing (see Section 2.7.5).

2.7.2 ANOVA *F* Test

Analysis of variance (ANOVA) can be used in a univariate way to test whether the mean expression levels of a particular gene differ significantly between the J populations, where $J > 2$. The ANOVA makes the following assumptions:

- the biological samples are independent,
- the gene expression is normally distributed with the same variance in each population.

The ANOVA F test statistic is based on the ratio of the variance between classes to the variance within classes[36] and is used to decide whether we can reject the null hypothesis of no difference between the J population means,

$$\begin{aligned} H_0&: \quad \mu_1 = \mu_2 = \cdots = \mu_J \\ H_a&: \quad \text{Not all population means are equal} \end{aligned} \tag{2.28}$$

Assuming that we compare J populations with n_j biological samples in each class j, $j = 1, \ldots, J$, with the total number of biological samples $N = \sum_{j=1}^{J} n_j$, the F test statistic is calculated as

$$F = \frac{MSTR}{MSE}, \tag{2.29}$$

where

- $MSTR$ is the mean square due to treatment,

$$MSTR = \frac{\sum_{j=1}^{J} n_j(\bar{x}_j - \bar{\bar{x}})^2}{J - 1}, \tag{2.30}$$

[36] If the null hypothesis is true, this ratio should be near 1. We reject the null hypothesis if the ratio is significantly greater than 1.

- *MSE* is the mean square due to error,

$$MSE = \frac{\sum_{j=1}^{J} (n_j - 1)s_j^2}{N - J}, \tag{2.31}$$

- $\bar{x}_j$ is the mean expression of the gene in class j,
- $\bar{\bar{x}}$ is the overall mean expression of the gene (across all classes),
- s_j^2 represents the variance of gene expression in class j.

Under H_0, the test statistic (2.29) follows an F distribution with $J - 1$ degrees of freedom in the numerator and $N - J$ degrees of freedom in the denominator. As for the t tests, the overall p-value of the F tests has to be adjusted for multiple comparisons.

2.7.3 SAM *t* Statistic

The *Significance Analysis of Microarrays* (SAM) software (Tusher et al. 2001; Chu et al. 2007) is a popular tool[37] for the univariate analysis of differential expression. To identify genes differentially expressed in a two-class experiment, it implements a modified t test statistic defined for a particular gene as the *relative difference d*,

$$d = \frac{\bar{x}_1 - \bar{x}_2}{s_p\sqrt{\frac{1}{n_1} + \frac{1}{n_2}} + s_0}. \tag{2.32}$$

It differs from (2.24), the statistic for the equal variances t test, only by the constant s_0 (called an exchangeability factor or fudge factor) added to the denominator. Note that the statistic (2.24) can be interpreted as the ratio of expression difference to variability, or more generally as the ratio of signal to noise. As some microarray experiments are still performed with very small sample sizes, their estimation of the standard deviation may be unreliable. Underestimation of the variability inflates the value of the statistic and may result in an increased number of false positives. The fudge factor s_0 reduces the correlation between the value of the d statistic and the estimated expression variability represented by s_p. The value of s_0 is expressed as a percentile of the pooled standard deviation values (s_p values) of all the genes. It is chosen for each data set as the percentile that minimizes the coefficient of variation of d, calculated as a function of s_p.

With the added fudge factor, the null distribution of the d statistic no longer follows a t distribution. To assign significance to d values calculated for genes, SAM estimates the empirical distribution of the d statistic by permuting class

[37]SAM software is free for academic use. It can be downloaded as an Excel add-in from http://www-stat.stanford.edu/~tibs/SAM/

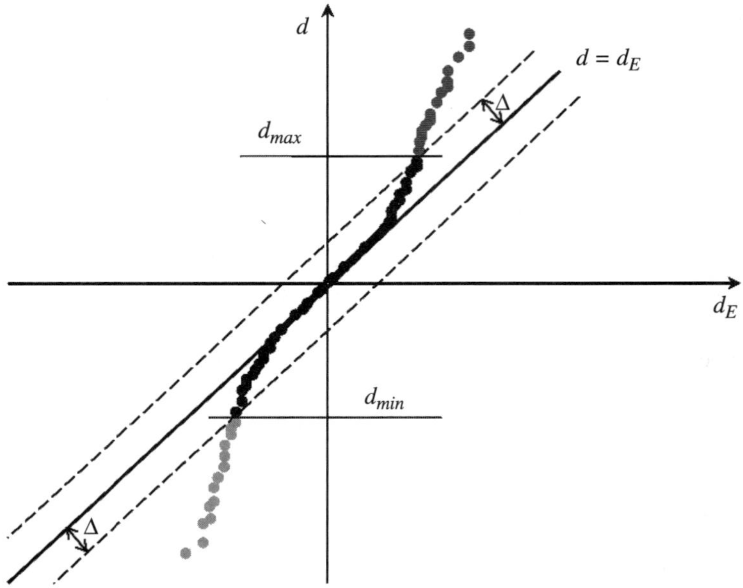

Figure 2.8: A scatterplot of the *observed relative difference d* versus the *expected relative difference* d_E. The dashed lines are at a threshold Δ distance from the $d = d_E$ identity line. The genes represented by the points that are outside of these threshold lines are deemed differentially expressed at the threshold level Δ. Depending on the sign of their relative difference *d*, SAM calls them either "significant positive genes" or "significant negative genes." (See color insert.)

labels. For each permutation,[38] a new *d* statistic value, say d_p, is calculated for each gene and the values are ranked. Then, the *expected relative difference* d_E is calculated for each rank as the mean of the d_p values for the rank over all permutations. A scatterplot of the observed relative differences *d* (ranked by their values) versus the expected relative difference d_E is used to identify differentially expressed genes (see Fig. 2.8). The genes represented by the points whose distance from the identity line $d = d_E$ is greater than some threshold distance Δ are called significant. The values of the threshold Δ are associated with estimates of the false discovery rate (FDR) and the user selects Δ based on the FDR level appropriate for the experiment.

To associate the threshold Δ with an estimated FDR, SAM uses the already available permutation data. First, the minimum positive *d* value (d_{max}) and the maximum negative *d* value (d_{min}) associated with the genes deemed significant are found (the horizontal cutoffs in Fig. 2.8). For each permutation, SAM determines the number of false positive genes by counting the genes with d_p values outside the

[38]Note that for very small experiments there are not enough distinct permutations to treat permutation-based estimates as reliable. An acceptable number of permutations should be at least 1000, preferably about 10,000 (Drăghici 2003).

(d_{min}, d_{max}) range. To estimate FDR for the threshold Δ, the median (or the 90th percentile)[39] of the number of false positive genes[40] across all permutations is divided by the number of genes deemed significant in the original data. For each gene called significant, SAM reports a q-value that represents the lowest FDR at which this particular gene would be called significant. For multiclass comparisons, SAM implements a similar approach based on an F-like test statistic with a fudge factor added to the denominator.

2.7.4 Limma

The *Linear Models for Microarray Data* (Limma) package (Smyth 2004) uses a modified t statistic to reduce the significance of genes with underestimated variance. The idea is similar to the SAM approach, but unlike SAM, Limma fits a linear model for each gene and then calculates a moderated t statistic by using the empirical Bayes approach. The variance for each gene is moderated by replacing it with the weighted average of the gene-specific variance and the estimated global variance (calculated by pooling all genes). This borrowing of information from other genes makes the analysis more stable, especially for experiments with a small number of biological samples. The moderated t statistic follows a t distribution with degrees of freedom estimated from the data. The p-value for each gene can be calculated and then adjusted for multiple testing. This approach can be extended to more than two classes, by using a similarly moderated F statistic. Limma is implemented as a part of the *Bioconductor* project (www.bioconductor.org).

2.7.5 Adjustment for Multiple Comparisons

A typical microarray study generates a gene expression matrix with tens of thousands of rows—probe sets representing genes. Assume we have 10,000 genes and we are performing 10,000 univariate tests. If the significance level $\alpha = 0.05$ is used, then for each of these tests we allow a five percent chance of making a Type I error. This means that we expect five percent of the 10,000 genes to be deemed significant (significantly differentially expressed) just by chance alone—amounting to 500 false positives. To control the overall probability of a Type I error, we have to apply a correction for multiple testing. Whether we are repeatedly performing a t test, an ANOVA F test, or any other (univariate or multivariate) test resulting in a p-value, we have to adjust the individual raw p-values for multiplicity in order to control the overall posterior false positive rate.

[39]SAM calculates both the median and the 90th percentile FDR and presents them on the Delta Table that can be used to select an appropriate threshold Δ. On the list of significant genes, the reported q-value corresponds to the median FDR.

[40]After calculating the "raw" median (and the 90th percentile) of the number of false positive genes, it is multiplied by π_0, the estimated proportion of genes that are truly not differentially expressed. SAM estimates π_0 from the permutation data and reports the median (and the 90th percentile) after this adjustment (Chu et al. 2007).

Exercise

If the reader is not convinced about the need for a multiplicity adjustment and is inclined to assume that a gene with consistently different expression levels in the two classes should be deemed differentially expressed regardless of whether the gene is tested as the only variable or in parallel with thousands of other genes, we suggest performing the following simple but enlightening Excel exercise described by Drăghici (Drăghici 2003):

- Using the RAND function in Excel, generate a random number for each cell in a spreadsheet with 10,000 rows (genes) and 20 columns (samples).
- Copy all the data and "Paste Special" as "Values" to another spreadsheet.
- Assume that the first ten columns represent Class A (say, Disease), and the last ten columns represent Class B (Control).
- In the 21st column, use Excel's TTEST function to calculate p-value for each gene (row).
- Sort the data in ascending order of the p-values.
- Check whether approximately 500 top genes have the p-value below 0.05.
- Recall that the data is randomly generated noise.

When performing multiple tests, rather than considering the significance level α of individual tests, we should use a procedure that controls one of the Type I error rates defined for testing multiple null hypotheses. Among the commonly used Type I error rates are the *family-wise error rate* (FWER) and the *false discovery rate* (FDR).

Family-wise error rate (FWER)

The family-wise error rate is defined as the probability of at least one Type I error (i.e., at least one false positive) over all tests. This probability for a single test is equal to the significance level α of the test. However, if we perform M independent tests, this probability is equal to $1 - (1 - \alpha)^M$, which for large M is close to 1.

False discovery rate (FDR)

The false discovery rate is the expected proportion of false positives among the rejected null hypotheses (i.e., among all genes reported as differentially expressed). When all null hypotheses are true (i.e., none of the tested genes is differentially expressed), FDR is equal to FWER, but otherwise is smaller (Benjamini and Hochberg 1995).

Generally, procedures controlling the FWER are more conservative than those controlling FDR (Dudoit and van der Laan 2008). The best known among the FWER-controlling procedures are: the classical single-step Bonferroni adjustment, the single-step Sidak procedure, and the step-down Holm procedure. The most popular among the FDR-controlling procedures is the step-up Benjamini and Hochberg procedure. The single-step procedures apply the same multiplicity adjustment to each individual α or raw p-value whereas adjustments made by the stepwise approaches depend on the rank of the gene among all tested genes and on the outcomes of the tests for other genes (Ge et al. 2003).

To avoid confusion with p used in this section for p-value, assume—for this section only—that M denotes the number of variables in the gene expression matrix as well as the number of multiple univariate tests. Thus, each gene and each test can be indexed by $m = 1, \ldots, M$. Assume also the following notation:

- α—the significance level for each of the M tests,
- $p(m)$—the unadjusted (raw) p-value calculated for test m, $m = 1, \ldots, M$,
- α^*—the modified significance level,
- $p^*(m)$—the adjusted p-value for test m.

2.7.5.1 Single-Step Bonferroni Procedure

This classical procedure (introduced by Bonferroni in 1936) compares the unadjusted p-value of each test, $p(m)$, with the modified significance level α^* defined as

$$\alpha^* = \frac{\alpha}{M}. \tag{2.33}$$

This is equivalent to comparing the adjusted p-value $p^*(m)$,

$$p^*(m) = \min\{p(m)M,\ 1\}, \tag{2.34}$$

to the original significance level α. Therefore, gene m is considered differentially expressed if $p^*(m) < \alpha$ or equivalently if $p(m) < \alpha^*$.

2.7.5.2 Single-Step Sidak Procedure

The Sidak procedure (Sidak 1967) defines the modified significance level α^* as

$$\alpha^* = 1 - (1 - \alpha)^{\frac{1}{M}}. \tag{2.35}$$

Instead of comparing each p-value $p(m)$ to α^*, the adjusted p-value may be calculated as

$$p^*(m) = 1 - [1 - p(m)]^M. \tag{2.36}$$

The adjustments made by the Sidak procedure are similar, though slightly less conservative, than the Bonferroni adjustments. Here *conservatism* means that if a gene is called differentially expressed by a conservative procedure, then its expression is more likely truly different between the classes (Drăghici 2003). However, the price to pay for this is an inflated level of false negatives.

2.7.5.3 Step-Down Holm Procedure

The step-down Holm adjustment (Holm 1979), also known as the step-down Bonferroni or sequential Bonferroni adjustment, is less conservative than the classical

Bonferroni adjustment. After the raw p-values are calculated, the data are sorted in ascending order of these p-values. Then we start with the smallest p-value and multiply it by the total number of genes M. For the subsequent p-values, we are less conservative and multiply them by $M-1$, $M-2$, and so on. The last (and the largest) p-value is multiplied by 1. However, there is an additional constraint requiring that each subsequently adjusted p-value cannot be less than the one preceding it in the ordered list (this guarantees the monotonicity of the adjusted p-values). Therefore, finding the first gene that is not differentially expressed (i.e., having the adjusted p-value greater than or equal to α) means that all the genes below this one on the sorted list are deemed also not differentially expressed.[41] To describe this procedure more formally, assume the following notation:

- m, $m = 1, \ldots, M$ is the index for the original gene order,
- k, $k = 1, \ldots, M$ is the index for the sorted unadjusted p-values,
- $O(k)$ is the function translating the rank of a gene, that is, its position k on the sorted list, into its original position m, thus $O(k) = m$,
- $p(O(k))$ represents sorted unadjusted p-values, thus $p(O(1)) \leq p(O(2)) \leq \cdots \leq p(O(M))$.

Adding the obvious constraint that no p-value may be greater than 1, the adjusted p-values are calculated using the following formula (Dudoit and van der Laan 2008):

$$p^*(O(k)) = \max_{l=1,\ldots,k} \{\min[p(O(l)) \cdot (M - l + 1), 1]\}, \quad k = 1, \ldots, M. \tag{2.37}$$

The same result is achieved by comparing the unadjusted p-values $p(O(k))$ to modified significance levels $\alpha^*(k)$, where

$$\alpha^*(k) = \frac{\alpha}{M - k + 1}, \quad k = 1, \ldots, M. \tag{2.38}$$

Starting from the top of the sorted list, the comparisons are made until the first gene with $p(O(k)) \geq \alpha^*(k)$ is found. This gene and all the genes below this one on the sorted list are reported as not differentially expressed.

2.7.5.4 Step-Up Benjamini and Hochberg Procedure

For microarray data sets it is not unusual for the FWER-controlling adjustments (especially the most conservative Bonferroni adjustment) to report zero or very few differentially expressed genes. Generally, these adjustments may result in too many false negatives when the number of tested hypotheses M is large. Instead of focusing on the *number* of false positives, the Benjamini and Hochberg procedure (Benjamini and Hochberg 1995) controls the *proportion* of false positives among the genes called differentially expressed (i.e., FDR). In general, the FDR-controlling approaches are

[41] In particular, this means that a gene with a given raw p-value cannot be called differentially expressed if there is a gene with a smaller raw p-value that is deemed not differentially expressed.

less stringent and may be more appropriate for some data sets with very large numbers of genes.

Similar to the Holm procedure, the Benjamini and Hochberg procedure starts with calculating the raw p-values and sorting the data in ascending order of these p-values. Then, we start at the *bottom* of the sorted list, that is, with the largest p-value. This p-value is multiplied by 1, or M/M. The second p-value from the bottom of the list is multiplied by $M/(M-1)$, the third by $M/(M-2)$ and so on. However, none of the adjusted p-values may be greater than the adjusted p-value below it. This means that when the first differentially expressed gene is identified, all the genes above this one on the sorted list are also called differentially expressed. Using the notation introduced in the subsection on the Holm procedure, the adjusted p-value of the Benjamini and Hochberg procedure is calculated in the following way (Dudoit and van der Laan 2008):

$$p^*(O(k)) = \min_{l=k,\dots,M}\left\{\min\left[p(O(l))\cdot\frac{M}{l},\ 1\right]\right\},\quad k=1,\dots,M. \tag{2.39}$$

Instead of adjusting the p-values, we may compare the sorted unadjusted p-values $p(O(k))$ to modified significance levels $\alpha^*(k)$ calculated as

$$\alpha^*(k) = k\frac{\alpha}{M},\quad k=1,\dots,M. \tag{2.40}$$

Again, the first gene from the bottom of the sorted list, for which $p(O(k)) < \alpha^*(k)$ will make all the genes above it reported as differentially expressed.

2.7.5.5 Permutation Based Multiplicity Adjustment

The permutation (or bootstrap) approach to evaluate statistical significance for multiple comparisons has recently become quite popular. This approach is based on generating the empirical null distribution of the test statistic by using permutations of the data. There are many possible scenarios (Westfall and Young 1993; Dudoit et al. 2002; Ge et al. 2003; Dudoit and van der Laan 2008). Some of them use permutations to estimate the unadjusted p-values and then calculate the adjusted ones using one of the multiplicity procedures. Some apply the permutation approach to estimate both the unadjusted and adjusted p-values. Still others first approximate the unadjusted p-values and then use permutations to find the adjusted ones.

Let us consider a basic permutation scenario. Assume that we use the t statistic (2.26) for a data set with M genes and N biological samples assigned to two differentiated classes ($J=2$). Permuting the class labels (equivalent to permuting the columns of the gene expression matrix) generates a new data set with the random assignment of the biological samples to classes, but preserves the size of each class and at least some relations between gene expression levels. For each permutation $b=1,\dots,B$, we use (2.26) to calculate the value of the t statistic $t_b(m)$ for each gene m, $m=1,\dots,M$. The permutation-based unadjusted p-value $\hat{p}(m)$ for gene m is then calculated as the proportion of the permutations for which the absolute value of the original test statistic $|t(m)|$ of the gene (for not permuted data) is less than $|t_b(m)|$,

$$\hat{p}(m) = \frac{1}{B}\sum_{b=1}^{B} \mathbf{1}_{|\,t(m)|<|\,t_b(m)\,|}, \quad m = 1, \ldots, M, \tag{2.41}$$

where

$$\mathbf{1}_{|\,t(m)\,|<|\,t_b(m)\,|} = \begin{Bmatrix} 1 & \text{when } |t(m)| < |t_b(m)| \\ 0 & \text{otherwise} \end{Bmatrix}.$$

By replacing $p(m)$ with $\hat{p}(m)$ in one of the previously described multiplicity adjustment procedures, we can calculate the adjusted p-value $p^*(m)$ for each gene m. Note that for reliable estimates, we should perform thousands of permutations. Note also that instead of permutations we may use bootstrap-based resampling, in which randomized data sets are generated by sampling with replacement.

Univariate analysis results in an ordered list of genes sorted by the significance of their differential expression. As the univariate approach neglects correlations between genes, such results are not very informative. A more advanced, multivariate analysis should follow. There are, however, studies reported in the literature that stop here. Often, they are studies with very few samples per class (which adds to doubts about their statistical validity and biomedical relevance). Nonetheless, whether our results are univariate lists of genes, clusters of genes with similar expression, or multivariate biomarkers (small sets of genes significantly differentiating the classes—more on this in Chapter 3), the next step is to look for the biological interpretation of the results. We will just mention here that biological interpretation of a univariate list of genes is usually done by first assigning annotations to each gene (or probe set; for the Affymetrix microarrays, extensive annotations of probe sets can be downloaded from www.affymetrix.com or from other sources like http://genecruiser.broadinstitute.org), and then by analyzing annotations such as the Gene Ontology (GO) terms (www.geneontology.org) or metabolic pathways in order to identify functional categories or pathways that are significantly affected by the expression changes of the genes from the list.

2.8 UNSUPERVISED LEARNING (*TAXONOMY-RELATED* ANALYSIS)[42]

Unsupervised learning methods, such as clustering, organize data into structures that may be useful for developing taxonomies. Clustering can be done on genes (probe sets), samples, or on both genes and samples simultaneously (two-way clustering).

[42]In the (domain-free) data mining language, methods such as clustering or principal component analysis are called *unsupervised learning*. This naming is based on *how* the methods work. Though this naming scheme is appropriate also in specific domains, in some domains it may be more convenient to name the methods according to the major scientific results they can provide. Such results for genomics and proteomics would be *new taxonomic knowledge*. For example, assume we know five leukemia subtypes; we cluster a large leukemia data set and results point to a possibility of having six well defined clusters. If five of them align with known five subtypes, the sixth may indicate a new unknown subtype.

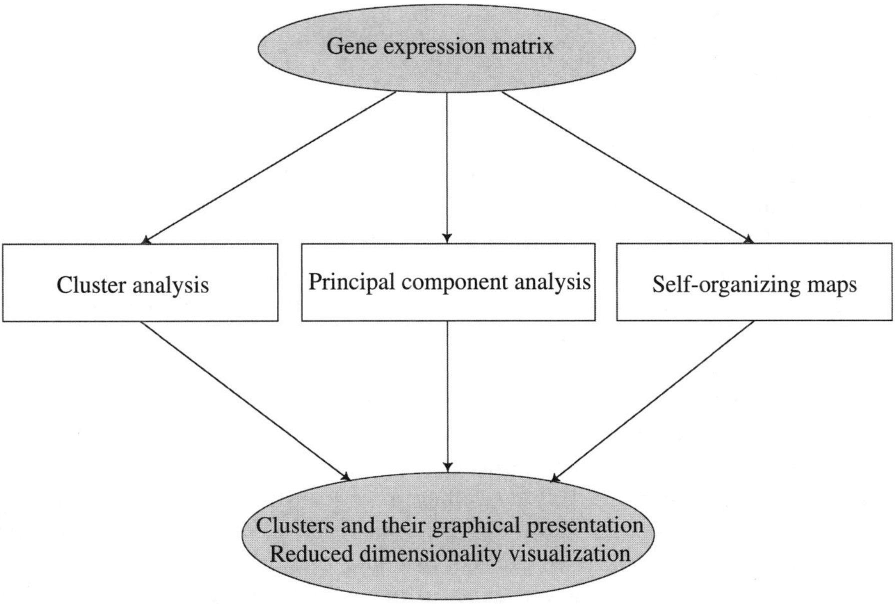

Figure 2.9: Unsupervised (taxonomy-related) analysis of gene expression data—an example.

Clustering samples may reveal new subtypes of a disease. Clustering genes may aim at identifying groups of genes performing similar functions or belonging to the same genetic pathway. The goal of two-way clustering (performing clustering simultaneously for genes and samples) is to identify subsets of genes that exhibit distinct expression patterns over only a subset of samples. Principal component analysis and self-organizing maps represent popular methods that can be used to reduce the dimensionality and visualize data in order to gain some information about grouping of samples or genes (see Fig. 2.9).

2.8.1 Cluster Analysis

Cluster analysis groups objects into categories, or *clusters*, based on some definition of object similarity. Objects assigned to a cluster should be more similar to each other than to objects in other clusters. Although clustering is sometimes presented as a way of finding "natural" groupings in the data, there is rarely a single way of grouping objects of a particular data set. Due to various possible similarity measures and various grouping algorithms, clustering may lead to very different results and is most often a tool to generate hypotheses about data structure rather than a tool to test such hypotheses (Fielding 2007). This is especially true for data sets with many variables.

A common mistake in early gene expression studies was to use clustering as a preprocessing step for the supervised analysis. The reasoning behind this approach was that grouping variables (genes) by their similarity and then selecting only some of them to represent the groups should be a valid approach to reducing the dimensionality of the supervised analysis. A serious problem with this approach is that we usually have no way of knowing whether a particular similarity among variables in the unsupervised environment can be translated to their similarity in supervised analysis. In other words, by the very definition of unsupervised learning, we do not know whether genes assigned to the same cluster carry similar discriminatory information that would be important in the differentiation of a particular set of phenotypic classes using supervised analysis.

Consider, for example, a gene expression matrix with biological samples representing patients assigned to a few subtypes of a disease. If the goal of a study was to build a classification system, but we did not know how to perform the supervised analysis with thousands of variables, it could seem tempting to cluster genes and use much fewer variables—only the ones selected to represent clusters. However, the genes of a cluster are very rarely perfectly or very highly correlated. Their common variation may be relatively small, often even 50 percent or less. May we assume that they represent similar discriminatory information for classification of a particular set of phenotypic classes under supervised analysis? May we choose one or a few of them to represent each cluster and discard the rest? If we are not convinced yet that we would be most likely discarding important discriminatory information, let us look at our supervised study. For the same data, we can define different sets of phenotypic classes to differentiate. For example, one supervised goal may be to build a classifier assigning new patients to one of the disease subtypes, another may be to identify a biomarker predicting relapse, and so on. Consequently, the same group of patients may be split differently into different phenotypic classes. Different goals of such supervised studies are associated with different discriminatory information carried by the same set of variables. With which set of phenotypic classes is our clustering of genes associated? Conceivably with none, since clustering does not take into account any metadata information assigning the patients to classes. Please read Section 3.2.3 of Chapter 3 for more on this subject. However, please also visit Chapter 4 to read about using cluster analysis in a supervised setting—at the stage when we *do know* that our variables are associated with the discrimination of a particular set of classes.

One may argue that if we have specific information about the genes clustered together, for example, that they belong to the same genetic pathway involved in differentiation of phenotypic classes we are interested in, then we should be able to use the clustering results to reduce the number of these genes. Yes, we can use biological knowledge about a specific supervised problem. However, this refers to the supervised environment of the study. Clustering (or any unsupervised analysis) by itself cannot supply such knowledge. One gene may belong to several pathways whose activities may be different for different conditions. Furthermore, if our goal is classification or biomarker discovery, it would be much better to incorporate this biological knowledge into supervised multivariate feature selection rather than use it with any unsupervised method.

For studies with unsupervised (taxonomy-related) goals, cluster analysis is the primary, and a very useful, tool with many excellent software implementations. Among methods used for clustering gene expression data are K-means clustering, hierarchical clustering and two-way clustering. The K-means clustering represents a partitioning approach, which divides a set of objects into a predetermined number of clusters. Hierarchical clustering provides a sequence of nested clusters. Because objects to be clustered (either genes or biological samples) are grouped by their similarity, a measure of similarity is crucial for cluster analysis.

2.8.1.1 Measures of Similarity or Distance

Many metrics (or measures) can be defined to describe *similarity* between two objects. Sometimes it may be more convenient to define them in terms of *distance* or *dissimilarity* since the increase in the distance between objects corresponds to their decreased similarity. Assume that we have n objects to cluster and each of them is represented by a v-dimensional vector of measurements. If we cluster genes, then $n = p$ and $v = N$, where p represents the number of rows (genes) and N the number of columns (biological samples) in the gene expression matrix. If we cluster samples, then $n = N$ and $v = p$. Denote the distance between two objects represented by v-dimensional vectors $\mathbf{a}_i$ and $\mathbf{a}_j$, $\mathbf{a}_i, \mathbf{a}_j \in \Re^v$, $i, j = 1, \ldots, n$,

$$\mathbf{a}_i = \begin{bmatrix} a_{1i} \\ a_{2i} \\ \vdots \\ a_{vi} \end{bmatrix}, \quad \mathbf{a}_j = \begin{bmatrix} a_{1j} \\ a_{2j} \\ \vdots \\ a_{vj} \end{bmatrix}, \tag{2.42}$$

as $d(\mathbf{a}_i, \mathbf{a}_j)$.

A well designed distance measure should satisfy most or preferably all of the following conditions.[43]

1. The distance in always non-negative,

$$d(\mathbf{a}_i, \mathbf{a}_j) \geq 0 \tag{2.43}$$

2. The distance can be equal to zero only when the two objects are identical, that is, represented by the same vector,

$$d(\mathbf{a}_i, \mathbf{a}_j) = 0 \quad \text{if and only if } \mathbf{a}_i = \mathbf{a}_j \tag{2.44}$$

3. The distance measure is symmetric,

$$d(\mathbf{a}_i, \mathbf{a}_j) = d(\mathbf{a}_j, \mathbf{a}_i) \tag{2.45}$$

4. The distances between any three objects $\mathbf{a}_i$, $\mathbf{a}_j$, $\mathbf{a}_\gamma$, $i, j, \gamma = 1, \ldots, n$ follow the triangle inequality,

$$d(\mathbf{a}_i, \mathbf{a}_j) \leq d(\mathbf{a}_i, \mathbf{a}_\gamma) + d(\mathbf{a}_\gamma, \mathbf{a}_j) \tag{2.46}$$

[43]These four conditions are properties of a *metric*: non-negativity, reflexivity, symmetry, and the triangle inequality. Although the terms *measure* and *metric* are sometimes used interchangeably, a distance measure is a metric only when it has all these four properties (Duda et al. 2001). For example, the Euclidean distance is a metric, but the correlation distance is not.

The following measures are among the most commonly used by clustering algorithms to determine distance (dissimilarity) or similarity between objects.

Euclidean Distance

The Euclidean distance (or dissimilarity measure) between two objects is defined as the geometric distance between the two points representing the objects in the v-dimensional space of their attributes,

$$d(\mathbf{a}_i, \mathbf{a}_j) = \sqrt{\sum_{k=1}^{v} (a_{ki} - a_{kj})^2}. \qquad (2.47)$$

Squared Euclidean Distance

The squared Euclidean distance,

$$d(\mathbf{a}_i, \mathbf{a}_j) = \sum_{k=1}^{v} (a_{ki} - a_{kj})^2, \qquad (2.48)$$

is sometimes used by the clustering algorithms that minimize the within-cluster sum of squares. However, if the data is not normalized, this distance measure tends to overemphasize the variables with large variances.

Manhattan Distance

The Manhattan (or city block) distance is the sum of absolute values of the geometric distances along each of the v dimensions,

$$d(\mathbf{a}_i, \mathbf{a}_j) = \sum_{k=1}^{v} \left| a_{ki} - a_{kj} \right|. \qquad (2.49)$$

Minkowski Distance

The Minkowski distance[44] generalizes both the Euclidean and Manhattan distances and is defined as

$$d(\mathbf{a}_i, \mathbf{a}_j) = \left[\sum_{k=1}^{v} \left| a_{ki} - a_{kj} \right|^m \right]^{\frac{1}{m}}. \qquad (2.50)$$

When $m = 1$, the Minkowski distance is equivalent to the Manhattan distance, and when $m = 2$, it is equivalent to the Euclidean one.

[44]The Minkowski distance $d(\mathbf{a}_i, \mathbf{a}_j)$ is also known as the L_m norm, $L_m(\mathbf{a}_i, \mathbf{a}_j)$. The Euclidean distance is then the L_2 norm, and the Manhattan distance the L_1 norm (Duda et al. 2001).

Chebyshev Distance

The Chebyshev distance is equal to the largest distance along any single dimension,

$$d(\mathbf{a}_i, \mathbf{a}_j) = \max_k |a_{ki} - a_{kj}|. \tag{2.51}$$

Mahalanobis Distance

If the variation in the data differs significantly between dimensions, the Euclidean or Manhattan distances may not constitute appropriate metrics of objects' similarity. The Mahalanobis distance takes into account the variations by defining the distance as

$$d(\mathbf{a}_i, \mathbf{a}_j) = \sqrt{(\mathbf{a}_i - \mathbf{a}_j)^T \mathbf{S}^{-1} (\mathbf{a}_i - \mathbf{a}_j)}, \tag{2.52}$$

where **S** is the variance-covariance matrix,[45] $\mathbf{S}^{-1}$ denotes the inverted matrix **S**, and the superscript *T* denotes transposition.[46] If **S** is the identity matrix, the Mahalanobis distance is equivalent to the Euclidean distance.

Correlation Distance

Pearson's correlation distance can be defined as

$$d(\mathbf{a}_i, \mathbf{a}_j) = \frac{1 - r(\mathbf{a}_i, \mathbf{a}_j)}{2}, \tag{2.53}$$

where

- $r(\mathbf{a}_i, \mathbf{a}_j)$ is the Pearson correlation coefficient between the vectors $\mathbf{a}_i$ and $\mathbf{a}_j$,

$$r(\mathbf{a}_i, \mathbf{a}_j) = r_{a_i a_j} = \frac{s_{a_i a_j}}{s_{a_i} s_{a_j}} = \frac{\sum_{k=1}^{v} (a_{ki} - \bar{a}_i)(a_{kj} - \bar{a}_j)}{\sqrt{\sum_{k=1}^{v} (a_{ki} - \bar{a}_i)^2 \sum_{k=1}^{v} (a_{kj} - \bar{a}_j)^2}} \tag{2.54}$$

- $s_{a_i a_j}$ is the covariance between $\mathbf{a}_i$ and $\mathbf{a}_j$,
- s_{a_i} and s_{a_j} are the standard deviations of $\mathbf{a}_i$ and $\mathbf{a}_j$ respectively,
- $\bar{a}_i$ and $\bar{a}_j$ are the mean values of $\mathbf{a}_i$ and $\mathbf{a}_j$,

$$\bar{a}_i = \frac{1}{v} \sum_{k=1}^{v} a_{ki}, \quad \bar{a}_j = \frac{1}{v} \sum_{k=1}^{v} a_{kj}.$$

[45]Diagonal elements of the matrix **S** represent the variances of the v variables, whereas the off-diagonal elements of **S** are covariances between the variables.

[46]The operation of *transposing* the column vector $(\mathbf{a}_i - \mathbf{a}_j)$ to the row vector $(\mathbf{a}_i - \mathbf{a}_j)^T$.

Note that the correlation coefficient $r(\mathbf{a}_i, \mathbf{a}_j)$ can be used as a measure of similarity between two objects whereas the distance (2.53) is a related measure of dissimilarity. As $r(\mathbf{a}_i, \mathbf{a}_j)$ assumes values between -1 and $+1$, the values of the correlation distance defined by (2.53) are between 0 and 1. The zero correlation distance corresponds to the perfect positive linear correlation between the two vectors, and the largest distance of one to perfect negative linear correlation. If we prefer to neglect the direction of the correlation and cluster objects according to the strength of their linear correlation only, we can replace the distance measure (2.53) with

$$d(\mathbf{a}_i, \mathbf{a}_j) = 1 - |r(\mathbf{a}_i, \mathbf{a}_j)|. \tag{2.55}$$

Each of these correlation distances is associated with the pattern of changes in the gene expression measurements (similar variation) rather than the magnitude of the expression values. Thus, the distance measures (2.53) and (2.55) are equal to zero not only for identical objects, but also for objects that are perfectly correlated. This is a departure from condition (2.44). However, this may be advantageous in clustering gene expression data since it leads to clustering genes that are co-expressed but have different expression levels together (Amaratunga and Cabrera 2004). For standardized vectors $\mathbf{a}_i$ and $\mathbf{a}_j$, the correlation distance (2.53) is proportional to the squared Euclidean distance between the vectors.

Cosine Similarity

The cosine measure of similarity is associated with the angle between two v-dimensional vectors and is defined as

$$\cos(\sphericalangle \mathbf{a}_i \mathbf{a}_j) = \frac{\sum_{k=1}^{v} a_{ki} a_{kj}}{\sqrt{\sum_{k=1}^{v} a_{ki}^2 \sum_{k=1}^{v} a_{kj}^2}}. \tag{2.56}$$

Note that for normalized data with zero means, the cosine similarity is equivalent to the correlation similarity $r(\mathbf{a}_i, \mathbf{a}_j)$. The distance measures based on the cosine similarity can be defined analogically to the correlation distances (2.53) or (2.55).

2.8.1.2 K-Means Clustering

The K-means clustering algorithm (Lloyd 1957; MacQueen 1967) is one of the simplest and most popular partitioning methods.[47] It starts with the random selection of K cluster centers, where K is the input parameter. These centers may be random points in v-dimensional space of v variables or randomly selected objects of the data. Using a similarity (or distance) measure between objects and the cluster centers, each object

[47] In the area of signal processing, K-means is known as *vector quantization*.

is assigned to the cluster with the closest center. Then, the cluster centers are redefined, usually by finding the mean vector of all points assigned to each cluster.[48] Objects are re-assigned according to their distance to these new class centers. This iterative process continues until there are no changes in the assignment of objects to clusters.[49] When using the *K*-means algorithm with squared Euclidean distances, the within-cluster variation will be minimized.[50] However, it can be a local minimum (Hastie et al. 2001). Since different starting points can result in a different partitioning of objects, it may be advisable to run the algorithm several times and select the solution with the smallest within-cluster variation.

The main disadvantage of the *K*-means method is the necessity to decide on the number of clusters. Although an additional loop of iterations can be added for different values of the parameter *K* (and, for example, the *K* value offering the smallest within-cluster variation selected), the solution is always a set of *K* clusters with not much information about the relation between clusters or between objects.

2.8.1.3 Hierarchical Clustering

Hierarchical clustering requires neither upfront determination of the number of clusters nor selection of the initial cluster centers. It, however, requires a measure of similarity (or a measure of distance) defined not only between objects, but also between groups of objects. Hierarchical clustering algorithms generate a tree-like diagram, called a *dendrogram*, representing a hierarchy of objects and clusters. All levels of nested clusters are represented simultaneously. This may be very valuable in gene expression studies interested in small clusters of similar genes (or samples) as well as in a few large groups of them. Commonly, the height of the dendrogram branches (the vertical lines in Fig. 2.10) joining two clusters is proportional to the distance between the clusters. To select a specific number of clusters, we may cut the dendrogram with a horizontal line crossing that particular number of branches. Sometimes, however, more customized cluster selection may be preferable that does not correspond to a single horizontal line (Chipman et al. 2003). Please note that the horizontal proximity of objects depends on their ordering within clusters and does not necessarily correspond to their similarity.

A dendrogram can be created using one of the two basic approaches: divisive (top-down) and agglomerative (bottom-up). The *divisive strategy* starts with a single cluster including all objects, splits it first into two clusters and then at each step splits recursively one of the clusters into two until each object is in its own cluster. The *agglomerative strategy* starts with n clusters (where n is the number of objects to be clustered), with each of them including one object. At each step, the two most similar clusters are joined into one, and this binary merging continues until all objects form a single cluster. Both strategies produce a hierarchy

[48]The *K*-medoids method is similar to *K*-means except that each cluster center has to be one of the cluster objects (the one with the smallest average distance to other objects of the cluster).

[49]Or, more generally, until some stopping criterion is achieved.

[50]For other distance measures the algorithm does not necessarily converge to this minimum (Theodoridis and Koutroumbas 2006).

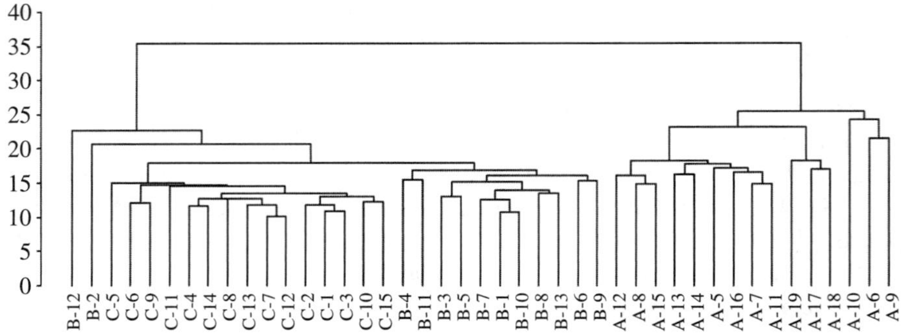

Figure 2.10: An example of a dendrogram resulting from clustering gene expression data with biological samples belonging to three classes (A, B, and C). The vertical lines represent distances (or dissimilarities) between samples or clusters. Please note that "rotating" the cluster including all samples of class C and all but two samples of class B would preserve all distances between samples or clusters; however, the two leftmost samples that belong to class B (samples B-2 and B-12) would be pictured next to other samples of class B.

with $n - 1$ levels. Each of the levels has a different number of clusters with the assignment of objects to clusters representing a "snapshot" of the ordered sequence of nested groupings.

By using a color scheme to code the expression values, we can create a *heat map* representation of the resulting gene expression matrix. This approach of coupling hierarchical clustering, and eventually two-way clustering, with heat map visualization of the results has become very popular in the area of gene expression analysis since Eisen's paper (Eisen et al. 1998). See Figure 2.11 for an example of a heat map showing the results of independent hierarchical clustering of genes and samples.

2.8.1.3.1 Agglomerative Clustering

Algorithms implementing agglomerative (or bottom-up) clustering[51] start with n clusters, with each of the n data set objects representing a separate cluster. At each step, the two closest clusters are identified and joined. To find out which clusters are closest, a measure of distance (or a measure of similarity) between any two clusters has to be defined. Assume we want to measure the distance $d(A, B)$ between clusters A and B, and that:

- objects $\mathbf{a}_\alpha$, $\alpha = 1, \ldots, n_A$ belong to the cluster A,
- objects $\mathbf{a}_\beta$, $\beta = 1, \ldots, n_B$ belong to the cluster B,
- n_A represents the number of objects in the cluster A,
- n_B represents the number of objects in the cluster B.

Commonly used distance measures are: single linkage, complete linkage, average linkage and centroid linkage.

[51]Also known as *agglomerative nesting*.

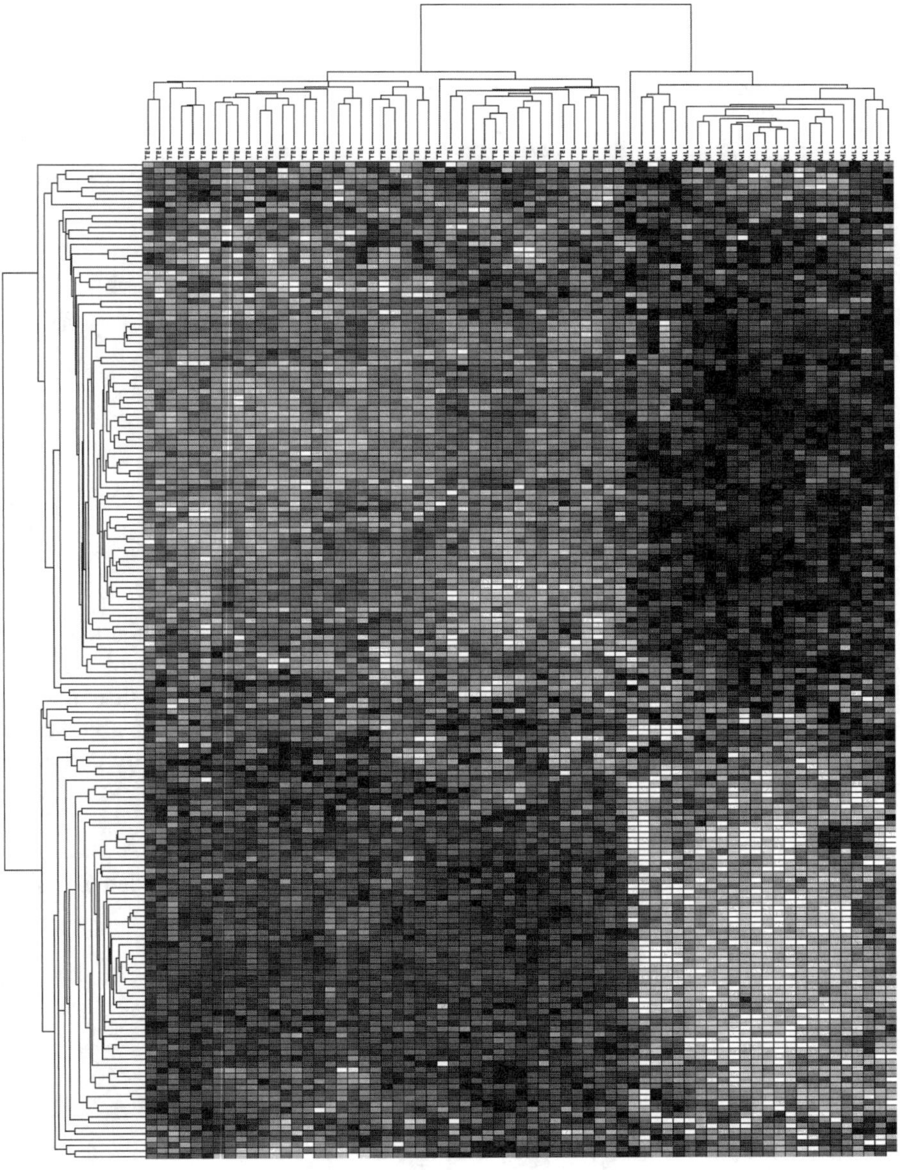

Figure 2.11: An example of a heat map showing the results of independent hierarchical clustering of genes (the dendrogram down the side of the image) and biological samples (the dendrogram across the top of the image). The expression levels are represented as color intensities or as shades of gray. When colors are used, red corresponds to higher expression levels and green to lower ones. When shades of gray are used, brighter spots represent higher expression levels. (See color insert.)

Single Linkage

The single linkage distance between two clusters A and B is defined as the distance between the *nearest neighbors* (see Fig. 2.12),

$$d(A, B) = \min_{\substack{\alpha=1,\ldots,n_A \\ \beta=1,\ldots,n_B}} d(\mathbf{a}_\alpha, \mathbf{a}_\beta). \tag{2.57}$$

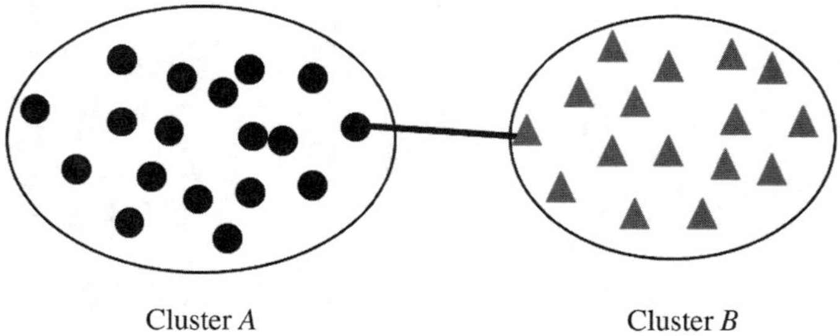

Figure 2.12: The single linkage (nearest neighbor) distance.

Since only one pair of points with a small distance is required for the distance between clusters to be declared small, the single linkage method has a tendency of "chaining"—combining long strings of clusters linked by a series of intermediate objects. This may result in only a few clusters that are quite heterogeneous. Furthermore, if the size of a cluster is defined by the largest distance between its objects, then single linkage tends to identify very large clusters (Hastie et al. 2001).

Complete Linkage

The complete linkage distance between two clusters A and B is defined as the distance between the *furthest neighbors* (see Fig. 2.13),

$$d(A, B) = \max_{\substack{\alpha=1,\ldots,n_A \\ \beta=1,\ldots,n_B}} d(\mathbf{a}_\alpha, \mathbf{a}_\beta). \tag{2.58}$$

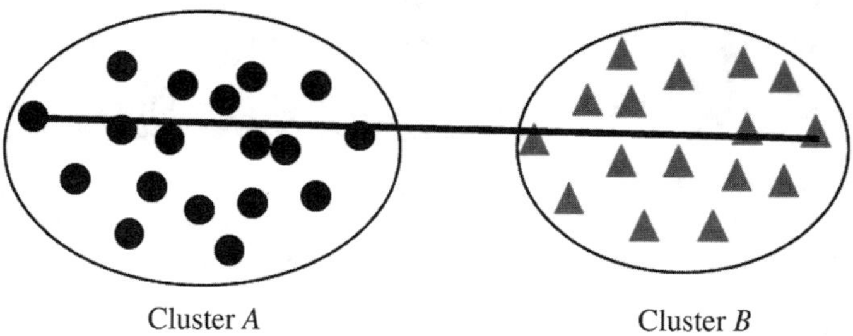

Figure 2.13: The complete linkage (furthest neighbor) distance.

The complete linkage approach tries to minimize the size of clusters, which usually results in many compact clusters. However, it may produce clusters that are so close to each other that some objects are closer to objects in another cluster than to some objects in its own cluster.

Average Linkage

The average linkage method uses the following formula to calculate the distance between clusters A and B (see Fig. 2.14),

$$d(A, B) = \frac{1}{n_A n_B} \sum_{\alpha=1}^{n_A} \sum_{\beta=1}^{n_B} d(\mathbf{a}_\alpha, \mathbf{a}_\beta). \tag{2.59}$$

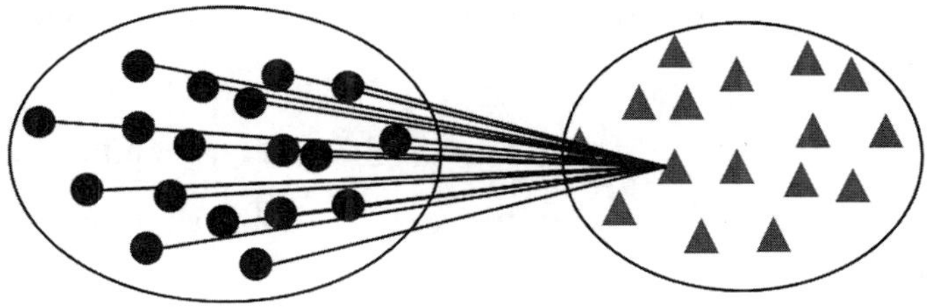

Figure 2.14: The average linkage is based on all pairwise distances between objects of the two clusters. Only distances from one of the Cluster B objects to all objects of Cluster A are shown.

The average linkage distance between two clusters is the arithmetic mean of distances between all pairs of objects in different clusters. This distance is also known as *UPGMA*—the Unweighted Pair-Group Method using Arithmetic mean.

Centroid Linkage

The centroid linkage distance is the distance between the centers of clusters A and B represented by their mean vectors (see Fig. 2.15),

$$d(A, B) = d(\mathbf{c}_A, \mathbf{c}_B), \tag{2.60}$$

where

$$\mathbf{c}_A = \frac{1}{n_A} \sum_{\alpha=1}^{n_A} \mathbf{a}_\alpha, \quad \mathbf{c}_B = \frac{1}{n_B} \sum_{\beta=1}^{n_B} \mathbf{a}_\beta. \tag{2.61}$$

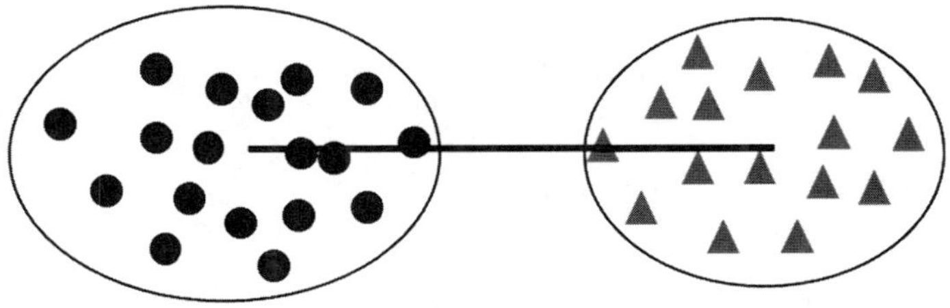

Figure 2.15: The centroid linkage distance between clusters.

The centroid linkage distance is also called *UPGMC*—the Unweighted Pair-Group Method using Centroids. Both centroid and average linkages are compromises between the extreme approaches of single and complete linkages.

Agglomerative clustering starts with calculating the distances between all possible pairs of the n initial clusters (objects $\mathbf{a}_i$, $i = 1, \ldots, n$) using one of the previously defined distance or similarity metrics. These distances can be represented by an $n \times n$ distance matrix $\mathbf{D}$ with the individual distances $d(\mathbf{a}_i, \mathbf{a}_j)$, i, $j = 1, \ldots, n$ between the initial n single-object clusters in the upper diagonal part of the matrix,

$$\mathbf{D} = \begin{bmatrix} 0 & d(\mathbf{a}_1, \mathbf{a}_2) & d(\mathbf{a}_1, \mathbf{a}_3) & \cdots & d(\mathbf{a}_1, \mathbf{a}_{n-1}) & d(\mathbf{a}_1, \mathbf{a}_n) \\ & 0 & d(\mathbf{a}_2, \mathbf{a}_3) & \cdots & d(\mathbf{a}_2, \mathbf{a}_{n-1}) & d(\mathbf{a}_2, \mathbf{a}_n) \\ & & 0 & \cdots & d(\mathbf{a}_3, \mathbf{a}_{n-1}) & d(\mathbf{a}_3, \mathbf{a}_n) \\ & & & \ddots & \vdots & \vdots \\ & & & & 0 & d(\mathbf{a}_{n-1}, \mathbf{a}_n) \\ & & & & & 0 \end{bmatrix}. \tag{2.62}$$

At each consecutive step, the two clusters with the smallest distance in the distance matrix $\mathbf{D}$ are identified and merged. The distances between the newly created cluster and all remaining ones are calculated. The two rows and columns representing the two merged clusters are replaced in the matrix with one column and one row of the distances calculated for the new cluster. Thus, at each step, a new distance matrix $\mathbf{D}$ is created and this process continues until all objects are assigned to a single cluster. Note that for each new (and smaller) distance matrix, we need to calculate distances only for the newly created cluster; all other distances remain the same. Furthermore, we do not need to use the original object data to calculate the distances for new clusters. At each stage, the current version of the distance matrix includes all distance information necessary for calculating these new distances. Assume that we are merging clusters A and B into a new cluster AB. The distance between the cluster AB and another cluster, say C, can be calculated using the distances $d(A, C)$ and $d(B, C)$ from the current matrix $\mathbf{D}$. For example, this distance $d(AB, C)$ can be calculated (Kaufman and Rousseeuw 1990):

- for the average linkage method as:

$$d(AB, C) = \frac{n_A}{n_{AB}} d(A, C) + \frac{n_B}{n_{AB}} d(B, C) \tag{2.63}$$

- for the single linkage method as:

$$d(AB, C) = \frac{1}{2}[d(A, C) + d(B, C)] - \frac{1}{2}|d(A, C) - d(B, C)| \tag{2.64}$$

- for the complete linkage method as:

$$d(AB, C) = \frac{1}{2}[d(A, C) + d(B, C)] + \frac{1}{2}|d(A, C) - d(B, C)| \qquad (2.65)$$

A hierarchy created by agglomerative clustering is monotone,[52] which means that the consecutive merging of clusters increases the distance between merged clusters.

2.8.1.3.2 Divisive Clustering

The divisive (or top-down) approach to hierarchical clustering is less often used for clustering gene expression data. However, if we are interested in the identification of a few large clusters, the divisive approach can provide better results than the agglomerative one. Top-down methods start with all objects assigned to a single cluster. The cluster is then partitioned into two clusters. At each consecutive step, one of the currently defined clusters is split into two. The process ends either when each cluster includes only one object or when a stopping criterion—such as a specified number of clusters—is achieved. To split a cluster into two child clusters, an iterative method such as the K-means method (with $K = 2$) is often used. However, if the K-means method is used, results will depend on the starting points selected at each step. Different runs may produce different results. Other approaches are, of course, possible. For example, to split a cluster, the object with the largest average distance from all other objects may be identified and used as the seed for the second cluster. Then, one by one, objects that are currently more similar to the second cluster than to their original one are moved. The process ends when there are no more such objects (MacNaughton-Smith et al. 1964).

2.8.1.3.3 Hybrid Hierarchical Clustering

Both agglomerative and divisive approaches to hierarchical clustering have their strengths and weaknesses. By the very nature of the hierarchical approach, mistakes made early in the clustering process cannot be corrected and their accumulation may lead to inferior results at later steps of the process. Therefore, agglomerative (bottom-up) clustering is recommended when we are interested in small clusters, but it is not particularly good at identifying large ones. On the other hand, the strength of the divisive (top-down) approach is in finding a few large clusters rather than many small ones. Chipman and Tibshirani proposed a *hybrid hierarchical clustering* that combines the strengths of these two methods (Chipman and Tibshirani 2006). They introduced the concept of *mutual clusters* as the central idea of the hybrid method. A mutual cluster is a group of objects that are closer to each other than to any object not in the group. This means that the maximal distance between objects of the mutual cluster is smaller than the minimal distance between any of these objects and any object outside the mutual cluster. Objects of the mutual cluster should never be separated. To accomplish this, the hybrid method starts with the identification of mutual clusters by performing preliminary agglomerative clustering using the

[52] There are exceptions. A hierarchy created with centroid linkage may sometimes be nonmonotone. However, software implementations usually correct this by adjusting branch heights (Chipman et al. 2003).

average linkage method,[53] and then uses this information in divisive clustering. The weakness of divisive clustering is in its tendency to break apart mutual clusters. To avoid this, the identified mutual clusters are replaced with their centroids and treated by divisive clustering as single objects. Once this top-down clustering of such modified data is completed, the mutual clusters are reinstated and further divided by performing either top-down or bottom-up clustering within each of them. By using divisive clustering to split the data into a few large clusters and agglomerative clustering to identify small mutual clusters, the hybrid method can produce effective clustering at both ends of the cluster size spectrum (Chipman and Tibshirani 2008).

2.8.1.4 Two-Way Clustering and Related Methods

By performing clustering simultaneously for genes and samples (rows and columns of the gene expression matrix), we may be able to identify subsets of genes with distinct expression patterns over only a subset of samples. Such blocks of coherent expression patterns may be important for pointing to biological processes associated with specific groups of genes and samples. This approach is called *two-way clustering* or *biclustering* and was described in Hartigan 1972. The *direct clustering* algorithm due to Hartigan—currently known as *block clustering*—reorders the rows and columns of the gene expression matrix in the search of blocks with homogeneous expression. It starts by treating the gene expression matrix as a single block. At consecutive steps, it splits one of the existing blocks (either by row or column) into two blocks in a way that results in the largest decrease in the total within-block variation. However, not all splits are allowed—only ones that result in hierarchical tree structures of both sample and gene clusters. The original Hartigan algorithm stops when the reduction in the total within-block variation is not greater than that expected by chance. A modified version of this algorithm (Tibshirani et al. 1999) implements a permutation-based estimation of the optimal number of blocks. Splitting continues until a large number of blocks is identified. Then pruning and permutation experiments are performed. A *gap* function, defined as the difference between the permutation-based estimation and the observed value of the within-block variation is calculated for each considered cardinality. Clustering based on the number of blocks that maximizes the gap function is deemed optimal.[54]

The preferred visualization of the results of this and similar two-way clustering methods is in the form of a heat map coupled with two dendrograms, one for samples and one for genes. Sometimes genes and samples are clustered separately and the results are shown together on the heat map in a way similar to that of two-way clustering (Eisen et al. 1998; Alizadeh et al. 2000).

Two-way clustering results in a single ordering of samples for all genes. Some methods investigate the possibility of more informative presentations allowing for the different ordering of samples for different subsets of genes or for overlapping clusters.

[53]Agglomerative clustering using the average, single, or complete linkage methods does not break mutual clusters (Chipman and Tibshirani 2006).

[54]Maximizing this gap function means minimizing the within-block variation when compared to the one expected by chance.

The *plaid model* (Lazzeroni and Owen 2002) can be seen as a method merging two-way clustering with analysis of variance (ANOVA). It identifies rectangular layers, each of them corresponding to a subset of genes with similar expression patterns over a subset of samples. The layers are similar to blocks in block clustering, but they may overlap. Furthermore, a gene may belong to many layers or it may belong to none of them.

Gene shaving (Tibshirani et al. 1999; Hastie et al. 2000) is another method related to two-way clustering. It looks usually for small clusters of highly correlated genes whose average expression has large variation across the samples. The gene shaving algorithm implements principal component analysis (refer to Section 2.8.2) to identify the direction of the most variation in the p-dimensional space of p genes. It starts with the entire gene expression matrix and finds the largest principal component, which can be called an *eigengene* (for it is a linear combination of the genes along the direction of the most variation in the data). Then genes are sorted according to their correlations with the eigengene and a fraction (for instance, 10 percent) of the least correlated genes is removed, or *shaved-off*. The procedure is then repeated in subsequent steps for gene expression submatrices including fewer and fewer genes, until only one gene remains. The result is a sequence of nested subsets of genes. Selection of one of these subsets is based on permutation experiments and the percent of variance explained by the vector of average expressions of the subset genes. The subset that maximizes the gap function—defined here as the difference between the observed variance and the variance expected by chance[55]—is selected as the first cluster of genes. The vector of the average expressions of the cluster genes is called a *supergene*. To identify the second and then subsequent optimal clusters of genes, the gene expression matrix is orthogonalized with respect to the identified supergene, that is, from each row of the matrix (representing a gene) the component correlated with the supergene is removed. The entire process is then repeated until the predefined number of optimal clusters is identified.

There are many biclustering algorithms (Madeira and Oliveira 2004) and new ones are being developed. Promising extensions include clustering 3D gene expression data (the gene expression matrix with the added temporal dimension). The *gene-sample-time*, or *GST*, microarray data may be used to identify clusters of genes with similar expression patterns over a subset of samples and across the series of time points (Zhang 2006).

A word of warning: given a large number of variables, two-way clustering (as well as independent clustering of genes and samples) will always find some patterns in data, even when the data is randomly generated noise. On a heat map, the patterns in the random data may even look as interesting as ones based on real gene expression data. Which of the gene data patterns are biologically relevant? This cannot be answered by cluster analysis. As Drăghici noted "*Contrary to the popular belief, clustering is not a goal in itself and, by itself, is seldom convincing.*" (Drăghici 2003). Biological knowledge has to drive the interpretation of clustering results by formulating hypotheses to be subsequently tested by approaches external

[55]By maximizing the gap function as defined, the cluster of genes for which the observed variance across samples most exceeds the one expected by chance is selected.

to cluster analysis. Furthermore, when interpreting a heat map we have to keep in mind that changes in the clustering algorithm or the distance measure may produce different results.

2.8.2 Principal Component Analysis

> *"Principal components are a sequence of projections of the data, mutually uncorrelated and ordered in variance."*
>
> —(Hastie et al. 2009)

Principal component analysis (PCA) is a technique allowing the reduction of the data dimensionality in such a way that the resulting lower-dimensional representation of the data preserves as much of its variation as possible. The technique dates back to Pearson and then Hotelling (Pearson 1901; Hotelling 1933) with some even earlier work on *singular value decomposition*, a related method.

As in cluster analysis, either genes or samples can be treated here as dimensions (variables). Let us choose one of these options and treat samples as objects and genes as variables. Therefore, our goal is to transform the p-dimensional space defined by p genes into a low dimensional space in order to facilitate the visualization of samples that may reveal their groupings. A data set of N biological samples and p variables can be represented by a p-dimensional cloud of N points. If we approximate the cloud by a p-dimensional hyperellipsoid, the directions of the principal components will be aligned with the axes of the hyperellipsoid.[56] Another interesting geometric interpretation of PCA has been given by Flury and Riedwyl (1988). If we enclose our N data points in a p-dimensional hypersphere (centered at the multivariate mean of the cloud of N points), then the principal components will correspond to such a set of orthogonal directions that successively minimize the Mahalanobis distance[57] between the data cloud and the hypersphere (and move across the origin of the hypersphere).

The first principal component (PC) identifies the direction with the most variation in the data. The second PC is orthogonal to the first and captures the direction in the data having the greatest remaining variation. The third PC is orthogonal to the first two PCs and corresponds to the direction with the greatest still remaining variation, and so on. Although up to p principal components can be defined, we hope that the first m of them, where $m \ll p$ will account for *almost all* of the variation in the p original variables (Jolliffe 2002). Usually, by transforming the original p-dimensional space into a space defined by the first few PCs, we can retain a significant portion of the total variation in the data. To facilitate visualization, it is preferable to use only the first two or three principal components. Since principal components are

[56]This geometric interpretation is statistically correct only if we assume that the original variables follow a multivariate normal distribution. However, the multivariate normality assumption is not necessary if we use the PCA technique as a descriptive tool rather than for statistical inference. Jolliffe notes that PCA "*can provide valuable descriptive information for a wide variety of data, whether the variables are continuous and normally distributed or not.*" (Jolliffe 2002).

[57]Mahalanobis distance has been defined in Section 2.8.1.

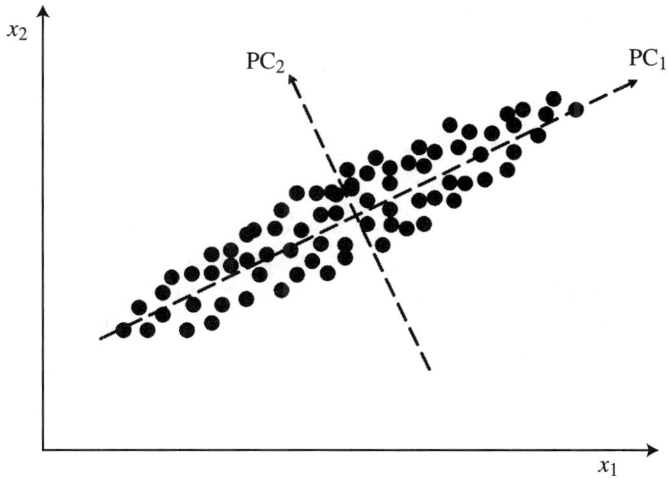

Figure 2.16: An illustration of principal component directions. The first principal component PC_1 identifies the direction of the most variation in the data. In this toy example, the original space is defined by only two variables. Hence, the second principal component PC_2, orthogonal to the first one, captures the remaining data variation. Each of the principal components can be represented by a linear function of original variables x_1 and x_2.

orthogonal to each other, they define a set of uncorrelated variables, each being a linear combination of the original variables (genes) (see Fig. 2.16).

To calculate principal components we can use either correlations or covariances. There are arguments for and against each of these approaches[58] and generally neither has a clear advantage over the other. For microarray data where all the gene expression variables are measured in the same units, it is more common to use the covariance matrix. Assuming, as before, that we treat our N biological samples as objects and the p genes as variables, each of the samples is represented by a p-dimensional vector $\mathbf{x}_i \in \Re^p$,

$$\mathbf{x}_i = \begin{bmatrix} x_{1i} \\ x_{2i} \\ \vdots \\ x_{pi} \end{bmatrix}, \quad i = 1, \ldots, N. \tag{2.66}$$

The general idea of PCA is to find the characteristic vectors of the covariance matrix, which are solutions to the following eigenproblem:

$$\mathbf{Se} = \lambda \mathbf{e} \tag{2.67}$$

where

[58]For example, if a covariance matrix is used when the original variables are of different types, with different units of measurement and very different variances, the first few PCs may be strongly influenced by the relative sizes of the variances, hardly giving us any new information.

- $\mathbf{S}$ is the [statistical] sample covariance matrix,[59]

$$\mathbf{S} = \frac{1}{N-1}\sum_{i=1}^{N}(\mathbf{x}_i - \overline{\mathbf{x}})^T(\mathbf{x}_i - \overline{\mathbf{x}}). \tag{2.68}$$

The $p \times p$ covariance matrix $\mathbf{S}$ is also called the variance-covariance matrix since its diagonal elements $s_{kk} = s_k^2$ represent the variances of variables x_k, $k = 1, \ldots, p$, and its off-diagonal elements s_{kl} are the covariances between variables x_k and x_l, $k \neq l$ and $k, l = 1, \ldots, p$. Since $s_{kl} = s_{lk}$ for any $k, l = 1, \ldots, p$, the matrix $\mathbf{S}$ is always symmetric,

$$\mathbf{S} = \begin{pmatrix} s_{11} & s_{12} & \cdots & s_{1p} \\ s_{12} & s_{22} & \cdots & s_{2p} \\ \vdots & \vdots & \ddots & \vdots \\ s_{1p} & s_{2p} & \cdots & s_{pp} \end{pmatrix}, \tag{2.69}$$

- $\overline{\mathbf{x}}$ is the mean vector of the N sample points

$$\overline{\mathbf{x}} = \frac{1}{N}\sum_{i=1}^{N}\mathbf{x}_i \tag{2.70}$$

- $\mathbf{e}$ is a normalized eigenvector of the matrix $\mathbf{S}$ and λ is its corresponding eigenvalue.[60] There are p eigenvalue-eigenvector pairs that are the solutions to (2.67). For small p, the eigenvalues can be found by expanding the characteristic equation[61]

$$\det|\mathbf{S} - \lambda\mathbf{I}| = 0 \tag{2.71}$$

and solving the resulting pth degree polynomial in λ whose roots are the eigenvalues (Anderson 2003). However, for a large number of variables p, this method would be computationally intensive. Due to the fact that the covariance matrix $\mathbf{S}$ is always symmetric, *singular value decomposition* (SVD), a very efficient numerical method, can be used in such situations to find eigenvalues and eigenvectors of the matrix $\mathbf{S}$ (Press et al. 2007).

[59] $\mathbf{S}$ is an estimate of the unknown population covariance matrix $\boldsymbol{\Sigma}$.

[60] Texts that discuss both population principal components and sample principal components usually use λ to denote population eigenvalues and l for sample eigenvalues. Here we discuss only the sample principal components, and λ is used to denote the *sample* eigenvalues.

[61] det $|\mathbf{A}|$ denotes the determinant of matrix $\mathbf{A}$, and $\mathbf{I}$ is the $p \times p$ identity matrix with 1s on the diagonal and 0s off the diagonal.

Solving the eigenproblem (2.67) gives us a set of p ordered eigenvalues λ_k, $k = 1, \ldots, p$,

$$\lambda_1 \geq \lambda_2 \geq \cdots \geq \lambda_p \geq 0 \tag{2.72}$$

and, paired with them, p normalized eigenvectors

$$\mathbf{e}_k = \begin{bmatrix} e_{1k} \\ e_{2k} \\ \vdots \\ e_{pk} \end{bmatrix}, \quad k = 1, \ldots, p. \tag{2.73}$$

Each eigenvalue-eigenvector pair (λ_k, $\mathbf{e}_k$) is associated with a principal component. The eigenvector $\mathbf{e}_k$ defines the direction of the kth principal component (PC_k) whereas the eigenvalue λ_k represents the amount of data variance along this principal component direction. Each principal component can be described as a linear combination of the original p variables,

$$\begin{aligned} PC_k &= e_{1k}x_1 + e_{2k}x_2 + \cdots + e_{pk}x_p \\ &= \mathbf{e}_k^T \mathbf{x} \end{aligned} \tag{2.74}$$

with the elements of the eigenvector $\mathbf{e}_k$ representing the weights (or loadings) of the original variables. The proportion of the total variance explained by the kth principal component can be calculated as

$$\gamma_k = \frac{\lambda_k}{\sum_{l=1}^{p} \lambda_l}, \quad k = 1, \ldots, p \tag{2.75}$$

and the cumulative proportion of total variance explained by the first m principal components, $m \leq p$, as

$$\begin{aligned} \Gamma_m &= \sum_{h=1}^{m} \gamma_h \\ &= \frac{\sum_{h=1}^{m} \lambda_h}{\sum_{l=1}^{p} \lambda_l}. \end{aligned} \tag{2.76}$$

The denominators of (2.75) and (2.76) represent the total variance in the data set, which is equal to the sum of all p eigenvalues, and also to the sum of the diagonal elements of the variance-covariance matrix **S**,

$$\begin{aligned}\text{Total variance} &= \sum_{k=1}^{p} \lambda_k \\ &= \sum_{k=1}^{p} s_{kk}.\end{aligned} \tag{2.77}$$

Please note that if the original variables are standardized to a mean of zero and standard deviation of one, then the total data variance will be equal to the number of variables p,

$$\begin{aligned}\text{Total variance (standardized)} &= \sum_{k=1}^{p} s_{kk} \\ &= \sum_{k=1}^{p} s_k^2 \\ &= \sum_{k=1}^{p} 1 \\ &= p.\end{aligned} \tag{2.78}$$

If, as stated earlier, our main goal for performing PCA is visualization of samples (or genes), then we would like to retain only two or three principal components. From (2.76) we can find out how much of the total data variance would be represented by either of these visualizations. This would give us some information on the validity of conclusions about sample grouping based on such visualizations. If we wanted to go beyond visualization, a valid question after identifying the PCs could concern the number of principal components to retain. The simplest way would be to decide what proportion of the explained variance is satisfactory for our particular study,[62] and use (2.76) to drive the selection. If, however, we want to discard only the uninformative PCs, we need to define some cut-off for their informativeness. One possible way to do this is called the *broken stick model* (Jolliffe 2002). Consider a unit length stick. If we break the stick, at random, into p pieces, then the expected length of the kth longest piece can be calculated as

$$\gamma_k^* = \frac{1}{p} \sum_{l=k}^{p} \frac{1}{l}. \tag{2.79}$$

We can compare the proportion of the variance explained by each principal component (2.75) with that expected by chance (2.79) and retain only the principal

[62]Often the cut-off is in the range of 0.7–0.9. It depends, however, on the data set and goals of the study. The number of PCs selected for different cut-off values may also be taken into account.

components for which the following inequality is true,

$$\gamma_k > \gamma_k^*, \quad k = 1, \ldots, p. \tag{2.80}$$

To project samples represented by the p-dimensional vectors $\mathbf{x}_i$, $i = 1, \ldots, N$, in the original space of p genes onto the space defined by the first m principal components, we will use the first m eigenvectors and for each sample will calculate the vector $\mathbf{w}_i$ of its new coordinates,

$$\mathbf{w}_i = \begin{bmatrix} w_{1i} \\ \vdots \\ w_{mi} \end{bmatrix} \tag{2.81}$$

as

$$\mathbf{w}_i = \mathbf{E}_m^T \mathbf{x}_i, \tag{2.82}$$

where $\mathbf{E}_m$ is a $p \times m$ matrix whose m columns are the eigenvectors associated with the first m principal components,

$$\mathbf{E}_m = \begin{bmatrix} e_{11} & \cdots & e_{1m} \\ e_{21} & \cdots & e_{2m} \\ \vdots & \ddots & \vdots \\ e_{p1} & \cdots & e_{pm} \end{bmatrix}. \tag{2.83}$$

We have to remember that PCA is an unsupervised technique that identifies the directions of the most data variation. These directions do not need to be in any way related to the discriminatory directions sought after by supervised classification problems. In particular, this means that principal component analysis should **not** be used as a preprocessing step for the supervised analysis. To read more on this subject, refer to Chapter 3.

2.8.3 Self-Organizing Maps

The *self-organizing map* (SOM), also known as *Kohonen network*, is the unsupervised artificial neural network (ANN) learning algorithm introduced by Teuvo Kohonen (Kohonen 1982a, 1982b). The algorithm is considered "*one of the most realistic models of the biological brain function*" (Kohonen 2001). It projects high-dimensional data usually onto a two-dimensional grid,[63] or map, in a way that preserves the topological relations between data objects and groups of these objects. The SOM is a clustering method (and a visualization tool) that groups objects into a predetermined rectangular $K_1 \times K_2$ grid of clusters. As the neighboring clusters

[63] One- or three-dimensional maps are also used as well as two-dimensional maps with topologies different from rectangular, for example, hexagonal ones.

are more similar to each other than to clusters that are farther away on the grid, the SOM clustering is more informative than K-means or even hierarchical clustering.[64] As with other clustering methods, we can use SOM to cluster either genes or biological samples. However, in gene expression analysis, we usually use this method to group genes into clusters of similar expression profiles.

Assume then that we treat the p genes of the gene expression matrix as objects and its N samples as variables. Therefore, each of the genes can be represented by an N-dimensional vector $\mathbf{x}_g \in \Re^N$ of the gene expression values for the N samples,

$$\mathbf{x}_g = \begin{bmatrix} x_{1g} \\ x_{2g} \\ \vdots \\ x_{Ng} \end{bmatrix}, \quad g = 1, \ldots, p. \tag{2.84}$$

The self-organizing map is a neural network (NN) with the input layer of N inputs representing the original N variables (here, biological samples), no hidden layers, and the output layer being a $K_1 \times K_2$ grid of neurons corresponding to cluster prototypes (Fig. 2.17).

Each input is connected to each neuron. Each of these $N \times K_1 \times K_2$ connections is characterized by a weight (Fig. 2.18), which is equivalent to saying that each neuron η_k, $k = 1, \ldots, K$, where $K = K_1 \times K_2$, has assigned to it an N-dimensional vector of weights $\mathbf{w}_k \in \Re^N$,

$$\mathbf{w}_k = \begin{bmatrix} w_{1k} \\ w_{2k} \\ \vdots \\ w_{Nk} \end{bmatrix}, \quad k = 1, \ldots, K. \tag{2.85}$$

The weight vectors $\mathbf{w}_k$ represent the cluster prototypes and are initialized to random values from the range of gene expression values in the data.[65] During the learning process, the network is presented with the p gene expression patterns as inputs, one at a time, preferably in a random order.

The neurons *compete* in the sense that the one whose weight vector $\mathbf{w}_k$ is most similar (according to the similarity or distance measure used) to the current input pattern $\mathbf{x}_g$ is declared the *winner* and its weights are adjusted towards the values of the vector $\mathbf{x}_g$. The magnitude of the adjustment depends on the *learning rate* α. Although this *competitive learning* is called the *winner-takes-all-approach*, the approach is complemented by the *network plasticity* characteristic resulting in

[64] K-means clustering may be misleading about relations between objects; it is quite possible that objects close to each other in the original space are assigned to different clusters and appear far from each other. Although hierarchical clustering is more informative, the horizontal dendrogram distances between objects of different clusters do not necessarily reflect their similarity (Drăghici 2003).

[65] Preferably using an appropriate distribution of the gene expression data.

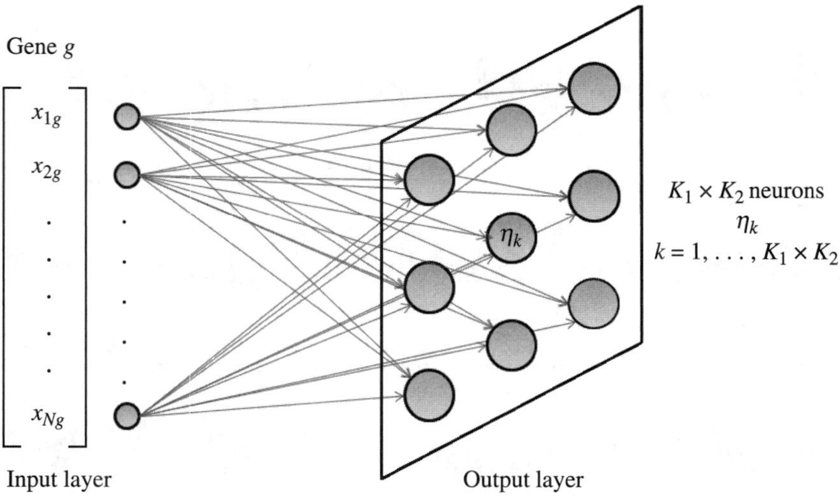

Figure 2.17: An example of the self-organizing map. The input layer has N nodes corresponding to N variables. When we cluster genes, each gene is represented by an N-dimensional vector of its expression values for the N samples of the data set. The output layer is a rectangular grid of $K_1 \times K_2$ neurons; in the example $K_1 = K_2 = 3$. Each input node is connected to each neuron.

the propagation of learning to a local neighborhood of the winning neuron. Thus, the "excitement" of the winner is shared with its neighbors; their weights are also adjusted, though to a lesser extent. To determine the winner's local neighborhood and the magnitude of each neighbor's weights adjustment, we define a neighborhood function $h(\eta_k, \eta_l)$, $k, l = 1, \ldots, K$ whose value decreases with the increase in the

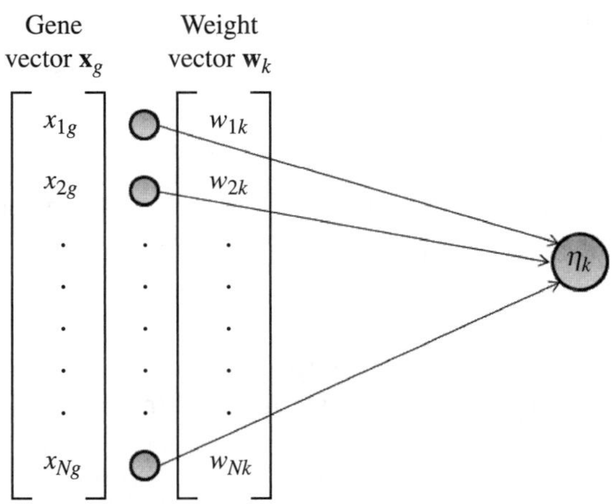

Figure 2.18: There are N connections to each neuron η_k, each with its own weight. Hence, an N-dimensional vector of weights, whose values are adjusted during the learning process, is associated with each neuron. The vector of weights $\mathbf{w}_k$ is a prototype of the gene expression profile for the cluster of genes assigned to the neuron η_k.

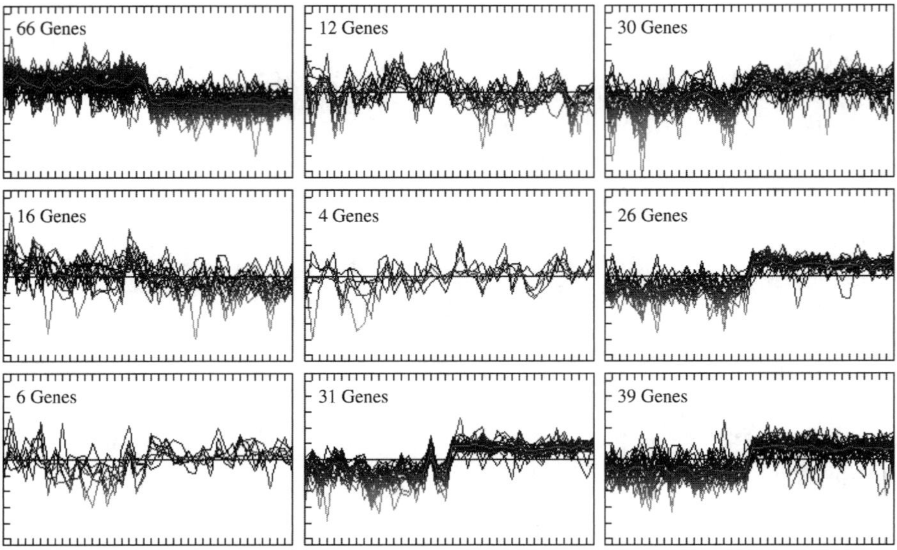

Figure 2.19: An example of graphical presentation of clustering genes by similarity of their expression profiles. Self-organizing map with 3 × 3 rectangular topology and correlation distance have been used. The image was obtained with *MultiExperiment Viewer* software (Saeed et al. 2003). (See color insert.)

distance[66] between neurons η_k and η_l. The first iteration[67] of the learning algorithm ends when all p genes are presented to the network. The learning rate α as well as the size of the local neighborhood are then reduced and the second iteration started. The process continues until convergence or until a specified number of iterations has been performed. When the learning process ends, each neuron is associated with the cluster prototype—the expression profile represented by its weight vector. Each gene is assigned to the neuron (cluster) with the most similar expression profile (Fig. 2.19).

Example of the SOM Algorithm

1. For the selected $K_1 \times K_2$ rectangular topology of the neural network:
   - Initialize the weight vectors of all $K = K_1 \times K_2$ neurons to random values

$$\mathbf{w}_k(t), \quad k = 1, \ldots, K, \tag{2.86}$$

   where t is the index of learning steps; it starts with $t = 1$ and then is increased after each consecutive step of weights adjustment.

[66]Note that we use two distance measures. One is the distance between neurons on the two-dimensional grid; this distance is used to define the local neighborhood. The other is the distance between two vectors in the N-dimensional space of the original variables; one vector represents the gene expression pattern of a data point, the other is the cluster prototype represented by the neuron's weight vector.

[67]In neural network terminology, the iteration—or a single pass through all input vectors—is called an *epoch*.

- Initialize the learning rate α.
- Initialize the neighborhood function

$$h(\eta_k, \eta_l), \quad k, l = 1, \ldots, K. \tag{2.87}$$

The value of the neighborhood function equals 1 if $k = l$, is between 0 and 1 for neurons η_l that are in the local neighborhood of the neuron η_k, and is zero otherwise.

- Specify the maximum number of iterations (epochs).

2. Repeat until convergence or until the maximum number of iterations is performed:

a) Loop over the data set of the p gene expression patterns. The steps are indexed by the consecutive values of t:

- present a single input pattern $\mathbf{x}_g(t)$ to the network,
- using the selected distance measure $d(\mathbf{x}_g, \mathbf{w}_k)$, identify the neuron most similar to $\mathbf{x}_g(t)$ and call it the winning neuron $\eta_c(t)$,

$$\eta_c(t): \quad c = \arg\min_k d(\mathbf{x}_g(t), \mathbf{w}_k(t)), \tag{2.88}$$

- assign the input gene expression pattern $\mathbf{x}_g(t)$ to the cluster represented by the winning neuron,
- adjust the weights of the winning neuron $\eta_c(t)$ and all neurons in its local neighborhood

$$\mathbf{w}_k(t+1) = \mathbf{w}_k(t) + \alpha \cdot h(\eta_c(t), \eta_k)[\mathbf{x}_g(t) - \mathbf{w}_k(t)], \quad \text{for } k = 1, \ldots, K. \tag{2.89}$$

The neighborhood function $h(\eta_c(t), \eta_k)$ assumes the following values:

$$h(\eta_c(t), \eta_k) = 1 \qquad \text{for the winning neuron } \eta_k = \eta_c(t),$$

$$0 < h(\eta_c(t), \eta_k) < 1 \qquad \text{for neurons in the current neighborhood of the neuron } \eta_c(t),$$

$$h(\eta_c(t), \eta_k) = 0 \qquad \text{otherwise.}$$

b) Decrease the value of the learning rate α.

c) Decrease the size of the local neighborhood by redefining the neighborhood function $h(\eta_k, \eta_l)$.

When neurons are initialized to random weights[68] (selected from the range of expression values represented in the data), it is possible that some neurons will

[68]Instead of using randomly selected weights, we may initialize neurons to the vectors representing points lying on the two-dimensional plane defined by the first two principal components of the data. This would allow for an interesting geometric interpretation of the learning process (Hastie et al. 2001).

never win, and the clusters associated with them will be empty. To avoid such situations, we may initialize the cluster prototypes to equal the gene expression vectors randomly selected from the data set. The neighborhood function should be initialized in such a way that defines a large local neighborhood that may even include the entire network.[69] This way initial learning is widely propagated to allow for the correct global mapping of the topological relationships between groups of expression patterns. Then the size of the local neighborhood is gradually decreased until it includes only the winner. This allows focusing on the local spatial resolution after the global relationships have been mapped. A simple definition of the neighborhood may be based on the Euclidean distance between the neurons on the grid. Note that if the initial local neighborhood is too small and includes only the winning neuron, then the spatial relations between clusters cannot be mapped and the SOM algorithm is reduced to K-means clustering.

As with the local neighborhood, the learning rate gradually decreases. The high initial rate results in crude clusters, which are then fine-tuned when the learning rate decreases. Commonly, either the Euclidean or the correlation distance is used to identify the neuron most similar to the presented pattern.

Although both SOM and PCA project data onto a low-dimensional space, the SOM algorithm—unlike PCA—allows for nonlinear projections. In that sense, SOM can be regarded as a nonlinear generalization of PCA (Oja et al. 2006).[70]

Running the SOM algorithm with different topologies of the network, different initializations of cluster prototypes, different learning rates and local neighborhood functions, different orders of presenting data points, and different distance measures may lead to different results. Consider, for example, the measure of distance between gene vectors and cluster prototype vectors. The Euclidean distance may tend to group together genes with similar expression levels whereas the correlation distance will group genes with similar shapes of their expression patterns across the samples. Sometimes the Chebyshev distance may provide even better separation of profiles by both the shape and the expression level (Drăghici 2003). To decide on the network topology, we may explore maps with different numbers of neurons. If we start with only a few neurons, the resulting clusters will most likely have a large within-cluster variation. Then, we may keep adding nodes until distinctive clusters with low variation are identified (Tamayo et al. 1999).

EXERCISES

2.1 Search public repositories for at least five gene expression data sets related to cancer or central nervous system (CNS) diseases. Limit your search to the *Homo sapiens* species, Affymetrix gene expression microarrays, and experiments submitted during the last five years. Do *not* consider experiments with the total number of biological samples

[69] Although, if the local neighborhood is too large for a given grid resolution, the cluster prototypes tend to be updated in blocks (Ripley 1996).

[70] *Principal surfaces* also generalize PCA. Thus, SOM can be seen as a discrete version of principal surfaces (Hastie et al. 2001).

less than twenty. Your selection should include experiments with two classes, three classes, and at least one with more than four classes. For each data set, report:

- accession number (e.g., *GSE13425*),
- year of submission,
- microarray type (e.g., *HG-U133 2.0 Plus*),
- total number of biological samples,
- number and names of the differentiated classes,
- number of biological samples in each class,
- brief description of main goals of the original study.

2.2 From *ArrayExpress* or *Gene Expression Omnibus*, select a gene expression data set with two differentiated classes. It has to include detection call information (Present, Absent, and Marginal). Furthermore, raw data (CEL files) should also be available.

a) Download the gene expression and detection call data.

b) In Excel, perform gene expression level quality assessment. Perform also any additional preprocessing that you determine necessary.

c) Using Excel, filter the data by the detection calls and by the range of expression values. Decide on filtering criteria that are appropriate for the selected data set.

d) Describe all steps of your experiment. Present and comment on the results. Include the Excel file showing the steps of your experiment.

Note:
When downloading data from *ArrayExpress*, select fields corresponding to gene expression signal, detection call and probe set name. This will give you a data matrix with a single probe set column and with two columns (signal and detection call) for each biological sample. Save the exported data as a .txt file and then open it in Excel. Commonly, class names are included in sample names. Alternatively, you may download a separate *sample annotation* file.

2.3 Perform basic exploratory analysis of the data set prepared in Exercise 2.2.

a) Download the *Significance Analysis of Microarrays* (SAM) software http://www-stat-class.stanford.edu/~tibs/clickwrap/sam.html (free for academic use).

b) Familiarize yourself with the software, implemented methods, and input parameters.

c) Analyze the data set prepared in Exercise 2.2 and identify a list of differentially expressed genes. Remember about appropriate selection of parameters resulting in a reasonable FDR rate.

d) Using the TTEST function in Excel, perform an ordinary t test for the same data set. Remember about adjusting for multiple testing—apply the following corrections:

- the single-step Bonferroni procedure,
- the step-down Holm procedure,
- the step-up Benjamini and Hochberg procedure.

e) Compare the results of the three methods of correction for multiple comparisons. Then, compare them with the SAM results. Present the results and their comparison in the form of Excel worksheets of the same Excel file.

f) Describe your experiments (including discussion of the selected values of SAM parameters), observations and conclusions.

2.4 Perform all steps of Exercises 2.2 and 2.3 for a gene expression data set with three differentiated classes. In addition, explain why "significant negative genes" are not reported for the *Multiclass* SAM experiment.

Note: Since there are more than two classes, the ANOVA F test (instead of the t test) should be performed. Unfortunately, Excel does not have a cell function for the ANOVA F test. Although it can be done using a one-cell formula, it would be more instructive if you add four columns to the spreadsheet and—for each probe set (row)—calculate:

- variance between classes (*MSTR*),
- variance within classes (*MSE*),
- the value of the F statistic (the ratio of these two variances),
- the p-value (using Excel's FDIST function).

2.5 Perform low-level preprocessing of raw expression data using the Affymetrix *Expression Console* software:

a) Download the raw expression data (CEL files) corresponding to the data set used in Exercise 2.2 (or 2.4). If necessary, download additional annotation information (assignment of samples to classes).

b) From the Affymetrix website, download and install *Expression Console* software.

c) Familiarize yourself with the software and configure it for your experiment:

- download the library and annotation files appropriate for the microarray type used in the experiment,
- create a new study and configure MAS5 algorithm to use the same target intensity value as the one used in the original gene expression data set.

d) Run the low-level analysis of the raw data. Export the resulting probe set level expression data and compare them with the data downloaded for Exercise 2.2 (or 2.4).

2.6 Repeat the experiments performed in Exercise 2.5 using the other low-level preprocessing algorithms implemented in the *Expression Console* software—RMA and PLIER. Compare results and discuss their differences. For example, visualize the relationship between expression levels of a selected pair of arrays using MA plots. Discuss the differences in the MA plots for the data preprocessed with the three algorithms.

2.7 Using the *self-organizing map* (SOM) algorithm, group genes into clusters of similar expression patterns:

a) Use a data set including differentially expressed genes identified by one of the univariate methods used in Exercise 2.2 (or 2.4).

b) Use any software package that implements the *self-organizing map* algorithm (e.g., *MultiExperiment Viewer*).

c) Depending on capabilities of the selected SOM implementation, perform experiments with different settings of SOM parameters. Parameters to consider include:

- topology of the neural network,
- distance metric,
- neighborhood definition,

- learning rate,
- initialization of the weight vectors,
- number of iterations.

d) Compare results of SOM clustering performed with several different sets of parameters.

CHAPTER 3

BIOMARKER DISCOVERY AND CLASSIFICATION

Supervised analysis of gene expression microarray data:

- Biomarker discovery
- Feature selection
- Classification

"The field of Statistics is constantly challenged by the problems that science and industry brings to its door".

—(Hastie et al. 2001)

3.1 OVERVIEW

Basic exploratory analysis and unsupervised (taxonomy-related) analysis of gene expression data (covered in Chapter 2) are well represented and usually well implemented in many software packages and programs. Generally, the implementation and use of unsupervised methods of data mining can be considered quite mature in the area of biomedical research. Unfortunately, the same cannot be said about supervised methods, especially for biomarker discovery. In this chapter, we will focus on this area (shown on Fig. 3.1 as Element D). Although many well established or newer methods (such as Discriminant Analysis, Support Vector Machines, Logistic Regression, or Decision Trees) are implemented and used, many publications and software applications seem to neglect or underplay the crucial step of feature selection. Figure 3.2 shows main elements of biomarker discovery, which include:

- feature selection,
- building classification model (depending on the feature selection approach, the first two steps can be separated or coupled together),
- model validation (preferably on an independent test set),

Data Mining for Genomics and Proteomics. By Darius M. Dziuda

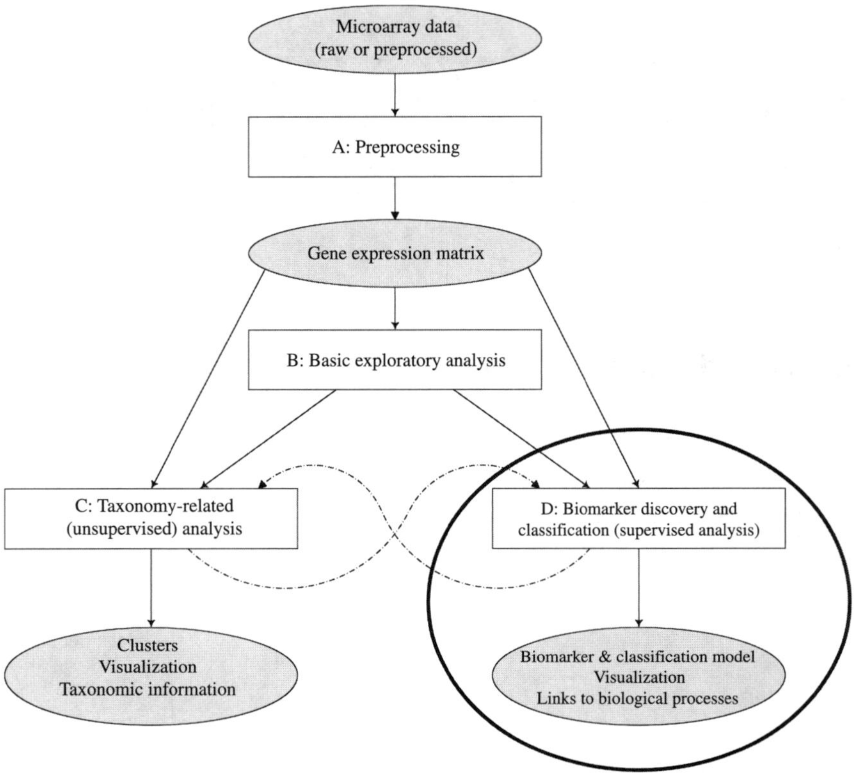

Figure 3.1: Elements of microarray gene expression data analysis—an example. The focus of this chapter is on the element D: Biomarker discovery and classification.

- implementation of the classification model (preferably including visualization of the discriminatory space),
- elucidation of biological processes underlying the class differentiation.

The sequence of the biomarker discovery steps shown in Figure 3.2 is only an example and simplification of the process. Although biomarker discovery can be performed in such a simple sequential way (e.g., in situations where discriminated classes are easily separable), usually the process includes iterations of some elements or their combinations. In this chapter, we will look at the elements of the biomarker discovery process and discuss their interactions. In Chapter 4, we will describe how to identify the *Informative Set of Genes* facilitating biological interpretation of class differences, and then show how to use this set in the biomarker discovery process for additional optimization that may lead to robust multivariate biomarkers with plausible biological interpretation.

Identification of biological processes associated with the class discrimination is sometimes considered a 'follow-up' step for the main goals of biomarker discovery. One may argue that positive validation of a biomarker on a large and *independent* test set is enough for its deployment, even if we do not understand the biology

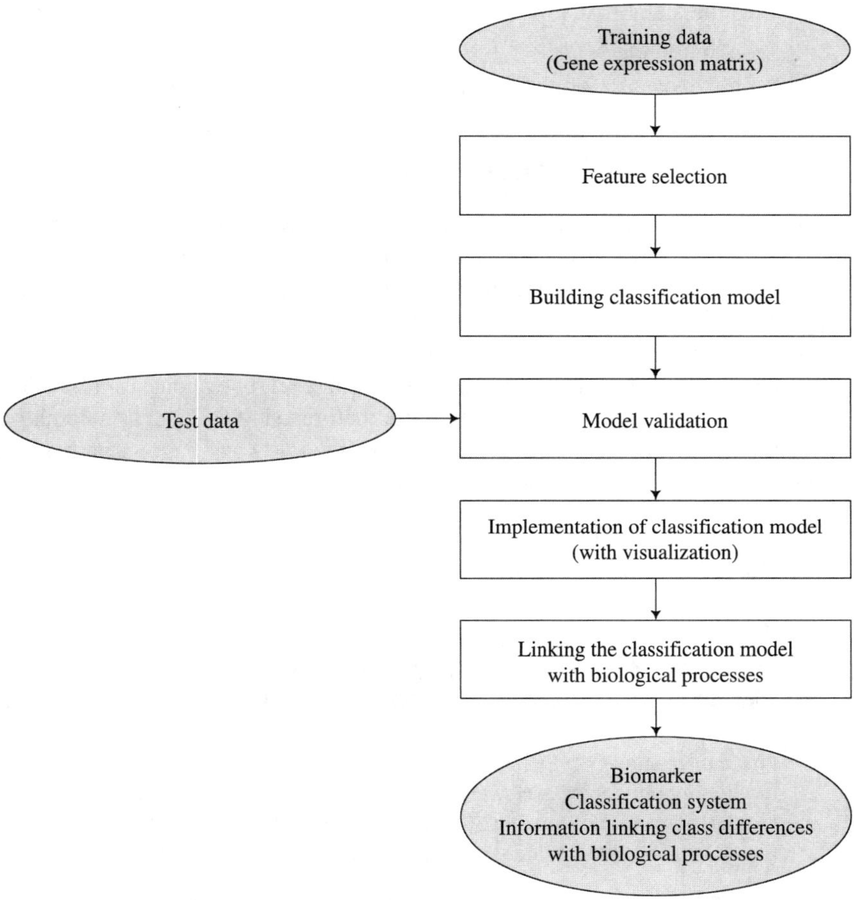

Figure 3.2: Elements of biomarker discovery. Please note that this is a simplification of the biomarker discovery process. Often, biomarker discovery includes various iterations and combinations of these elements. In Chapter 4, we will describe how to combine these elements into a process allowing the identification of robust and interpretable multivariate biomarkers.

underlying the differentiation (Baker 2005). This opinion is easier to accept for some types of biomarkers, for example, diagnostic ones. However, for some other types, such as biomarkers predicting drug efficacy, it seems to be crucial to understand the relationship between the biomarker and the mechanisms of disease. Especially valuable are study outcomes, in which a new biomarker leads to new knowledge about the biology of the disease.

The main goals of biomarker discovery are:

- to identify small subsets of variables that can be used for the efficient classification of new samples,
- to provide fast and cost-effective classifiers that can be easily implemented into clinical practice,

- to link the identified biomarkers and the class differences with underlying biological processes,
- to facilitate visualization of the discriminatory space and classification results.

3.1.1 Gene Expression Matrix . . . Again

We will again start with the data—this time preprocessed, scaled, quality-controlled, and noise-filtered gene expression matrix (Table 3.1). As before, its N columns represent biological samples and p rows represent genes (probe sets, variables). The minimum meta-data information includes assignment of samples to classes.

Although the data is said to be quality-controlled, it is a good practice (especially if this is our first encounter with a particular data set) to evaluate (or re-evaluate) the quality of the data and experimental design before using the data as a training set for supervised learning. One should remember the GIGO (garbage-in garbage-out) rule and re-examine such elements of the data quality as:

- Quality of the individual measurements.
- Quality of the assignment of samples to classes.

  Note:

  If you are involved in data mining projects for genomics or proteomics, you may be approached by a researcher with a problem like this:

  > I have interesting data and want to find markers differentiating these two (or more) diseases (or phenotypes). However, I do not have all the diagnoses (or the certainty of the diagnoses is questionable). Can I cluster my samples first and then, once I have the classes identified, build the classification system?

  What would be your response? Try to answer this question before reading the footnote.[1]

- Number of samples in each class (is it sufficient for statistical reasoning?).
- Homogeneity (or reasonable heterogeneity) of the classes.
- Randomness of sample selection and independence of the samples.
- How well do the samples represent populations we want to differentiate? This is very important for the results to be generalizable—applicable to the investigated populations rather than only to the training set.

[1]The answer is NO. Unsupervised methods may not be used to generate high-dimensional training data sets. Even if clustering results would align with our tentative diagnoses or phenotypes, the clustering may be driven by variables that are not related to the classes we want to differentiate. For high-dimensional data, interpretation of clustering results is difficult and far from the quality level required for good training data. Generally, selection of the unsupervised or supervised approach depends on the goal of a study. If the goal is to expand taxonomic knowledge (e.g., to identify a new subtype of a disease), unsupervised methods should be used. If, however, the goal is to build a classification model or to identify a biomarker significantly differentiating known classes, then we need supervised learning—if we do not have good quality training data, we are risking researching factors that have nothing to do with what we are looking for. Refer to the Section 3.2.3 for more on this topic.

TABLE 3.1: Gene Expression Matrix, *J* Classes, *N* Samples, *p* Variables

	Class 1						Class 2		***	Class *J*		
	Sample 1	Sample 2	Sample 3	Sample 4	Sample 5	***	***	***	***	***	Sample $N-1$	Sample N
1053_at	6.5248	5.7101	6.2165	5.9984	6.2434						6.8442	6.9399
1316_at	5.8610	5.4029	6.5561	6.0639	5.5503						5.4944	5.1969
1494_f_at	5.5814	6.0217	6.1789	6.6498	5.9189						5.9639	6.4087
1729_at	4.2750	4.5192	5.0687	4.1301	4.5680						6.0076	6.4880
200595_s_at	9.5081	8.6934	7.9957	8.6349	9.0861						9.1352	8.8701
200600_at	9.7710	8.9780	9.1160	10.3930	9.9285						10.0923	10.5569
***												
***												
***												
81811_at	6.4634	6.9962	6.9654	6.8983	6.4436						6.3183	4.6599
89948_at	8.2129	8.1692	8.0640	7.8943	6.5841						6.1843	5.7522
91617_at	6.0796	5.6061	2.8094	4.6165	5.1416						6.5173	6.0670
91682_at	7.6769	6.2937	7.0522	7.1586	6.5065						5.4893	6.6147
91684_g_at	6.1999	6.3268	6.5926	5.3147	5.9832						6.0896	4.9364
91952_at	7.2515	8.6567	7.1778	7.2326	6.2506						6.5232	6.6493

3.1.2 Biomarker Discovery

In the context of gene expression analysis, biomarker discovery means identification of an optimal subset of variables that significantly differentiates the classes[2] and can be used for accurate prediction of the class membership. Although it may happen that a biomarker consists of just one variable, most often finding a good biomarker means searching for a set of variables, which together—as a set—can separate the classes.[3] A heuristic process or algorithm leading to such an optimal subset of variables is called *feature selection.*

In the context of supervised gene expression analysis, we introduce the following definitions:

Feature = variable

We will use the term *feature* as a synonym of original input *variable* (for instance, a probe set representing a gene and associated with a row in the gene expression matrix). More generally, the term could refer to original variables, to their combinations, or to variables constructed from the original variables. In the realm of biomedical applications it is, however, advantageous to create classifiers that directly use some of the original variables. When original variables are defining the dimensions of the discriminatory space, more straightforward biological interpretation of classification results is possible. Therefore, by ***feature selection*** we will mean selection of an optimal subset of the original variables.

Sample = biological sample

Since bioinformatics is an interdisciplinary area of research, we may—or sometimes have to—decide on which naming conventions to use when they are different in the overlapping areas included in bioinformatics. We will use the term *sample* as corresponding to a *biological sample* rather than in its statistical meaning. *Statistical sample* will then correspond to a group of *biological samples* selected from and representing—in a training data set—one of the populations we are to investigate.

Training Data—data representing biological samples (e.g., patients, tissues, objects, observations) characterized by the measured expression level of some number of variables (probe sets, genes,[4] exons) and with known, confirmed and considered highly accurate assignments to phenotypic classes (diseases, disease states, prognoses, responses to treatment, etc.). After

[2]Phenotypes, or more generally response variables, may be quantitative or qualitative. We are focusing here on qualitative phenotypes—discrimination between a set of classes, or categories (classification rather than regression). The main reason for this focus is the fact that in biomedical research common response variables are—or may be considered as—categorical (such as a set of differentiated diagnoses, a set of possible outcomes, a set of considered treatment protocols, etc.).

[3]In the context of data mining, the term *biomarker* is often used as a synonym for *multivariate biomarker* (a possibly small set of variables with possibly large joint discriminatory power).

[4]In this chapter, we focus on gene expression studies. However, all considerations of this chapter are equally applicable to protein expression studies, where variables represent expression level of proteins.

low-level preprocessing of the gene expression data, the training data set is represented by a gene expression matrix.

Test Data—data that is independent from the training data (or at least not used for training the system), and is used to estimate efficiency of the classification system, in particular the level of compromise between overfitting and generalization.

Overfitting—the ability of the classification system to perfectly or near perfectly classify training samples (by the means of reclassification or internal cross-validation) whereas performing poorly when classifying new samples.

Generalization—the ability of the classification system to correctly classify samples that were not included in the training data set. Generalization can be estimated best by evaluating performance of the classification system measured on an independent test data.

Biomarker *(more general definition)*—a biochemical characteristic that can be used to diagnose a disease, predict outcome, select treatment, assess efficacy or toxicity of a drug candidate, or—more generally—to predict class membership.

Multivariate biomarker (*more context-specific definition)*—a set of genes (or variables representing genes) with satisfactory discriminatory power that can significantly separate the differentiated classes and can be used to create a classification system of high sensitivity and high specificity. In other words, it is a set of genes whose joint expression pattern is predictive of class membership (is capable of highly accurate assignment of unknown samples to their true classes).

Optimal multivariate biomarker—a parsimonious multivariate biomarker that provides the best compromise between *overfitting* and *generalization*. As gene expression analysis deals with thousands of variables, the exhaustive search for *the best* subset of genes is intractable. Heuristic methods need to be used to identify the optimal biomarker—a possibly small set of genes with satisfactory and possibly large discriminatory power.

Classification versus prediction

Similarly as with the term *sample*, the terms *prediction* and *classification* are often used in biomedical research differently than in statistics. In statistics, they are associated with different types of response variables, continuous for prediction and categorical for classification. In biomedical research, they are often used interchangeably. For instance, assigning a biological sample to one of differentiated classes may be called *outcome prediction*; an estimate of generalization of a classification system may be called *predictive* accuracy of the classification system.

A common misconception in early biomarker discovery studies was the assumption that each gene selected into a multivariate biomarker has to be individually correlated with the disease or phenotype. Some of the researchers used to the *one-gene-at-a-time* approach attempted to project this univariate bias onto studies trying to identify a biomarker consisting of a set of genes. We hope that confusing a union of univariately identified variables for a truly multivariate biomarker is now a relic of the past. For a long time, the multivariate approach has been successfully applied in the fields of machine learning and artificial intelligence. Though propagated for many years by some bioinformatics researchers, the multivariate paradigm has only recently become a mainstream and leading approach in biomedical studies based on gene or protein expression data.

As biomarker discovery is applied to a rapidly increasing number of research areas, there is no clear taxonomy of biomarkers. The following are just a few examples of biomarker types:

- Diagnostic biomarkers—indicate the presence of a disease or the presence of a specific state or subtype of the disease.
- Prognostic biomarkers—indicate the probability of specific outcomes of a treatment.
- Biomarkers for personalized medicine (for instance, biomarkers for therapy selection or for minimizing the risk of adverse drug reactions)—indicate the probability of a specific outcome for the considered therapy options or medications.
- Toxicity biomarkers—indicate the level of a drug toxicity (often used during the drug discovery process and during clinical trials).
- Efficacy biomarkers—used in drug discovery to select most promising compounds.
- Pharmacodynamic and pharmacogenomic biomarkers—indicate the relationship between response to the drug and its dose; used to determine the dose associated with an optimal response.

Genomic, proteomic, and metabolomic biomarkers have a great potential for supporting personalized medicine. We will mention just two goals of biomarker discovery for personalized biomedicine—therapy selection and minimizing the risk of adverse drug reactions. Tailoring therapy to the condition of a patient is important in many disease areas, but especially in cancer treatment. Over-treatment of patients that have a low risk of relapse may cause therapy-induced conditions, such as leukemia. Although the outcomes for at least some types of cancer improved with currently available therapies, there are still significant challenges in predicting which of available treatment options will be most appropriate for a particular patient and which patient would benefit from less invasive or less toxic treatment (Carroll et al. 2003). Selection of the treatment protocol that is most appropriate for a patient may be improved by the identification of characteristic expression patterns associated with different responses to a variety of treatments. To identify such patterns, large repositories of expression data for patients with known diagnosis, treatment, and outcome parameters are necessary. Multivariate feature selection algorithms can be used to identify genomic or proteomic biomarkers with high classification efficiency.

A small size of multivariate biomarkers is important for their easy clinical implementations that may utilize such techniques as RT-PCR (Dziuda and Czar 2006). Biomarker discovery can also target minimizing the risk of adverse drug reactions. It is estimated that adverse drug reactions are causing, or contributing to, about 100,000 deaths annually in the United States (Ingelman-Sundberg 2008). Some of these adverse drug reactions (such as errors in drug administration) are preventable, but others are considered unpreventable by the current state of biomedical knowledge. It is estimated, that the latter are the fourth to sixth leading cause of death in the United States (Bates 1998). It is quite possible that data mining of genomic, proteomic, or metabolomic expression data may lead to the identification of predictors of currently nonpreventable adverse drug reactions. When identified, such predictors could be incorporated into diagnostic test profiles along with other types of predictors, such as genotyping markers (based on gene variant associations). It would be advisable to focus first on the most perilous adverse reactions, such as those related to antibiotics and allergic responses. Many lives would be saved if we could identify expression profiles of patients that are likely to develop adverse reactions to the most commonly administered antibiotics. Furthermore, it is conceivable that biological processes underlying particular reactions are common for groups of medications; it would be very important to identify profiles predicting such reactions (Dziuda and Czar 2006).

Typical (quality-controlled and noise-filtered) gene expression data sets include 5000–20,000 variables. Due to such a large number of variables, p, an exhaustive search that would guarantee finding the best subset of variables cannot be implemented, as the order of the search space is $O(2^p)$. Generally, many problems related to feature selection have been shown to be NP-hard (Amaldi and Kann 1998). Due to these difficulties, known since Bellman (Bellman 1961) as *the curse of dimensionality*, many studies reported in the literature pretty much neglect the feature selection step and apply more or less arbitrary selection of features that are subsequently used to build classification models. The usual approach is to find an ordered list of variables (using simple univariate methods like t-test, ANOVA F-test or their derivatives) and then use some number of the variables from the top of the list. Such a univariate approach not only neglects correlations between variables but also usually results in removing from consideration important discriminatory information. In some studies, the multivariate approach is applied to some number of genes from the top of a univariately-identified list. Although better than the univariate-only approach, multivariate methods are applied here after the harm was (most likely) already done.[5] Due to their limitations, both approaches (the univariate one and the multivariate approach with a strong univariate bias) should be avoided whenever possible.

A more sophisticated, truly multivariate approach to feature selection is necessary. The feature selection process is essentially a heuristic search, which—for

[5]Consider the following example. Assume that multivariate analysis is applied to 100 genes from the top of the univariate list (for instance, ANOVA F-test results sorted in ascending order of p-values). This top ranking sublist may be dominated by one or a few sets of highly correlated (highly redundant) genes. In an extreme (but not unrealistic) situation, most or all of the 100 genes may belong to two groups of highly correlated genes (one group with high expression in one class and low in the other, and the other group with expression low and high, respectively), thus each of these groups may provide about the same amount of discriminatory information as a single most discriminatory gene in the group. See also Section 3.2.2.

very large numbers of variables—may implement either sequential or random search methods. The sequential searches may implement forward or backward stepwise selection, or combine both of them in a hybrid selection. For example, the sequential hybrid selection may start with an empty set and then at each step may add or remove one or more variables to maximize some predefined metric of discriminatory power. The metric (or the *goodness* of the current subset) may be based on a measure of class separation, on information content of the subset, on the amount of explained variation, on an estimate of misclassification error of the classification system, etc. A random search starts with a randomly selected variable and then either follows the sequential search (more common) or generates the next subset also in a random manner.

Evaluating subsets with a well-defined metric of discriminatory power (such as the Lawley-Hotelling trace criterion) is preferable. Evaluating them exclusively on the basis of statistical significance of class separation is not recommended. A very small p-value associated with a set of variables tells us that the set differentiates the classes at a level unlikely to occur by chance, but this level of discrimination may still be far from satisfactory for efficient classification. However, assigning the significance to calculated values of discriminatory power is recommended. As distributions of some metrics may be unknown, the statistical significance is calculated for them either from distributions approximating the unknown ones, or with the use of bootstrap or permutation-based estimation of the unknown distributions.

The search continues until a stopping criterion (or one of multiple stopping criteria) is satisfied. Examples of stopping criteria include:

- a predetermined level of discriminatory power is achieved,
- a cut-off size of the marker (number of included variables) is reached,
- maximum number of iterations has been performed,
- incremental increase of the discriminatory power is below the specified *epsilon* level.

The feature selection process may be independent of the learning algorithm of the classification system (filter model), or may be related to the learning algorithm (wrapper and embedded models). The latter tends to provide more accurate classification systems as the selected subset of features is usually better suited to the predetermined classification algorithm (Liu and Yu 2005).

After the feature selection process ends, we are presented with one or more potential biomarkers (some algorithms report all the best subsets identified for each considered cardinality). Generally, large biomarkers tend to *overfit* the training data (they perfectly, or near perfectly, classify the training samples but perform poorly in classifying novel data). On the other hand, too small subsets may not have enough discriminatory power at all (and would perform poorly on both training and novel cases). The biomarker and model selection should be based on a compromise between the model's *overfitting* and *generalization* (the ability to properly classify new samples from the targeted populations). In other words, the model should be selected in such a way that it is "*not so simple that it cannot explain the differences between the categories, yet not so complex as to give poor classification of novel patterns*" (Duda et al. 2001).

Preferable Size of a Biomarker

Truly multivariate biomarkers should preferably consist of no more than ten variables. If differentiated classes are known to be heterogeneous, more variables may be necessary to construct a biomarker (assuming that an efficient biomarker can be identified in spite of the heterogeneity). In any case, if the resulting biomarker includes more than 20 variables, quality of the training data set, homogeneity of the classes (or reasonable heterogeneity), experimental design, and assumptions of the study need to be re-evaluated. As a rule of thumb, a red flag should be raised if a study reports a multivariate biomarker with more than 30 variables; most likely there are problems with the approaches applied.

The search for a multivariate biomarker can be seen as an optimization problem in a high-dimensional space. The selected biomarker may be associated with a local maximum (or optimum) in this space. A local maximum that is also the global one may seem preferable, though due to the mentioned curse of dimensionality we have no tractable method of verification. We can, however, check whether we are not trapped in a particularly inefficient local maximum by performing multiple repetitions of the feature selection process using subsamples of the training set and eventually with additional randomness introduced into the process itself. Another face of the curse of dimensionality is the sparsity of data points in a high-dimensional space. This means that good separation of the classes may be found even for the data that is simply random noise (especially for the data with a small number of samples). Therefore multiple searches and analysis of their results may be necessary for identification of a potentially stable biomarker. Furthermore, it is extremely important to properly validate the biomarker, preferably on an independent test data set.

3.1.3 Classification Systems

A classification system, or classifier, is the result of applying a machine learning algorithm to identify the relationship between patterns of variables and classes represented in the training data set. The classes correspond to populations we want to differentiate. Parameters of the classifier are *learned* from the training data, but the goal is to design a *generalizable* classifier—one capable of accurate prediction of class membership for new data points.

The following methods are among commonly used learning algorithms:

- Discriminant Analysis,
- Support Vector Machines,
- Random Forests and other tree-based classifiers,
- *k*-Nearest Neighbors,
- Artificial Neural Networks.

Some of them are classical, well-established methods (like *Fisher's Linear Discriminant Analysis*) and some represent the newest and promising trends (like *Random Forests* or other ensemble-based classifiers). Some of the classical methods, especially discriminant analysis (LDA—linear, or QDA—quadratic) have a very good record of accomplishment. Hastie et al. describe this as follows: "*Both LDA and QDA perform well on an amazingly large and diverse set of classification tasks.*" They recommend that discriminant analysis should always be available in spite of "*whatever exotic tools are the rage of the day*" (Hastie et al. 2009). We need to realize, however, that feature selection is usually more important than the selection of classification algorithm. This is especially true for training sets with thousands of variables and much fewer biological samples. Once a good quality (and hopefully parsimonious) biomarker is identified, many classification methods may yield similarly good classification results.

Later in this chapter, three learning algorithms will be described in detail—linear discriminant analysis, support vector machines and random forests. Linear discriminant analysis represents classical parametric methods that should be in the portfolio of any data miner and bioinformatician. Support vector machines are newer but already well-established nonparametric methods for designing both linear and nonlinear classifiers. Random forests represent recent ensemble-based approaches. In addition, two other learning algorithms will be described—k-nearest neighbors and artificial neural networks, useful in some situations, even if they are usually not our first choice in biomarker discovery.

3.1.3.1 Parametric and Nonparametric Learning Algorithms

Learning algorithms may be parametric or nonparametric. Parametric methods make some assumptions about the distribution of variables in the differentiated populations, about relations between some parameters of the populations, about independence of biological samples, etc. For example, the assumptions under which the linear discriminant analysis is performed include independence, multivariate normality, and equality of class covariance matrices. Nonparametric methods make no such assumptions. Random forests and other decision-tree-based algorithms are examples of nonparametric methods. As usual, there are advantages and disadvantages associated with either approach. Good performance of linear discriminant analysis may be due to the fact that the data often can only support simple boundaries between classes (linear decision boundaries that are the result of the equal covariance matrices assumption) or the fact that the estimates based on the normality assumption are stable (Hastie et al. 2009). On the other hand, some ensemble classifiers—such as random forests—can get away with sampling from the training set with replacement because they make no assumptions about independence of biological samples or about the form of underlying distributions. This may result in a relatively large pool of out-of-bag samples whose classification yields a good estimation of the classifier generalization (see Section 3.5 on random forests and 3.6 on ensemble classifiers).

3.1.3.2 Terms Associated with Common Assumptions Underlying Parametric Learning Algorithms

Multivariate Normal Distribution

Measurable biological variables are often normally distributed. As they can be considered the sum of many independent random effects, the central limit theorem leads to their normal distribution (Anderson 2003). The probability density function of the univariate normal distribution with the population mean μ and standard deviation σ is

$$f(x) = \frac{1}{\sqrt{2\pi\sigma^2}} e^{-\frac{1}{2}\left(\frac{x-\mu}{\sigma}\right)^2} \tag{3.1}$$

Multivariate and parametric learning algorithms often require the assumption that the p variables selected for consideration follow a multivariate normal distribution. This assumption is based on the direct generalization of the central limit theorem to multidimensional inputs (Morrison 2005). The multivariate normal distribution of p variables ($p \geq 2$) has the following density function

$$f(\mathbf{x}) = \frac{1}{(2\pi)^{p/2}|\boldsymbol{\Sigma}|^{1/2}} e^{-\frac{1}{2}(\mathbf{x}-\boldsymbol{\mu})^T\boldsymbol{\Sigma}^{-1}(\mathbf{x}-\boldsymbol{\mu})} \tag{3.2}$$

where

- $\mathbf{x}$ is a vector of values of p variables x_k, $k = 1, \ldots, p$ representing a biological sample,

$$\mathbf{x} = \begin{bmatrix} x_1 \\ x_2 \\ \vdots \\ x_p \end{bmatrix}, \tag{3.3}$$

- $\boldsymbol{\Sigma}$ is a $p \times p$ variance–covariance matrix,

$$\boldsymbol{\Sigma} = \begin{pmatrix} \sigma_{11} & \sigma_{12} & \cdots & \sigma_{1p} \\ \sigma_{12} & \sigma_{22} & \cdots & \sigma_{2p} \\ \vdots & \vdots & \ddots & \vdots \\ \sigma_{1p} & \sigma_{2p} & \cdots & \sigma_{pp} \end{pmatrix} \tag{3.4}$$

  whose diagonal elements $\sigma_{kk} = \sigma_k^2$ represent the variances of the variables x_k whereas the off-diagonal elements σ_{kl} are the covariances between variables x_k and x_l, $k \neq l$ and $k, l = 1, \ldots, p$. Since $\sigma_{kl} = \sigma_{lk}$ for any $k, l = 1, \ldots, p$, the matrix $\boldsymbol{\Sigma}$ is always symmetric (Srivastava 2002),
- $|\Sigma|$ denotes determinant of the matrix $\boldsymbol{\Sigma}$,
- $\boldsymbol{\Sigma}^{-1}$ is the inverse of matrix $\boldsymbol{\Sigma}$,
- $(\mathbf{x} - \boldsymbol{\mu})^T$ denotes the transpose of $\mathbf{x} - \boldsymbol{\mu}$,
- $\boldsymbol{\mu}$ is the p-dimensional vector of mean values of the p variables in the population,

$$\boldsymbol{\mu} = \begin{bmatrix} \mu_1 \\ \mu_2 \\ \vdots \\ \mu_p \end{bmatrix}. \tag{3.5}$$

Note that

$$\left(\frac{x-\mu}{\sigma}\right)^2 = (x-\mu)(\sigma^2)^{-1}(x-\mu) \tag{3.6}$$

in the exponent of (3.1) represents the squared distance between x and μ measured in standard deviation units. In (3.2), the term

$$(\mathbf{x}-\boldsymbol{\mu})^T\boldsymbol{\Sigma}^{-1}(\mathbf{x}-\boldsymbol{\mu}) \tag{3.7}$$

represents analogical distance between vectors **x** and **μ** in the p-dimensional space (Johnson and Wichern 2007). As probabilities in the univariate case are represented by areas under the univariate normal density curve, probabilities in the multivariate case are represented by volumes under the surface defined by (3.2) (see Fig. 3.3). The assumption of multivariate normality means that:

- each of the p variables is normally distributed,
- all linear combinations of the variables are normally distributed,
- contours of constant density for the multivariate normal distribution are ellipsoids.

With thousands of variables in a typical gene expression data set, it would be quite impractical to test for the normality of *all* their combinations. There are multivariate normality tests described in the literature (see for example Mardia 1970; Cox and Small 1978; Smith and Jain 1988; Srivastava 2002; Szekely and Rizzo 2005). One of better known tests uses Mardia's statistic (Mardia 1970) based on multivariate measures of skewness and kurtosis. However, "*it has proved difficult to construct a 'good' overall test of joint normality in more than two dimensions because of the large number of things that can go wrong*" (Johnson and Wichern 2007). This is

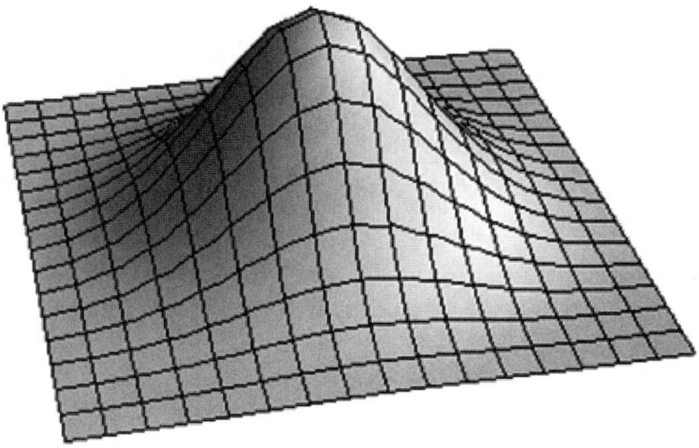

Figure 3.3: Bivariate ($p = 2$) normal density function—the simplest example of the multivariate normal distribution.

especially true for data sets with thousands of variables. Common approaches are then limited to some of the following:

- testing for the univariate normality of each variable,
- testing for the bivariate normality of a few pairs of variables (elliptical scatter plots are expected),
- testing for gross outliers,
- preprocessing and/or transforming the data in a way that should increase chances for multivariate normality.

One may ask, "What are advantages of making the multivariate normality assumption if we cannot test it in practice?" And additionally, "What is the value of a statistical procedure when the data may violate its assumption(s)?" To answer these questions, let us indicate the following:

(i) Multivariate methods based on normal distribution allow for exact mathematical solutions, usually based on standard linear algebra operations. As a result of this, many test statistics have a known distribution or one that can be approximated by a known distribution (Anderson 2003).

(ii) Most of these methods are relatively robust to violations of the normality assumption (unless departures from the multivariate normality are extensive and severe).

(iii) For gene expression data sets with a reasonable number of biological samples per class, univariate and bivariate normality tests are usually sufficient for unraveling serious violations of multivariate normality.

The univariate normal distribution of each variable is the necessary, but not satisfactory, condition of multivariate normality. However, normality of all p variables increases the likelihood of their multivariate normal distribution (Tabachnick and Fidell 2007). Screening for univariate normality may involve statistical or graphical methods. For example, skewness and kurtosis measures, reported by statistical packages, near zero indicate normally distributed variables.[6] *Skewness* measures the symmetry, or the lack thereof, of a distribution. A variable is skewed when its mean value is not in the center of the distribution. Negative skewness means that the left tail of the density function is more pronounced than the right tail (and the reverse is true for positive skewness). *Kurtosis* measures the degree of peakedness of a distribution. Positive kurtosis corresponds to a distribution with a high peak, and negative kurtosis to a flat-topped one. Among the graphical methods commonly used to assess univariate normality are normal probability and quantile–quantile plots. They rank order values of the variable and plot the observed scores against expected scores calculated from the normal distribution. For normally distributed variables, all the points of the plot are close to the diagonal straight line between (0, 0) and (1, 1).

[6]Statistical packages report *excess kurtosis*, which equals to *proper kurtosis* minus 3. Proper kurtosis (defined as the normalized fourth central moment of a distribution) is equal to 3 for a normal distribution.

Linearity

The linearity assumption means that the relationship between two variables can be approximated by a straight line. Many multivariate methods make this assumption. These are often algorithms that utilize the Pearson correlation in their statistical analysis. The Pearson correlation coefficient r assesses the strength and direction of the linear relationship between two variables. As such, r is incapable of capturing the nonlinear element of a relationship. Bivariate scatterplots can be used to test for a linear relationship. When both variables are normally distributed and linearly related, the scatterplots are ellipsoidal. This assessment is, however, unreliable for data sets with small number of biological samples, and impractical for large number of variables (Tabachnick and Fidell 2007).

Homoscedasticity

In the case of multivariate differentiation of J classes, homoscedasticity means homogeneity of the J variance–covariance matrices $\mathbf{\Sigma}_j, j = 1, \ldots, J$,

$$\mathbf{\Sigma}_1 = \mathbf{\Sigma}_2 = \cdots = \mathbf{\Sigma}_J. \tag{3.8}$$

This implies that the variances σ_k^2 of p variables x_k as well as the covariances σ_{kl} for all possible pairs of variables x_k and x_l, where $k, l = 1, \ldots, p, k \neq l$, are equal across the J classes. Box's M test (Morrison 2005) or Bartlett's test (Tabachnick and Fidell 2007) may be used to screen for violations of the homogeneity of variance–covariance matrices. With a small number of biological samples, it may be preferable to perform Box's M test using a higher than conventional alpha level, for example $\alpha = 0.10$ (Warner 2008).

Multicollinearity and Singularity

Multicollinearity and singularity refer to the degree of variable redundancy. Multicollinearity exists when two or more variables are highly correlated, and singularity when the variables are totally redundant. With singularity, correlation or variance–covariance matrices cannot be inverted; with multicollinearity, results of the matrix inversion are unstable. Many software packages screen for singularity and multicollinearity by calculating the squared multiple correlation R^2 between each individual variable and all other variables taken together. The R^2 coefficient can be interpreted as the proportion of total variation in the individual variable explained by the rest of variables.

Outliers

Some parametric learning algorithms are very sensitive to outliers. Identification of multivariate outliers is a challenging task, especially for the data with more than two classes. One of the methods of screening for multivariate outliers is to calculate Mahalanobis distance (3.37) between each biological sample in the training data set and the centroid of the sample's class. A test statistic with a χ^2 distribution can be used to determine the probability of the sample being a *typical* member of the class (Huberty and Olejnik 2006).

3.1.3.3 Visualization of Classification Results

Well-implemented classification models should include visualization of the discriminatory space and classification results. A model based on a multivariate biomarker of

m variables may perform classification in the m-dimensional space defined by the selected variables. For $m > 3$, direct visualization of classification results is not possible, and we need to apply some dimension reduction techniques. For example, in linear discriminant analysis, the m-dimensional hyperspace can be transformed to a space with $J - 1$ dimensions, where J is the number of differentiated classes. This means that for experiments comparing simultaneously less than five classes, the entire discriminatory information can be graphically presented in three or fewer dimensions.

3.1.4 Validation of the Classification Model

The search for the optimal biomarker and design of a classifier cannot be deemed complete until we estimate the prediction error of the classification model. The main utilitarian goal of biomarker discovery is a generalizable classification model—one that would have a low misclassification error rate when applied to new samples from the discriminated populations. We will discuss the following validation and cross-validation methods used in data mining, although not all of them are recommended for studies based on typical gene expression data.

- Reclassification
- Leave-One-Out cross-validation
- *K*-Fold cross-validation
- Holdout method of validation
- Validation based on classification of out-of-bag samples
- Validation on an independent test data set.

3.1.4.1 Reclassification

Reclassification means estimating the misclassification error rate of a classifier by classifying the very samples that were used to train the classifier. Although there are indications that for data mining projects based on training sets with large N/p ratios (where N is the number of cases and p the number of variables), reclassification may yield quite reliable results (Huberty and Olejnik 2006), **reclassification should never be used with typical gene expression data sets** that have relatively small numbers of biological samples and large numbers of variables.

3.1.4.2 Leave-One-Out and K-Fold Cross-Validation

The Leave-One-Out and K-Fold schemas are classical cross-validation methods utilizing the training data to estimate predictive ability of the classification model. In fact, the Leave-One-Out method is a special case of K-Fold cross-validation, but usually they are performed and reported independently.

Leave-One-Out Cross-Validation

The Leave-One-Out cross-validation method (also known as *jackknifed* classification) sets aside one sample from the training data set of N samples and builds a classification model based on the remaining $N - 1$ samples. The singled-out sample is then classified by this new model. This process is repeated N times, each time a different sample is taken out. This way each sample is classified by a model trained without this sample. The Leave-One-Out classification efficiency of the model is usually reported as the proportion of correctly classified samples.

K-Fold Cross-Validation

The K-Fold cross-validation method randomly divides the training set into K non-overlapping subsets of approximately equal size. A classification model is trained on $K - 1$ subsets and then used to classify samples from the remaining subset. This process is repeated K times, so each sample is classified by a model built without this sample. Commonly used values of K are 5 or 10. When $K = N$, we have the Leave-One-Out cross-validation.

3.1.4.3 External and Internal Cross-Validation

The Leave-One-Out and K-Fold schemas can be implemented as an external or internal cross-validation (Ambroise and McLachlan 2002). The difference is illustrated below.

- ***Internal cross-validation*** deals with a single biomarker identified from the entire training data set. After the feature selection process gives us an optimal biomarker, we split the training set by applying either the Leave-One-Out or K-Fold schema. We then build a number of classifiers (N for Leave-One-Out or K for K-Fold); each of them is trained on a current subset of training samples and then used to classify the remaining samples. However, each classifier uses *the same set of variables*—the optimal biomarker already having been identified from the entire training data set.
- For ***external cross-validation***, we use the schemas independently of the main feature selection process. This can be done either before or after our optimal biomarker is identified. As with the internal cross-validation, we are splitting the training set, but now we use each of the N or K training subsets to perform *independent feature selection*, and identify a biomarker for each of them. Then we use each biomarker to build a classification system and classify the samples not included in the training subset used for the biomarker identification.

In internal cross-validation, each of the N or K classifiers is built from a subset of variables identified using the entire training data set. Therefore the left-out samples classified at the cross-validation step cannot be treated as an independent test set. In external cross-validation, the left-out samples are never "seen" either by each of N or K feature selection processes or by the resulting classifiers. Thus they can much better emulate an independent test set.

In business data mining, where we deal with small numbers of variables and huge numbers of cases, $p \ll N$, there may be plausible arguments for using either of these approaches. However, **internal cross-validation should not be used for gene expression data sets** with large numbers of variables and relatively small numbers of biological samples ($p \gg N$). For such data, there could be many subsets of variables that yield excellent internal cross-validation results (zero or near zero misclassification error rates) but at the same time overfit the training data and be of little use in the classification of new samples. We have to be vigilant when reading publications or software documentations; the term *cross-validation* is quite often used without clear indication as to whether it refers to external or internal cross-validation. If the feature selection step is not performed independently for each of the N or K training subsets, we should not assign a lot of weight to the reported results and should treat them as overly optimistic. In the $p \gg N$ situations, the internal cross-validation may report very low (or even zero) misclassification error rates even for the data that are random noise.

3.1.4.4 Holdout Method of Validation

If an independent test data set is not available, we may withhold from the analysis a number of randomly selected samples, train the model on the remaining data, and use the withheld data as a test set to validate the classification model. This approach (or a similar one where the data is split into three parts—training, test, and validation sets) is known as the *holdout* method and is popular and very useful in data mining domains generating data sets with a large number of cases and a small number of variables. When applied to typical gene expression data, this approach has, however, the following disadvantages (Theodoridis and Koutroumbas 2006; Huberty and Olejnik 2006).

- A basic requirement of the holdout method is the large number of biological samples in each class.
- Our training set would be smaller and—unless we have a very large number of samples—the quality of the identified biomarker and classification model would be worse than when developed from the whole training set.
- The model that is validated is not the one that would be used in practice, that is, the one based on the entire training set.
- A decision on the size of the withheld test set is problematic: if the test set is relatively large, the estimation of the classification error will be better, but the model itself is likely to be poor; selecting a small test set will result in a better quality model, but the estimation of its generalization will be highly variable.

Furthermore, such a holdout test set—even if not used in the biomarker discovery process before the validation step—is not as independent as test data generated in a different lab, a different institution, and maybe a different country.

3.1.4.5 Ensemble-Based Validation (Using Out-of-Bag Samples)

One of the recent trends in model validation is to utilize ensemble classifiers. Even if we are not interested in the practical use of an ensemble classifier (for it usually does not implement a parsimonious biomarker, which is one of the main goals and requirements of biomarker discovery), we may use its out-of-bag samples for validation purposes. Consider, for example, a study based on a relatively small training data set, say, a couple of dozen biological samples per class. As this training set is too small for the holdout method of validation, we utilize the entire training set to perform biomarker discovery. Assume that we identified a promising biomarker and used it to build a classification model. How can we validate that model if we do not have an independent test data set? We can sample the training set to create a number of derivative training sets. Depending on the learning algorithm we use, the sampling may be with or without replacement. Assume for simplicity that we use a nonparametric learning algorithm and sample with replacement. Applying the *bagging* (bootstrap aggregating) approach, we generate many bootstrap training sets and use them to build an ensemble of classifiers. Please note that, like in external cross-validation, feature selection has to be performed independently for each of the bootstrap training sets. Each of these training sets leaves out a number of samples that are not used during feature selection and training a classifier. These *out-of-bag samples* can be used to estimate the misclassification error rate of a particular classifier. Although sets of variables used by the ensemble's classifiers are most likely different from our biomarker, averaging the out-of-bag misclassification error rate of all ensemble classifiers can serve as a reasonable estimate of the predictive ability of our original classifier. Even as such estimate is only an indirect indication of the generalization ability of our optimal biomarker (built from the entire training data set), it may often be the best estimate we can get for a typical gene expression study when we do not have an independent test set. For more information on bagging and ensemble classifiers, see Sections 3.5 and 3.6. Please note that ensembles of classifiers built on randomized or perturbed versions of training data may be used beyond validation. In Chapter 4, a method utilizing the distribution of genes among a large number of classifiers for optimization of the biomarker discovery process will be presented.

3.1.4.6 Validation on an Independent Data Set

We should remember that the ultimate estimation of the model generalization (its predictive ability for the correct assignment of new samples from the populations targeted by the classification model) is its validation on a possibly large and *independent* test data set. We should always try to find an independent data set to validate our final multivariate biomarker. If such a set is not available, we recommend using the ensemble-based validation (and eventually the external K-Fold cross-validation if, for whatever reason, the ensemble-based validation cannot be performed).

3.1.5 Reporting Validation Results

To report the estimated efficiency of the classification system we may calculate its *accuracy* as the proportion of test samples that are classified correctly into their true classes. Subtracting this proportion from one will result in the *misclassification rate*. However, these measures are sufficient only when the differentiated classes are of similar size (*a priori* probabilities are similar) and when the gravity of misclassification is similar for each class. Consider a simple situation when we differentiate two subtypes of a disease Δ, denoting the subtypes δ_1 and δ_2. If only one percent of the population of patients with the disease Δ suffer from the subtype δ_2, then a dummy "classifier" that is always assigning patients to class δ_1 will have the accuracy of 0.99, or 99 percent (and a misclassification error rate of 0.01, or 1 percent). However, such a classification system will misclassify 100 percent of the subtype δ_2 patients. Proper evaluation of the classifier would have to take into account the

TABLE 3.2: Confusion Matrix for a Binary Classifier

		Predicted Class	
		Disease (positive)	No Disease (negative)
True Class	Disease (positive)	*True Positive (TP)*	*False Negative (FN)*
	No Disease (negative)	*False Positive (FP)*	*True Negative (TN)*

The rows represent true classes of the test cases. The columns represent classes predicted for them by the classifier.

costs of different kinds of misclassifications. To present validation results in a more informative way, more detailed measures of the estimated classification efficiency are often reported. They are usually based on the presentation of the validation results in the form of a *contingency table*, or *confusion matrix*. Cells of the confusion matrix represent combinations of the predicted and actual classes. Each of them can be associated with the cost or benefit of making that particular decision, which allows for the estimation of the overall loss (Hand et al. 2001) or for selection of the optimal model (Larose 2006).

3.1.5.1 Binary Classifiers

Let us consider first the simplest, but quite frequent, situation of differentiating only two classes. Outcomes of such a binary classification may be (more or less arbitrarily) labeled as positive or negative. For instance, if we differentiate the *disease* class versus the *nondisease* class, each classified patient will be diagnosed as either disease-positive or disease-negative (assuming that the classes are mutually exclusive and no other outcomes are possible). Validation results can be presented in the 2×2 confusion matrix (see Table 3.2),

where

True Positive (TP)—the number of positive test cases that are correctly classified as positive (also known as *hits*).

True Negative (TN)—the number of negative test cases that are correctly classified as negative (correct rejections).

False Positive (FP)—the number of negative test cases that are incorrectly classified as positive (false alarms).

False Negative (FN)—the number of positive test cases that are incorrectly classified as negative (misses).

Denote the number of positive cases in the test data set as $P = TP + FN$, and the number of negative cases in the test data set as $N = TN + FP$. Based on the confusion matrix, we can define the following partial measures of performance of a binary classifier:

Sensitivity—the proportion of positive test cases that are correctly classified as positive (the probability of predicting disease when the actual class is disease; sensitivity is also called *true positive rate—TPR*, or *hit rate*),

$$Sensitivity = TPR = \frac{TP}{TP + FN} = \frac{TP}{P} \tag{3.9}$$

Specificity—the proportion of negative test cases that are correctly classified as negative (the probability of predicting "no disease" when the actual class is "no disease," specificity is also called *true negative rate—TNR*),

$$Specificity = TNR = \frac{TN}{TN + FP} = \frac{TN}{N} \tag{3.10}$$

False Positive Rate (FPR)—the proportion of negative test cases that are incorrectly classified as positive,

$$\begin{aligned} FPR &= \frac{FP}{FP + TN} = \frac{FP}{N} \\ &= 1 - Specificity \end{aligned} \tag{3.11}$$

False Negative Rate (FNR)—the proportion of positive test cases that are incorrectly classified as negative,

$$\begin{aligned} FNR &= \frac{FN}{FN + TP} = \frac{FN}{P} \\ &= 1 - Sensitivity \end{aligned} \tag{3.12}$$

False Discovery Rate (FDR)—the proportion of test cases classified as positive that are false positive,

$$FDR = \frac{FP}{FP + TP} \tag{3.13}$$

Accuracy—the proportion of all test cases that are correctly classified,

$$Accuracy = \frac{TP + TN}{TP + TN + FP + FN} = \frac{TP + TN}{P + N} \tag{3.14}$$

Misclassification Rate—the proportion of all test cases that are incorrectly classified,

$$\begin{aligned} Misclassification\ Rate &= \frac{FP + FN}{TP + TN + FP + FN} = \frac{FP + FN}{P + N} \\ &= 1 - Accuracy. \end{aligned} \tag{3.15}$$

In addition, we can define the following two measures of predictive value of the particular classification result:

Positive Predictive Value (PPV)—the proportion of test cases classified as positive that are true positive (for instance, the *PPV* can be interpreted as the probability of disease given that the patient was classified into the disease-positive class),

$$\begin{aligned} PPV = Precision &= \frac{TP}{TP + FP} \\ &= 1 - FDR \end{aligned} \tag{3.16}$$

Negative Predictive Value (NPV)—the proportion of test cases classified as negative that are true negative (for instance, $1 - NPV$ can be interpreted as the probability of disease given that the patient was classified into the disease-negative class),

$$NPV = \frac{TN}{TN + FN}. \tag{3.17}$$

Let us revisit our example of the dummy classifier, which by assigning every case to the subtype δ_1 had been achieving 99 percent accuracy. If we choose to label the subtype δ_1 of the disease Δ as the negative outcome and the subtype δ_2 as the positive one, the dummy classifier will have specificity of 100 percent and sensitivity of 0 percent—clearly unacceptable (especially when the subtype δ_2 of the disease is much more dangerous than δ_1). Proper evaluation of a classifier should take into account the possibly different costs of different misclassification errors. These costs may also be incorporated into the training process, either by assigning weights to training cases or by generating a training set with a different proportion of class sizes (Witten and Frank 2005). Generally, we are searching for the tradeoff between sensitivity and specificity that is most appropriate for our study. To visually assess this tradeoff for different values of the parameters of a binary classifier, we may use the *Receiver Operator Characteristic* (ROC) curve,[7] which usually plots sensitivity versus (1 – specificity). The area under the ROC curve (AUC) may be used to compare different classification models.

3.1.5.2 Multiclass Classifiers

For multiclass classifiers, the confusion matrix is of size $J \times J$, where J is the number of differentiated classes (Table 3.3). The cells along the main diagonal of the confusion matrix represent correct classifications, the cells off the diagonal—misclassifications.

Assuming the following notation,

- C_{ij} —number of test cases with the true class i that are classified into the class j, where $i, j = 1, \ldots, J$.

[7]The name of the ROC method reflects its origin in signal detection theory.

TABLE 3.3: Confusion Matrix for a Multiclass Classifier

		Predicted Class				
		Class 1	Class 2	...	Class J
True Class	Class 1	C_{11}	C_{12}	...	C_{1J}
	Class 2	C_{21}	C_{22}	...	C_{2J}
	...	...	...	...	...
	Class J	C_{J1}	C_{J2}	...	C_{JJ}

The rows of the confusion matrix represent true classes. The columns represent the predicted classes.

- N_i —number of test cases in class i (i.e., whose true class is i); this is the row i total,

$$N_i = \sum_{j=1}^{J} C_{ij}$$

- P_j —number of test cases classified into class j; this is the column j total,

$$P_j = \sum_{i=1}^{J} C_{ij}$$

- N —total number of cases in the test set,

$$N = \sum_{i=1}^{J} N_i = \sum_{j=1}^{J} P_j = \sum_{i=1}^{J} \sum_{j=1}^{J} C_{ij},$$

we will define sensitivity and specificity for each of the differentiated classes:

Sensitivity for the Class k—the proportion of test cases in class k that are correctly classified into the class k, where $k = 1, \ldots, J$,

$$Sensitivity(k) = \frac{C_{kk}}{N_k} \tag{3.18}$$

Specificity for the Class k—the proportion of test cases not in class k (i.e., whose true class is different from k) that are classified into a non-k class,

$$Specificity(k) = \frac{N - P_k}{N - N_k}. \tag{3.19}$$

We can also calculate the overall accuracy and misclassification rate of the multiclass classifier:

$$Accuracy = \frac{\sum_{k=1}^{J} C_{kk}}{N}, \tag{3.20}$$

$$Misclassification\,Rate = 1 - Accuracy. \tag{3.21}$$

3.1.6 Identifying Biological Processes Underlying the Class Differentiation

There are many possible approaches to this task. The most common one (though with arguably inferior ability to perform the task), frequently reported in the literature is based on an ordered univariate list of features cut by either some statistical significance level (hopefully adjusted for the number of variables) or including just some number of features from the top of the list. When it results in new biological knowledge or biologically valid and verifiable conclusions, shortcomings of the process are not important. Often, however, that is not the case. Other approaches combine several (usually also univariate) methods generating independent lists of genes, and then look at their overlapping set. This method has a better chance of finding some relevant links to biological processes. Recent and most promising approaches start from the multivariate biomarker and extend the small set of variables represented in the marker (using either correlation-based or other multivariate methods) into a larger informative set. After either a list of genes or the informative set of genes is identified, available gene annotation information (such as the Gene Ontology or metabolic pathway annotations) is used to link the class differentiation with underlying biological processes.

3.2 FEATURE SELECTION

"The features selected matter more than the classifier used."

—(Guyon et al. 2002)

3.2.1 Introduction

The human brain is well adapted to the processes of instant feature selection and classification when they are related to perceptual problems, such as recognizing a face in a crowd. From a large amount of visual information, we can efficiently remove irrelevant and redundant information. Focusing on a few important characteristics, we can easily and instantly recognize a face or person we know. Unfortunately, these abilities do not extend to problems with many numerical variables, whose visualization would require a multidimensional representation with the number of dimensions being greater than three. Michio Kaku provides an admirable explanation of this "*accident of evolution*" by pointing out that "*Lions and tigers do not lunge at us through the fourth dimension*" (Kaku 1994).

Nevertheless, a lesson from the evolution of our brain's perceptual abilities is that for efficient classification it is crucial to know how to select this small set of few important features. Understanding that "*less is more*" (Liu and Motoda 2007b) is the essence of feature selection.

For typical gene expression microarray data sets, the number of variables that pass the quality control assessment is in the thousands. Simple calculations can reveal that the exhaustive search for the best multivariate marker (the global optimum) consisting of a few, say up to 10, genes would require inordinate amount of time.[8] This is why one may easily find examples of gene expression studies that are limited to a univariate approach, or to multivariate analysis dealing only with a more or less arbitrarily selected group of genes (for example, the ones with top univariate scores). Both approaches may seriously limit the ability to find efficient biomarkers as they leave out genes that do not have high univariate significance, but may carry important discriminatory information. Furthermore, a multivariate approach without a clear definition of the discriminatory power of a set of genes may lead to unnecessarily large and suboptimal markers and to overfitting.

The goal of feature selection is to find a small subset of variables[9] that can significantly separate the differentiated populations (represented by the classes of samples in the training data set). The small size of a biomarker is very important for its fast and cost-effective clinical validation and implementation. Therefore, we are looking for a set with as few variables as possible and with as high discriminatory power as possible. To find such a parsimonious biomarker, we need efficient heuristic methods for

(i) the removal of irrelevant and redundant variables,

(ii) finding an optimal set of variables, with the optimization meaning minimizing the size of the set and maximizing its discriminatory power.

Heuristic searches result in local optima, which may or may not lead to efficient classifiers. However, there are indications that heuristic searches are less prone to overfitting than the exhaustive search, especially for data sets with a small number of samples (Liu and Motoda 2007b). Hence, even if we had computational abilities to find the global optimum, there might be no reason to do so.

Applying a single search method with a single starting point (for example, the starting point for a heuristic forward search would be the variable selected into a set first) would result in a single set of variables being deemed as optimal. This could be a sufficiently good result for typical business data sets with many more records than variables. For gene or protein expression data—which may have thousands of variables and only dozens of biological samples—this may lead to overfitting

[8]For example, the exhaustive search of all subsets of 10 variables out of only 1000 variables would require the processing of $2.6 * 10^{23}$ subsets. Assuming an optimistic processing time of only 10^{-6} seconds per subset, it would still take $2.6 * 10^{17}$ seconds, which is comparable to the age of the Universe.

[9]As mentioned before, the term *feature selection* can be, generally, understood as the search for a subset of original variables or as the search for a subset of features that are combinations of the original variables. However, for biomedical studies, the former is preferable since biomarkers consisting of original variables have the advantage of more straightforward biological interpretation.

and the identification of a biomarker that perfectly separates classes of training samples but cannot be generalized to the populations represented by those samples. Validating the biomarker on an independent test data set may yield a more realistic estimate of its generalization. If the biomarker is positively validated, we may end up with a good quality classification system, but if the biomarker cannot efficiently classify independent data, we end up with nothing.

Another approach, gaining popularity, is to perform many searches with different starting points and, eventually, with different algorithms. Sometimes, even different versions of the training data set are used (for instance, the same raw data can be processed with different methods at such stages as low-level preprocessing, normalization, or noise filtering). As a result of such parallel searches, a number of potential biomarkers (often dozens or even hundreds of them) would be identified. There could be many possible ways of analyzing or combining these biomarkers in order to build an efficient classification system. One could, for example, select one of these biomarkers based on the results of its validation on an independent data set. Such selection may indicate improved generalization, but the resulting classifier may not necessarily provide a satisfactory level of efficient classification of unknown samples from the researched populations. This is because of an increased risk of overfitting both the training and test data sets (unless we could validate each potential biomarker on a different test set, which would, however, be extremely unrealistic and wasteful). Better generalization and more stable solutions can usually be achieved when one builds an ensemble classifier. Either all or a subset of the identified biomarkers are used to classify new samples and the classification results are based on a voting schema. Some algorithms can achieve good generalization as well as good classification efficiency even in the situation where individual biomarkers are rather weak (see bagging, boosting, and random forests in Section 3.5). However, there is a disadvantage to this approach—ensemble classifiers use many (often hundreds or even thousands of) variables, which seriously limit their clinical applications. Our goal of finding a parsimonious multivariate biomarker is rarely achieved with this approach. A more promising way of utilizing the identified potential biomarkers is to use the ensemble's classifiers to vote for features (rather than for classes). For instance, we may select a small set of features based on the classifiers' performance and the frequency of the variables used. This approach may have many flavors and should, at least potentially, lead to better and more stable solutions. A novel method that uses an intermediate set of genes called *The Informative Set of Genes* will be introduced in Chapter 4.

3.2.2 Univariate Versus Multivariate Approaches

"A variable useless by itself can be useful together with others."

—(Guyon and Elisseeff 2003)

The main difference between the two approaches is that the univariate approach does not take into account the correlations and interactions between variables and the multivariate one does. The univariate approach evaluates each variable independently of the other variables (for instance, it takes one gene at a time and determines how the gene expression discriminates the researched classes) and the result is a list of variables ordered according to some univariate measure of class separation. As such, the univariate approach is appropriate only when either the variables are uncorrelated or we—for whatever reason—are interested in researching individual genes in isolation from others. Neither of these situations applies to biomarker discovery based on gene or protein expression data. Co-regulations and interactions among genes (or proteins) are important and should not be neglected. Genes that are far from the top of a univariately ordered list of genes and whose p-value (or other metric of statistical significance) makes them univariately insignificant, may carry important discriminatory information when their expression pattern is combined with expression patterns of some other genes. Removing them from biomarker discovery considerations may seriously limit the study's prospects for identification of efficient biomarkers. Furthermore, such an approach may fail to discover new biomedical knowledge that could be extracted from the training data had the multivariate approach been applied. Let us look at the following toy example (Fig. 3.4).

There are only two variables in the example and neither of them can significantly separate the two classes. But their combination results in a perfect class separation. That is easy to spot in a graphical presentation of such simple two-dimensional data. However, when we have thousands of variables, perhaps more than two classes to separate and a very likely possibility that a good separation may be achieved by combining more than two variables, identification of such variable combinations is not only a nontrivial task, but definitely one that could not be achieved by the univariate approach. Another disadvantage of limiting a study to some number of genes from the top of their univariately ranked list is that the group of selected genes may be

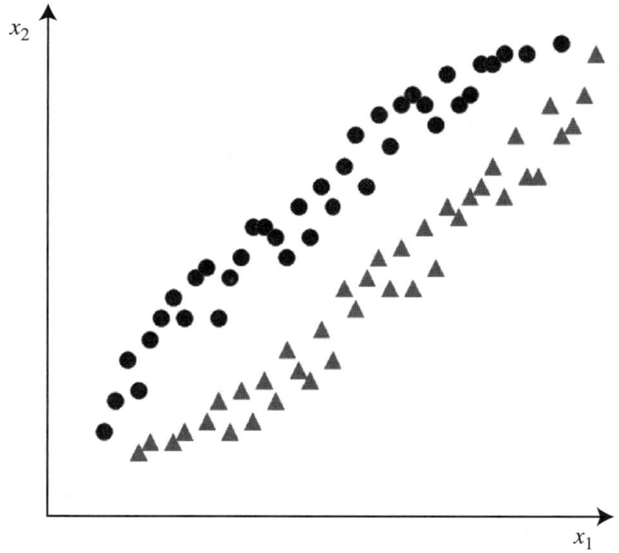

Figure 3.4: A data set with two classes of samples. There are only two variables. Neither of them is univariately significant for the class discrimination. However, as a set of two variables they can perfectly separate the classes. (See color insert.)

redundant or even highly redundant. Consider another simple—but not unrealistic—example. A group of 100 genes from the top of a univariate list (ordered by their p-values representing their significance of two class discrimination) may be dominated by two subgroups—one with genes over-expressed in one class and under-expressed in the other, and the other with genes under-expressed and over-expressed in the respective classes. Each of such subgroups may represent about the same discriminatory information as a single most discriminatory gene in the subgroup, and the entire selection of the top genes may have insufficient discriminatory power for significant class separation. One may argue that it may happen that a univariately identified group of genes results in an efficient biomarker and a good classification system. Yes, it may happen. Any method can be justified if the goal of a study is achieved and the results have sound biological interpretation. It may happen that a randomly selected single gene is all that is needed. However, what are the chances of such an event? Taking chances is not what we recommend here for biomarker discovery studies.

It may be a marginal note for this section, but it is worth mentioning that correlation between two variables does not necessarily mean they cannot be selected together to an efficient biomarker. Only perfectly correlated variables carry exactly the same discriminatory information. Variables that are less than perfectly correlated may share a lot of common variance, but it is still possible (and happening in practice) that one of them is the best complement to a set of variables already including the other, which means that it could increase discriminatory power of the set by more than any other variable that could be added to the set. However, identification and evaluation of such complementarity is possible only with the use of multivariate approaches.

As stated, multivariate approaches evaluate sets of variables, take into account interactions between variables and are capable of finding a small set of variables with high discriminatory power (assuming such a set exists). Although properly designed multivariate methods tend to select into a multivariate biomarker variables that are quasi-orthogonal, neither the orthogonality nor, more generally, correlations between variables are criteria for making selections. The primary criterion is to increase discriminatory power of the set of variables. That is why the inclusion of the two correlated variables described earlier is entirely possible as long as the addition of an even more highly correlated variable maximizes the class separation ability of the set. A multivariate biomarker consists of a set of variables, which complement themselves in a way that maximizes their joint discriminatory power. Usually, some of the variables in the set are univariately significant and some are not.

3.2.3 Supervised Versus Unsupervised Methods

Some studies try to overcome the curse of dimensionality by preceding biomarker discovery with unsupervised methods in order to reduce the number of features (with *features* meaning either original variables or their combinations). Generally, this approach yields inferior results and in extreme situations may lead to the worst possible solutions (Hand et al. 2001). The goal of supervised feature selection is to find a subset of the variables that maximizes some criterion of class separation, for

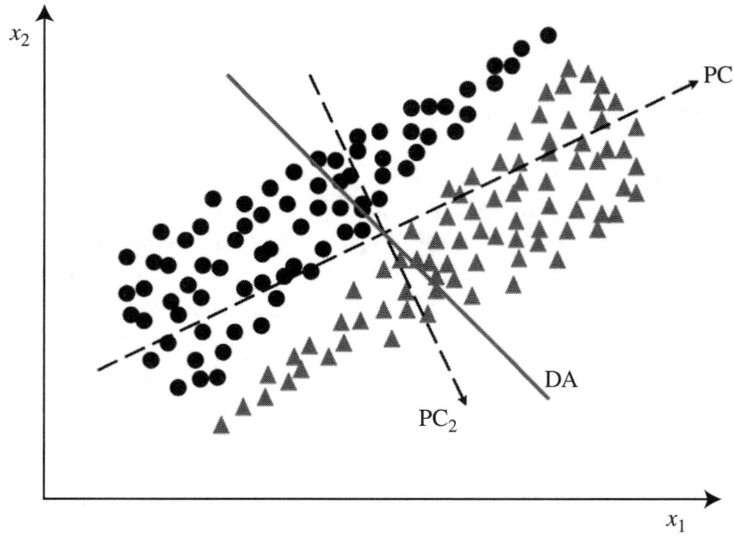

Figure 3.5: A data set with two classes of samples. The direction of the first principal component, PC_1, is aligned with the direction of the most variation in the data. This direction is very different from the direction of DA, which best separates the classes (and can be found by supervised methods such as discriminant analysis). As this is a toy example with only two dimensions, adding the second principal component, PC_2, will preserve the entire variation in the data. However, this would neither decrease the dimensionality nor identify the most discriminatory direction. (See color insert.)

instance the ratio of the variation between classes to the variation within classes, or the margin of a separating hyperplane. The goal of such unsupervised methods like principal component analysis (PCA) is to identify linear combinations of the original variables that explain most of the variation in the data. These goals, and the selection criteria associated with them, are very different. Consider another toy example presented in Figure 3.5. With only two variables ($p = 2$), we can easily visualize the variable space. Linear discriminant analysis[10]—a supervised learning algorithm—would identify DA as the direction best separating the two classes ($J = 2$) of samples. DA is the direction that maximizes the ratio of the variation between classes to the variation within classes.[11] PCA finds the direction PC_1. This would be the best single dimension to project the data if we were interested in preserving most of the variation in the data rather than in class separation. The DA and PC_1 directions are very different. If we use PCA to decrease dimensionality of the variable space, almost all of the discriminatory information would be removed from the data represented in the one-dimensional space of PC_1.

[10]Linear discriminant analysis is presented in Section 3.3.

[11]For this example, DA would also be the direction of the orientation vector orthogonal to the optimal separating hyperplane of a support vector machine—another supervised learning algorithm, described in Section 3.4.

In the realm of unsupervised analysis of typical gene expression data with thousands of variables there is no way to know how much, or how little, of discriminatory information is preserved in a space defined by the first few principal components. It is even possible that class separation can be aligned with directions of the last few principal components (Jolliffe 2002). To be sure we are not discarding important discriminatory information, we would need to use all principal components with nonzero eigenvalues. This generally results in no dimensionality reduction. Therefore, using PCA as a preprocessing step for biomarker discovery is not recommended.

Nevertheless, assume for a moment that we have strong indications that, for a particular data set, the first few principal components are aligned in the directions of good class separation. As principal components are orthogonal to each other, using a few of them would simplify the analysis and decrease dimensionality. Or, would it really? Remember that each of the principal components is a linear combination of all p original variables. This means that no progress would have been made towards identification of a parsimonious biomarker consisting of a few original (and biologically interpretable) variables.

Another approach to unsupervised preprocessing is to cluster variables by similarity of their expression patterns, and then limit biomarker discovery to variables chosen to represent the clusters (either original variables or new features representing cluster centroids). For instance, each cluster may be represented by the variable closest to its center. This may sound plausible, but the question is whether the clustering results would have anything to do with differentiating the classes. Genes are selected to a cluster on the basis of their expression similarity. Even with moderately sized data sets (e.g., with the length of the gene expression vectors of fifty or so), it may be hard to find many genes with near perfect or very high correlation between their expression patterns. Therefore, clusters often include genes whose expressions have a relatively small amount of common variance (even as small as 50 percent or less). Whether such genes carry similar discriminatory information is anybody's guess, and it cannot be determined within the unsupervised approach.

Consider the following intuitive explanation. Assume we have a data set of N samples. Assume further that this data set can be used as a training set in different studies. The studies may be interested in the classification of very different sets of phenotypic classes to which the N samples may be assigned. For example, one study may differentiate between disease states represented in the data, another study may split the N samples into classes representing different responses to radiation and aim at predicting radiation-sensitive patients, yet another study may be interested in differentiation between groups with different risks of relapse, and so on. What would be the reason to assume that the clustering results are aligned with a particular classification problem? Would expression similarities of genes within a cluster have anything to do with the differentiation of a particular set of phenotypes? Which of them, if any? By its very definition, unsupervised learning does not take into account the metadata information assigning samples to phenotypic classes. Similar to preprocessing with PCA, the problem with this clustering-based feature selection is that we do not know how much of the important discriminatory information this approach removes from the data. In conclusion, it is generally recommended that the feature selection process be kept within the realm of supervised analysis.

3.2.4 Taxonomy of Feature Selection Methods

Though the feature selection problem has been a domain of artificial intelligence and data mining, latest technological advances in biomedical research resulting in data sets with huge number of variables and small number of biological samples—a challenging combination rarely encountered in typical business data mining applications—has given rise to a new wave of data mining as well as bioinformatics publications and reviews on feature selection approaches and algorithms (Guyon and Elisseeff 2003; Liu and Yu 2005; Guyon et al. 2006; Liu and Motoda 2007a; Saeys et al. 2007).

Common taxonomy of feature selection methods divides them according to several criteria:

- search models defined by the relationship (or the lack thereof) between the feature selection search and the classification algorithm (filter, wrapper, hybrid, and embedded models),
- learning approach (supervised and unsupervised methods),
- whether interactions between variables are taken into account (univariate and multivariate methods),
- search strategy (exhaustive, complete, sequential, random, and hybrid searches).

As explained in the previous two sections, some of these categories—univariate and unsupervised methods—are inappropriate for biomarker discovery based on gene or protein expression data. We will exclude them from the scope of the *feature selection* term used in this book.

In addition to the above criteria, each feature selection search may be characterized by:

- subset evaluation criteria,
- search stopping criteria,
- methods of validation of search results.

3.2.4.1 *Filters, Wrappers, Hybrid, and Embedded Models*

The term *filter* is somewhat misleading for it implies univariate filtering, whereas filter models can be either univariate or multivariate (Saeys et al. 2007). We will classify the univariate models as belonging to variable *filtering* methods (described in Chapter 2) rather than to variable or feature *selection* methods. In the context of biomarker discovery we will consider only multivariate feature selection models.

3.2.4.1.1 Filter Models

Filter models are defined as the ones that perform feature selection independently of any classification algorithm. Their selection is based on the intrinsic characteristics of the training data and they evaluate each of the subsets under consideration using a metric independent of the algorithm that will be used to build a classification system (Fig. 3.6).

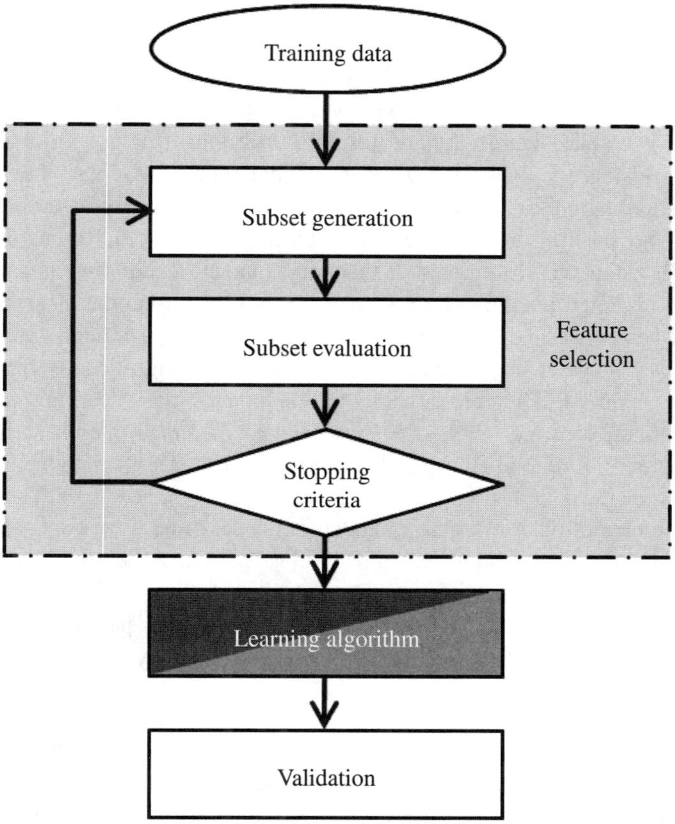

Figure 3.6: Filter model of feature selection.

Multivariate filter models may be further classified either as truly multivariate or as single-variable-centered multivariate models, which take into account interactions between variables only to some degree. Examples of single-variable-centered multivariate filter models include *Shrunken Centroid* filters, *Correlation-Based Feature Selection* and the *Fast Correlation-Based Filter*. Although they include elements of the multivariate approach, the relevance of each variable is determined univariately, by individual evaluation of the variable's prediction ability or its correlation to the class variable. Another example of a multivariate filter model is the *Markov Blanket Filter*. Though multivariate by itself, it is often preceded by univariate filtering making the combination univariately-biased. In practice, gene expression microarray studies limited to filter models of feature selection often report biomarkers consisting of dozens or even more than a hundred variables, which is not what we are looking for in biomarker discovery. An example of a truly multivariate filter model could be a heuristic sequential search, which evaluates subsets of variables using some metric of class separation that does not suffer from a univariate bias. It may be used independently or as the first stage of a hybrid feature selection (described later).

Shrunken Centroid Filters The *Shrunken Centroid* filter method (Tibshirani et al. 2002) starts by calculating average gene expression vectors for training samples assigned to each class (class centroids) and for all training samples (the global centroid). Elements of these vectors correspond to the average expression of a gene for each class (gene class centroids) and for all the samples (the gene global centroid). Class centroids are then shrunk toward the global centroid. Genes whose new class centroids are not significantly different from their global centroid are deemed irrelevant and removed from the analysis. The significance of the difference is evaluated by a t-test like statistic. The amount of shrinkage (and corresponding number of eliminated genes) is determined by minimizing the K-Fold cross-validation error. Note that filtering of genes is based on the individual evaluation of each gene. Genes retained after shrinkage have at least one of their class centroids significantly different from their global centroid. They are considered relevant for classification. An extension of the shrunken centroid filter, the *Uncorrelated Shrunken Centroid* method has been proposed (Yeung and Bumgarner 2003). It adds a step that evaluates the correlations between the relevant genes and removes highly correlated ones. The correlation is calculated for each pair of genes included in the set of relevant genes. If the correlation coefficient is greater than some threshold value, one of the genes is removed (the one with the smaller relative difference between its class centroids and its global centroid). For this version of the shrunken centroid method, both parameters (the amount of shrinkage and the correlation threshold) are selected by minimizing the cross-validation error.

Correlation-Based Feature Selection (CFS) The *Correlation-based Feature Selection* is based on the claim that "*A good feature subset is one that contains features highly correlated with (predictive of) the class, yet uncorrelated with (not predictive of) each other*" (Hall 1999). The CFS filter method performs a heuristic search in the feature space. It usually starts with the empty set and at each step adds the variable that has the highest correlation with the class but is not highly correlated with the variables already included in the set (Witten and Frank 2005). Formally, this forward search is driven by the maximization of the "merit" criterion defined for a set $\boldsymbol{X}$ of p features as

$$Merit(\boldsymbol{X}) = \frac{p\bar{r}_{cx}}{\sqrt{p + p(p-1)\bar{r}_{xx}}}, \tag{3.22}$$

where $\bar{r}_{cx}$ is the average value of correlations between individual features included in the set and the class variable C, and $\bar{r}_{xx}$ is the average feature to feature correlation (Hall and Smith 1999; Wang et al. 2005). To calculate the correlation between numeric features and labels of a nominal class variable, the features may first be discretized by splitting their ranges of values in a way that minimizes the class information entropy of the resulting intervals (Hall 1999). Focusing on variables with individual predictive ability, the CFS filter may miss variables that by themselves are not highly correlated with classes but are important for discrimination as complements to other features. Therefore, although multivariate in nature (evaluation of a set of variables), this method has a univariate bias.

Markov Blanket Filter (MBF) *Markov Blanket Filter* (Koller and Sahami 1996) is a method of eliminating redundant features from a set of features. The filtering process may start from either the set of all features or (more often) from a set including only features deemed as relevant, for example, ones individually correlated with the class variable. Similar to the CFS filter, feature relevance for class prediction may be measured by an entropy-based information gain (Xing et al. 2001; Xing 2003). For classification problems, class labels are values of a nominal class variable C. For each feature X_k included in a set of features $\boldsymbol{X} = (X_1, \ldots, X_p)$, a *Markov blanket* $\boldsymbol{M}$ is defined (Koller and Sahami 1996) as a subset of $\boldsymbol{X}$ such that:

- $\boldsymbol{M}$ does not include X_k,
- the class variable C is conditionally independent of the feature X_k given $\boldsymbol{M}$,
- $\boldsymbol{M}$ includes information that the feature X_k has about C,
- $\boldsymbol{M}$ includes information that the feature X_k has about all other features.

The MBF method performs backward elimination whereby a feature for which a Markov blanket exists within the remaining set of features is removed at each step. When applied to the set of all features, it should remove both irrelevant and redundant features. This sounds good intuitively. However, in practice it may be hard to find a Markov blanket for a feature. Furthermore, usually not many features have a Markov blanket of a limited size. Often then, a set of features highly correlated with the feature X_k is used as a surrogate of its Markov blanket (Koller and Sahami 1996; Xing et al. 2001). The original idea for using the MBF method was as a preprocessing step to reduce the feature space for a learning algorithm or for a wrapper method of feature selection (Koller and Sahami 1996). A variation of MBF is implemented in the Fast Correlation-Based Filter.

Fast Correlation-Based Filter (FCBF) The *Fast Correlation-Based Filter* method (Yu and Liu 2004) defines an approximate Markov blanket in terms of the information gain-based correlation between two variables. Assuming the following notation,

- $corr(X_k, C)$ – correlation between variable X_k and the class variable C,
- $corr(X_k, X_l)$ – correlation between variables X_k and X_l, where $X_k, X_l \in \boldsymbol{X}, k \neq l$,

X_k forms a single-variable approximate Markov blanket for X_l if and only if $corr(X_k, C) \geq corr(X_l, C)$ and $corr(X_k, X_l) \geq corr(X_l, C)$. The FCBF algorithm first selects a subset of relevant features based on their individual correlations with the class variable C. Then it processes the relevant features in descending order of their correlations with C. Each feature whose correlation with C is not greater than its correlation with the current feature is removed (i.e., the current feature forms an approximate Markov blanket for the feature to be removed).

3.2.4.1.2 Wrapper Models

Wrapper models (John et al. 1994) use a predetermined classification algorithm to evaluate relative usefulness of subsets of features. The process of searching for an optimal subset is *wrapped* around the classifier. By incorporating classifiers into the feature selection process (thus tailoring the selection to the classification learning

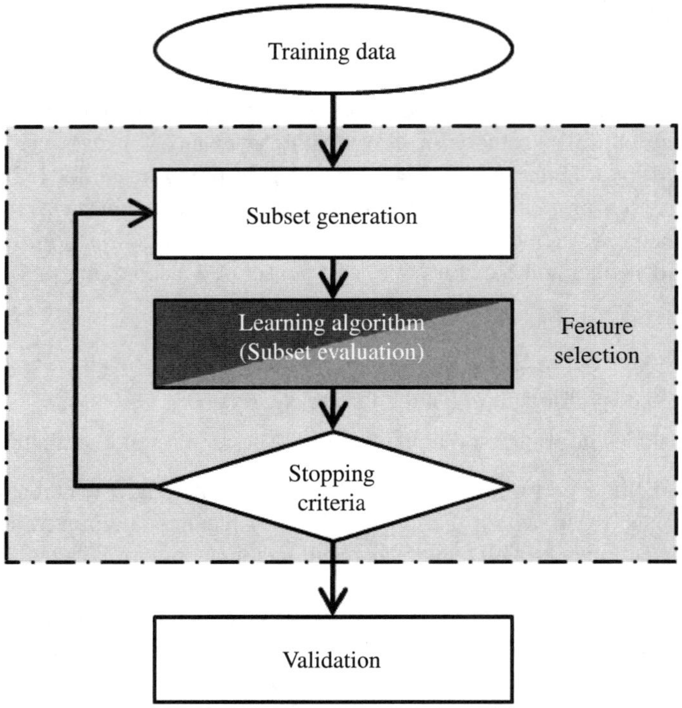

Figure 3.7: Wrapper model of feature selection.

algorithm), wrapper models tend to provide more accurate classification systems, though they are usually more computationally expensive than filter models (Fig. 3.7).

As typical wrapper models utilize internal cross-validation for subset evaluation (more precisely, to estimate the performance of the classifier built on the evaluated subset of variables), they may be prone to overfitting when applied to data sets with a large number of variables.

3.2.4.1.3 Hybrid Models

Hybrid models try to combine the advantages of filter and wrapper models (Das 2001; Xing et al. 2001; Liu and Yu 2005). Usually, they split the process of feature selection into two stages. At the first stage, a classifier-independent metric is used to identify the optimal subset for each of the considered cardinalities (subset sizes). At the second stage, the cross-validation performance of classifiers based on optimal subsets of different sizes is used to select one of these subsets.

3.2.4.1.4 Embedded Models

Embedded feature selection models (Fig. 3.8) have some similarity to wrappers in that the feature selection is associated with the learning algorithm. However, unlike in wrappers, the feature selection is *embedded* within the learning algorithm and is a part of the training process.

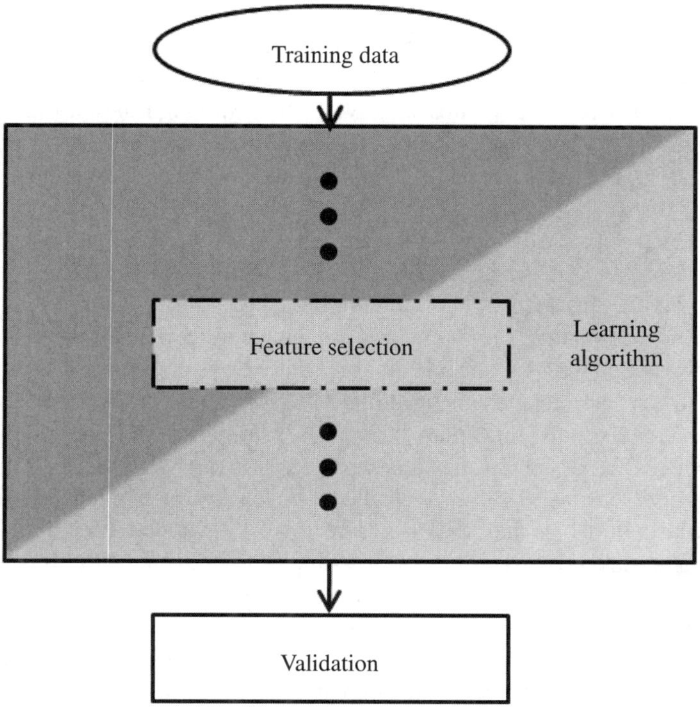

Figure 3.8: Embedded model of feature selection.

Embedded models can be implemented as extensions of learning algorithms. They usually calculate a weight or some other measure of multivariate importance for each variable in a subset. The variable with the least multivariate importance is eliminated and the classifier rebuilt on a smaller subset. The weights are then recalculated and the process repeated until optimal classifier performance and biomarker size is achieved. Sometimes, for computational reasons, more than one feature may be eliminated at a time. Examples of embedded models include random forest feature selection and Recursive Feature Elimination of Support Vector Machines (described in Sections 3.4 and 3.5 of this chapter).

3.2.4.2 Strategy: Exhaustive, Complete, Sequential, Random, and Hybrid Searches

An *exhaustive search* evaluates all possible subsets of variables and selects the subset that is the best according to some criteria. For a data set with p variables, the order of the search space is $O(2^p)$. This makes the exhaustive search infeasible for data sets with a number of variables much larger than 40 (Hastie et al. 2009). A *complete search* guarantees to find the best subset (according to the applied evaluation criteria) without evaluating all possible subsets. An example of a complete search is the *branch and bound search* with a monotonic criterion function (Hand et al. 2001). The complete search is more tractable than the exhaustive one, but still impractical for large numbers of variables. Neither of them can be used for typical gene or protein expression data sets with thousands of variables.

Heuristic searches can efficiently deal with large numbers of variables by traversing some good path through subsets of variables. Although they do not guarantee identification of the best subset, properly designed heuristic searches are capable of finding local optima resulting in efficient and parsimonious biomarkers. There are indications that such local optima may constitute better solutions to the gene expression biomarker search problem as they are more stable and less prone to overfitting than global optima (Guyon et al. 2002; Liu and Motoda 2007b). Heuristic searches are often classified as sequential or random ones.

Heuristic sequential searches could implement a stepwise forward or a stepwise backward selection strategy, or could combine these two strategies in a stepwise hybrid selection that at each step considers adding as well as removing variables. *Sequential forward selection* usually starts with the empty set and at each step adds one variable—the one which, according to set evaluation criteria, is the best addition to the current set. *Sequential backward selection* (called also backward elimination) starts with the set of all p variables and at each step removes the variable with the least multivariate importance. Both algorithms are simple to implement as well as fast, as the order of the search space is usually less than $O(p^2)$. Often called *greedy* or *hill-climbing* search strategies, they result in nested subsets of variables (a subset of m variables is included in a subset of $m + 1$ variables). In some situations, optimal subsets identified by backward elimination are better than the ones found by forward selection (for forward selection cannot evaluate the importance of a variable in the context of variables that are not included in the current subset). Backward elimination cannot, however, be implemented with set evaluation criteria that require the number of samples to be greater than the number of variables in the current subset. Class separation metrics based on the ratio of the between class variation to the within class variation belong to such criteria. These metrics cannot be calculated for the set of all p variables in a typical gene expression data set. A *stepwise hybrid selection* strategy may lead to the best results since at each step it considers both adding and removing variables until the class separation cannot be further improved for the current subset size. This strategy results in subsets that are not necessarily nested (an optimal subset of m variables does not have to be included in an optimal subset of $m + 1$ variables). An example of this strategy is presented in the Discriminant Analysis section of this chapter. All the sequential searches described here can be extended to versions that consider adding (or removing) more than one variable at a time.

Heuristic random searches may start with a randomly selected variable and then follow the sequential stepwise forward or hybrid strategy, or they may randomly generate a set of variables at each step. Either of these randomization approaches may be used to avoid trapping in inefficient local optima. Adding a random starting point to stepwise hybrid selection adds another level of hybridization to the strategy (which already combines forward and backward selections). Another level of randomization may be added when the feature selection process is performed a number of times with bootstrap versions of the training data set. Such versions of the training set may be created by random selections, with replacement, of N training samples from the original training data set. An example of this approach is described in the Random Forests section of this chapter. A *modified bagging* schema that generates bootstrap training sets by stratified random sampling of the training data set without replacement is presented in Section 3.6.

3.2.4.3 Subset Evaluation Criteria

Each subset of variables considered during the feature selection process has to be evaluated by comparing it to other subsets in order to decide on the path leading to optimal results. Depending on the search model, the criteria may be based on the characteristics of the training data or on a measure of performance of the selected classification algorithm. Examples of the former criteria are metrics of class separation or metrics of information gain. The latter criteria usually depend on classifier efficiency estimations based on a cross-validation method.

3.2.4.4 Search-Stopping Criteria

With the exception of some search procedures that may run till the search completes (such as the exhaustive search or full backward elimination—neither of which is practical for gene expression data sets), the feature selection process needs to be stopped at some point. Stopping criteria may be based on the size of a subset, the achieved value of the evaluation criterion, the number of search iterations, or finding a local optimum in the search space. For example, a search may be stopped:

- when the class separation metric reaches the predefined target value,
- when the subset size reaches the predefined maximum number of variables in the biomarker,
- when a satisfactory level of predictive accuracy of classification is achieved,
- when subsequent subsets do not offer a better solution.

Usually, we want to achieve a compromise between competing criteria (such as the minimal subset size and the maximal class separation) in order to avoid or minimize the danger of overfitting. This compromise may be incorporated into the stopping criteria defined by the algorithm. Alternatively, the search process may result in a sequence of optimal biomarkers—one biomarker for each of the considered subset sizes. The latter is preferable since, for each data set, we may decide on our optimal set of variables by taking into consideration additional information specific to the data set and the study. Please recall that the sequence of optimal biomarkers does not necessarily mean nested subsets of variables. With a well designed search, the optimal subset of $m + 1$ variables does not have to include the optimal subset of m variables.

3.2.5 Feature Selection for Multiclass Discrimination

Some feature selection methods are capable of direct handling of multiclass discrimination problems. Examples of such methods include those using subset evaluation criteria based on the ratio of the between class variation to the within class variation (such as linear discriminant analysis presented in Section 3.3), decision trees, or specialized zero-norm minimization methods for multiclass Support Vector

Machines (Weston et al. 2003). For methods that cannot directly handle multiclass data, the problem has to be treated as several two-class problems (for instance, decomposing the multiclass problem into either one-versus-the-rest comparisons or into pairwise comparisons). It may be argued that multiclass feature selection should be less prone to overfitting since it may be less likely that a random biomarker will provide a good separation of several classes. In practice, however, this may hold true only in some situations, for instance when we have only a few classes with a similar number of samples. A phenomenon often observed in multiclass discrimination is that one or two of the classes dominate the discrimination (they are easily separable) and render the remaining classes quite indistinguishable. A preferable way of dealing with such a situation is to design a multistage classification schema. This should be a data-driven schema. At the first stage, the highly overlapping classes should be treated as a single class and then this class and the easily separable classes should be discriminated. The second stage of the schema would consider only the classes that were overlapping at the first stage. In the absence of the class or classes originally dominating the discrimination, these overlapping classes may either be separable now, or—if we still have more than just few classes—at least one of them would dominate discrimination and the remaining ones that still heavily overlap will again be treated as a single class. This scenario will be continued. At each consecutive stage, we will consider fewer and fewer classes until all of the remaining classes can be significantly separated. For each of the stages, we will identify a single optimal biomarker and then use all these optimal biomarkers to build the multistage classification system. Each new sample will be processed through one or more stages, until it is classified into one of the original phenotypic classes.

3.2.6 Regularization and Feature Selection

Regularization can be seen as a way of controlling the complexity of a classification model in order to avoid (or minimize the danger of) overfitting the training data. In situations where the model is fitted by the explicit minimization of some error, or loss, function (that measures or estimates the discrepancy between the expected and observed outputs from the model), regularization may be implemented by adding a complexity penalty term to the error function. Models that are too complex will be penalized via increased total error. With a proper balance between the error and penalty terms, simpler and more generalizable classifiers can be designed. Common forms of explicit regularization include L_1 regularization (when the penalty term uses the sum of the absolute values of model coefficients) and L_2 regularization (when the sum of squares of the coefficients is used). Both the L_1 and L_2 penalties control model complexity by shrinking its coefficients towards zero. However, L_1 regularization may shrink some coefficients to exactly zero. Hence, if the model coefficients are weights associated with the variables, the L_1 regularization performs feature selection (Perkins et al. 2003; Zou and Hastie 2007).

Generally, any approach resulting in a more generalizable model can be considered a form of regularization. Hence, feature selection driven by a combined criterion of maximizing the goodness of fit and simultaneously minimizing the number of variables (selected into a multivariate biomarker) plays a role of regularization (Guyon and Elisseeff 2003).

Many learning algorithms may be seen as regularization methods. For example, shrinking class covariance matrices of quadratic discriminant analysis towards a common covariance matrix of linear discriminant analysis results in regularized discriminant analysis (Friedman 1989; Hastie et al. 2009). In particular, when all class covariance matrix estimates are replaced by the pooled estimate, decision boundaries become linear, and the resulting linear classifier can be considered a maximally regularized version of a quadratic one. Methods that further regularize linear discriminant analysis include shrinking the covariance matrix estimate towards its diagonal. This leads to diagonal linear discriminant analysis. Note, however, that using a diagonal covariance matrix is equivalent to making the assumption that genes are independent variables, which is unrealistic. For more information on linear discriminant analysis, see Section 3.3.

Soft-margin support vector machines, described in Section 3.4.2, employ regularization to find a compromise between maximizing the margin of class separation and minimizing the number of misclassified training points. Depending on its definition, the soft-margin optimization problem can be converted into the *loss* + *penalty* form that implements either the L_2 regularization or the L_1 regularization (Zou and Hastie 2007; Hastie et al. 2009).

Another approach to regularization is to design a classifier via exploiting information from many models built on perturbed versions of the training data (see descriptions of such methods as *bootstrap* and *bagging* later in this chapter). The method presented in Chapter 4 of optimizing feature selection by utilizing the *Informative Set of Genes* and information acquired from ensembles of classifiers built with the use of the *modified bagging schema* can also be considered a regularization approach leading to generalizable and more stable classifiers.

3.2.7 Stability of Biomarkers

When feature selection is performed on different versions of the training data set or by using different feature selection algorithms or their settings, the resulting multivariate biomarkers are most likely different.[12] However, the differences among such parsimonious sets of variables do not necessarily mean that the solution is unstable. The stability of biomarkers should not be mistaken with the equality of such sets. As long as these biomarkers consist of genes that point to a common set of biological processes underlying class differences, they may represent a stable solution. Hence, the stability of biomarkers can be defined in terms of the equivalence of biological processes represented by sets of genes selected into the biomarkers. Since the identification of these biological processes is often a nontrivial task that is undertaken after biomarker discovery, we can consider stability in terms of biomarkers consisting of genes representing the same *primary expression patterns* associated with class

[12]This is especially true when we deal with a large number of variables and possibly complex dependencies among them.

differences. For more information on primary expression patterns and on the process of optimizing feature selection to obtain stable biomarkers, see Chapter 4.

Other notions of stability may be based on the repeatability of classification results. For example, the stability of a learning algorithm may be measured by the degree of agreement between predictions generated by classifiers built on different training sets selected from the same populations. The agreement may be defined as the probability that the classifiers assign a test sample to the same class. Please note that this does not have to be the true class of the test sample. Thus, the stability of learning algorithms has to be considered together with their accuracy (Turney 1995).

3.3 DISCRIMINANT ANALYSIS

> *"Both LDA and QDA perform well on amazingly large and diverse set of classification tasks... It seems that whatever exotic tools are the rage of the day, we should always have available these two simple tools."*
>
> —(Hastie et al. 2009)

3.3.1 Introduction

Discriminant analysis (DA) for the differentiation between two classes was introduced by British statistician and geneticist Sir Ronald A. Fisher in the 1930s, with some earlier works related to the subject by Karl Pearson and Prasanta C. Mahalanobis. In the 1940s, discriminant analysis was extended to multiclass differentiation by C. R. Rao. Initially, discriminant analysis was applied to biological and medical problems, but was later extended to other areas, such as business. Some statistical texts divide discriminant analysis into two types—predictive DA and descriptive DA, with the predictive DA focusing on predicting class membership and the descriptive[13] DA focusing on the interpretation of class differences (more precisely, on the interpretation of the linear discriminant functions associated with class differences). With this distinction, methods and goals of descriptive DA are closely aligned with those of multivariate analysis of variance (MANOVA). This distinction of the two types of discriminant analysis may work well for some research areas. We feel, however, that in biomarker discovery it is best not to adhere to this distinction. Our goals in biomarker discovery are twofold. First, to identify a multivariate biomarker that can be used to efficiently predict class membership of new cases from targeted populations. Second, to interpret the biomarker and class differentiation in a way that can elucidate biological processes underlying class differences (it may mean either linking the biomarker with known biological processes, or discovering new biomedical knowledge, or both). Although it may be true that, in some situations (for instance, early diagnosis of a specific cancer for which there is no efficient diagnostic biomarker), positive validation of a new biomarker may be sufficient for its deployment even if we do not (yet) understand the biology underlying the differentiation, we should always assume that

[13]The term *descriptive discriminant analysis* is sometimes used in a way that includes Multivariate Analysis of Variance (MANOVA), and sometimes studies interested in group differences describe their approach as including MANOVA along with the descriptive discriminant analysis.

understanding how the biomarker reflects the mechanisms of disease is a crucial part of our studies.

Biomedical researchers may also be confused by different terminologies used in the two types of discriminant analysis. For example, the meaning of independent and dependent variables (or predictor and response variables) are reversed in these two approaches. Without necessarily implying causality, it is natural for many biomarker discovery studies to treat the characteristics of biological samples (such as the expression level of genes, proteins, or other variables representing or associated with genes or proteins) as independent variables and the differentiated classes as categories (or levels) of the dependent variable. It may, however, depend on whether our study is an experimental or observational one. In the experimental study, it would be more natural to consider variables manipulated by researchers as independent ones. To avoid such terminological confusions and to avoid switching the meaning of variables in the middle of a study (when, for instance, we start with MANOVA and descriptive DA, and then follow up with predictive DA), we will try to avoid using this taxonomy of discriminant analyses and, whenever possible, avoid describing study variables as independent and dependent.

Please recall from Section 3.1, that our use of the term *sample* corresponds to a biological sample rather than to its statistical meaning. Instead of describing variables as independent and dependent (or predictor and response), we will be saying that each sample is represented by some (usually large) number of *variables* (sample characteristics, which we want to use as a basis for classification and interpretation) and that each sample is assigned to one of some number of *classes* (targeted populations) we want to differentiate.

Yet another source of confusion when dealing with gene or protein expression data sets may be the ratio of samples to variables. Data miners experienced in analysis of very large business-type data sets may be puzzled as to why they should not assign much importance to results of the internal cross-validation methods, or even why some well-tested and often recommended (in typical data mining applications) approaches, such as dividing the data into training and test sets, should be avoided. This confusion is somewhat related to the theoretical distinction between the goals and methods of descriptive DA and predictive DA. An extension of this distinction facilitates the distinction between feature selection driven by group separation and feature selection driven by estimated classification accuracy. Estimation of classification accuracy by internal cross-validation or the holdout method can be reliable when applied to typical business problems where we may have thousands (or even hundreds of thousands) of records and a few dozen (or few hundred) variables. For example, practical advice from methodologists is that the holdout method can give us a reliable estimate of the expected classification accuracy only when the ratio of records (for the smallest class) to variables is at least five (Huberty and Olejnik 2006). Some suggest minimum ratios of 10 or even 20.

Even when we have a relatively large biomedical data set, say, with a few hundred biological samples (and if we forget for a moment about the required ratio of records to variables), where we may be tempted to try splitting the data set into a training set and a test set (or even training, validation and test sets), we are still far from a typical situation when mining business data. Splitting business data sets into training and test sets, with each of them containing a few (or even many) thousand of records, does not compromise the quality of the classification model. Splitting our gene expression data set will, however, most likely result in a classification model of lesser quality than one built on the entire data set.

These are the reasons why we will not only combine both goals of discriminant analysis (prediction and interpretation) by combining our biomarker identification process (feature selection) with the process leading to the informative set of variables (the set to be used to link class differences with biological phenomena), but we will also drive both these processes by a metric of group separation. This approach is championed by texts targeting the practical use of multivariate analysis; for example in Tabachnick and Fidell (2007) we can read (p. 23): "*If groups differ significantly on a set of variables in MANOVA, the set of variables significantly predicts group membership in discriminant analysis.*"

Let us note, however, that the theoretical differences between the group separation and classification accuracy criteria for feature selection disappear when the analysis is done under the assumptions of multivariate normality, equal covariance matrices, and equal prior probabilities (Huberty and Olejnik 2006). These assumptions are commonly made for practical applications of linear discriminant analysis. For data sets with thousands of variables, an efficient way of checking some of them is not an easy task (please refer to Section 3.1.3 of this chapter). Fortunately, discriminant analysis is quite robust to departures from these assumptions. Usually, then, we make the assumptions and consider what would happen to the results if they were not met. It is also common practice to preprocess the data in a way that will increase the chances that these assumptions are met.

The main assumptions of discriminant analysis are:

- The independence of biological samples.
- Multivariate normality. Under this assumption, each variable, as well as linear combinations of variables, follow normal distributions. With thousands of variables in typical gene expression data, it is impractical to test the normality of all possible variable combinations. Thus, practical considerations are often limited to checking on the univariate normality of the variables (this is a necessary but not satisfactory condition for multivariate normality) or to preprocessing the data in a way that should increase the chance of multivariate normality. It is worth noting, however, that discriminant analysis is quite robust to the violation of this assumption.
- Homogeneity of class covariances. This assumption results in linear discriminant analysis (LDA). If class covariance matrices are not equal, we have quadratic discriminant analysis (QDA).
- No singularities or multicollinearities. Singularity refers to completely redundant variables and multicollinearity to highly correlated ones. With singularity, covariance matrices cannot be inverted. With multicollinearity, the results of matrix inversion are unstable.
- No extreme outliers that have a large impact on the group means and increase variability.

3.3.2 Learning Algorithm

Let the training data set consist of N data points (biological samples) and p variables $x_1, \ldots, x_p$ (for instance, expression levels of p genes). Each of the N training data points is assigned to one of the J classes (populations) that we want to differentiate. Let each class j, where $j = 1, \ldots, J$, include n_j data points, so $N = \sum_{j=1}^{J} n_j$. Thus, a training data point belonging to class j can be represented by a p-dimensional vector $\mathbf{x}_{ji} \in \Re^p$,

$$\mathbf{x}_{ji} = \begin{bmatrix} x_{1ji} \\ x_{2ji} \\ \vdots \\ x_{pji} \end{bmatrix}, \tag{3.23}$$

where $i = 1, \ldots, n_j$ is used as a separate index for biological samples from each class j.

Our gene expression matrix $\mathbf{X}$ (see Section 3.1.1) can be then presented in the form

$$\mathbf{X} = \left(\begin{array}{ccc|ccc|c|ccc} x_{111} & \cdots & x_{11n_1} & x_{121} & \cdots & x_{12n_2} & \cdots & x_{1J1} & \cdots & x_{1Jn_J} \\ x_{211} & \cdots & x_{21n_1} & x_{221} & \cdots & x_{22n_2} & \cdots & x_{2J1} & \cdots & x_{2Jn_J} \\ \vdots & \ddots & \vdots & \vdots & \ddots & \vdots & \ddots & \vdots & \ddots & \vdots \\ x_{p11} & \cdots & x_{p1n_1} & x_{p21} & \cdots & x_{p2n_2} & \cdots & x_{pJ1} & \cdots & x_{pJn_J} \end{array} \right). \tag{3.24}$$

Let $\bar{\mathbf{x}}_j$ denote the mean vector for training data points belonging to class j,[14]

$$\bar{\mathbf{x}}_j = \frac{1}{n_j} \sum_{i=1}^{n_j} \mathbf{x}_{ji} \tag{3.25}$$

and $\bar{\bar{\mathbf{x}}}$ the mean vector for all N training data points (overall mean),

$$\bar{\bar{\mathbf{x}}} = \frac{1}{N} \sum_{j=1}^{J} \sum_{i=1}^{n_j} \mathbf{x}_{ji} = \frac{1}{N} \sum_{j=1}^{J} n_j \bar{\mathbf{x}}_j \tag{3.26}$$

so that $\bar{\mathbf{x}}_j$ is an unbiased estimator of the population mean vector $\boldsymbol{\mu}_j$ for class j, $j = 1, \ldots, J$, and $\bar{\bar{\mathbf{x}}}$ is the estimator of the overall mean vector $\boldsymbol{\mu}$ for all classes together.

The estimated covariance matrix[15] for the class j is[16]

$$\mathbf{S}_j = \frac{1}{n_j - 1} \sum_{i=1}^{n_j} (\mathbf{x}_{ji} - \bar{\mathbf{x}}_j)(\mathbf{x}_{ji} - \bar{\mathbf{x}}_j)^T. \tag{3.27}$$

[14]Training data points belonging to class j represent a statistical sample selected from this class (population j).

[15]This matrix is also called the variance–covariance matrix for its main diagonal elements are variances and the off-diagonal elements are covariances (with covariance being the common variance shared between two variables).

[16]The superscript ^{T} means transposition of a matrix, so a $k \times l$ matrix $\mathbf{A}$ becomes an $l \times k$ matrix $\mathbf{A}^T$ when transposed.

Under the assumptions that population observations (biological samples) are independent and normally distributed within classes (multivariate normal distribution in the p-dimensional space of the p variables) with the common covariance matrix $\mathbf{\Sigma}$, the pooled estimate of $\mathbf{\Sigma}$ is

$$\mathbf{S} = \frac{1}{N-J}\sum_{j=1}^{J}\sum_{i=1}^{n_j}(\mathbf{x}_{ji} - \bar{\mathbf{x}}_j)(\mathbf{x}_{ji} - \bar{\mathbf{x}}_j)^T \tag{3.28}$$

or

$$\mathbf{S} = \frac{1}{N-J}\sum_{j=1}^{J}(n_j - 1)\mathbf{S}_j. \tag{3.29}$$

For better stability of results, the assumption of equal covariance matrices is recommended for data sets with small n_j to p ratios (ratios of the number of biological samples in a class to the number of variables), even with covariance heterogeneity (Huberty and Olejnik 2006). Gene and protein expression data sets definitely qualify for having small values of these ratios. Hastie et al (2009) describe this stability in terms of bias-variance tradeoff: "*we can put up with the bias of a linear decision boundary because it can be estimated with much lower variance than more exotic alternatives.*"

Let us now define the *discriminatory power* of a set of p variables as a metric of class separation in the p-dimensional space of the variables. The metric can be interpreted as the value of the test statistic measuring departure from the null hypothesis H_0 of the equality of the J class centroids $\boldsymbol{\mu}_j$, $j = 1, \ldots, J$ (the p-dimensional vectors representing the population means),

$$\begin{aligned} H_0&: \quad \boldsymbol{\mu}_1 = \boldsymbol{\mu}_2 = \cdots = \boldsymbol{\mu}_J \\ H_a&: \quad \boldsymbol{\mu}_j \neq \boldsymbol{\mu}_k \text{ for some } j \neq k \end{aligned} \tag{3.30}$$

Our preferred metric of the multivariate discriminatory power of a set of p variables is the *Lawley–Hotelling trace statistic* T^2 (Lawley 1938; Hotelling 1951). There are other statistics that could be used in this role (for instance, the Wilks Λ criterion, or Roy's maximum root test), but the T^2 trace criterion has interesting features allowing for easy interpretation of the results.

The T^2 discriminatory power of a set of p variables is defined as

$$T^2 = T^2(x_1, \ldots, x_p) = tr(\mathbf{HE}^{-1}), \tag{3.31}$$

where

- $\mathbf{H}$ is the $p \times p$ "hypothesis" matrix describing variability between (or among) the classes,

$$\mathbf{H} = \sum_{j=1}^{J} n_j(\bar{\mathbf{x}}_j - \bar{\bar{\mathbf{x}}})(\bar{\mathbf{x}}_j - \bar{\bar{\mathbf{x}}})^T \tag{3.32}$$

- $\mathbf{E}$ is the $p \times p$ "error" matrix describing variability within classes[17],

$$\mathbf{E} = \sum_{j=1}^{J} \sum_{i=1}^{n_j} (\mathbf{x}_{ji} - \bar{\mathbf{x}}_j)(\mathbf{x}_{ji} - \bar{\mathbf{x}}_j)^T \tag{3.33}$$

- $tr(\mathbf{A})$ denotes the trace of a square matrix $\mathbf{A}$, or the sum of its diagonal elements. For the $k \times k$ matrix $\mathbf{A}$, $tr(\mathbf{A}) = \sum_{i=1}^{k} a_{ii}$.
- $\mathbf{E}^{-1}$ denotes an inverse of matrix $\mathbf{E}$, that is, $\mathbf{E}^{-1} \times \mathbf{E} = \mathbf{I}_{p \times p}$, where $\mathbf{I}_{p \times p}$ is the $p \times p$ identity matrix—a matrix with 1's on the diagonal and 0's off the diagonal.

Maximizing the T^2 metric of discriminatory power means that the variation between classes is maximized in relation to the variation within classes. In other words, maximal separation of classes does not necessarily mean maximal distances between the class centroids; rather, it means minimal overlaps between the classes (Hastie et al. 2001).

The exact distribution of T^2 is complicated (Anderson 2003; Morrison 2005) but there are several T^2 approximations using either the χ^2 or the F distribution. We can use, for example, the following F approximation of T^2 (Rencher 2002; Huberty and Olejnik 2006):

$$F = \frac{t(N - J - p - 1) + 2}{t^2 b} tr(\mathbf{HE}^{-1}), \tag{3.34}$$

which (under the assumption that the null hypothesis is true) has an F distribution with the following degrees of freedom for numerator (df_N) and denominator (df_D):

$$df_N = bt, \tag{3.35}$$
$$df_D = t(N - J - p - 1) + 2, \tag{3.36}$$

where

$$t = \min(p, J - 1),$$
$$b = \max(p, J - 1).$$

Let us take a look at some properties of the T^2 metric (Ahrens and Läuter 1974):

- The value of the T^2 metric tells us how well a set of p variables can separate the J classes. We always have $T^2 \geq 0$, with $T^2 = 0$ meaning that the classes are undistinguishable by the p variables. Increased values of T^2 correspond to increased class separability.

[17]Note that $\mathbf{S} = \frac{1}{N - J}\mathbf{E}$. The $\mathbf{E}$ matrix may be called the error sum-of-squares and cross-products matrix, and $\mathbf{H}$—the matrix for the hypothesis or the between-class sum-of-squares and cross-products matrix (Huberty and Olejnik 2006; Meyers et al. 2006).

- T^2 values can be used for direct comparison of discriminatory power of data sets with different numbers of variables and different numbers of biological samples.
- T^2 metric is monotonic, which means that adding a variable to a set cannot decrease the value of T^2,

$$T^2(x_1) \leq T^2(x_1, x_2) \leq \cdots \leq T^2(x_1, \ldots, x_p).$$

- T^2 is invariant for linear and regular transformations of the p variables.

We can use the T^2 metric to measure discriminatory power of a set of p variables and to answer the question about how well the set of p variables can separate the J classes. However, T^2 can be calculated only when $p < N - J - 1$ (where $N - J$ is the error-term degrees of freedom). Can we then use it for the microarray gene expression data set with, say, $p = 5000$ probe sets, $N = 100$ samples and $J = 3$ classes? Definitely yes, but not with all the 5000 variables at once. As our goal is biomarker discovery, we want to identify a small set of variables that (i) sufficiently separates the classes, and (ii) can be used for efficient classification of new cases. We may definitely use T^2 for the evaluation of the discriminatory power of small sets of variables. Furthermore, we can use the T^2 metric directly in the process of selecting the optimal multivariate biomarker for—as stated earlier—we want this process to be driven by a metric of class separation. We will cover this feature selection method in the next section. Here, assume that we already have an optimal multivariate biomarker consisting of p variables and that p is less than $N - J - 1$. In a real study, after the well performed (and successful) step of feature selection, we would have a biomarker with much fewer variables than $N - J - 1$ (which equals 96 for our hypothetical study). Assume then that our optimal biomarker is a set of, say, $p =$ 10 variables.

Now, we would like to build a classification system based on the multivariate biomarker. We can use the classifier to further validate the biomarker (preferably with a totally independent test data set) and then to classify new samples. Using the above example, the classification of external samples can be performed in the ten-dimensional space of the $p = 10$ biomarker variables (probe sets). We can calculate the centroid of each class, using (3.25), and classify a new sample based on some measure of the distances between the point representing the sample and each of the J centroids. Mahalanobis distance is often used since it takes into account correlations between variables. Euclidean distance would be appropriate only when variables are not correlated. Although after properly performed biomarker discovery we could expect that the variables in a multivariate biomarker are not highly correlated (if they were, they could be redundant and as such unlikely to be selected together to the multivariate biomarker), they are not necessarily orthogonal or even quasi-orthogonal. Mahalanobis distance D_j^* between the point $\mathbf{x} = [x_1, \ldots, x_p]^T$ representing the sample to classify and the centroid of class j, $\overline{\mathbf{x}}_j$, is equal to

$$D_j^* = \sqrt{(\mathbf{x} - \overline{\mathbf{x}}_j)^T \mathbf{S}^{-1} (\mathbf{x} - \overline{\mathbf{x}}_j)}. \tag{3.37}$$

The sample would be classified[18] into the class corresponding to the smallest D_j^*.

A graphical presentation of the classification results is usually very important for perception of the classification system by end users. For biomarkers with the number of variables $p > 3$ neither the entire p-dimensional discriminatory space nor the full classification results can be presented graphically. To facilitate visualization of classification models, we will decrease the dimensionality of the discriminatory space by solving the generalized eigenproblem (Duda et al. 2001; Rencher 2002; Huberty and Olejnik 2006; Johnson and Wichern 2007)

$$\mathbf{Hv} = \lambda \mathbf{Ev}. \tag{3.38}$$

This problem has $t = \min(p, J-1)$ nonzero eigenvalues $\lambda_1, \lambda_2, \ldots, \lambda_t$ such that

$$\lambda_1 \geq \lambda_2 \geq \cdots \geq \lambda_t. \tag{3.39}$$

Associated with the t eigenvalues are t normalized eigenvectors $\mathbf{v}_1, \mathbf{v}_2, \ldots, \mathbf{v}_t$. When $J \leq p$, the solution to this problem allows for the transformation of the p-dimensional space of the original biomarker into t-dimensional space having t linear discriminant functions. Because T^2 is invariant to such linear transformations and because we use linear classifiers, the resulting t-dimensional space represents the same discriminatory information as the original p-dimensional space. The t discriminant functions are linear combinations of the p biomarker variables. For example, the first discriminant function f_1, associated with the largest eigenvalue λ_1 and its p-dimensional eigenvector $\mathbf{v}_1$,

$$\mathbf{v}_1 = \begin{bmatrix} v_{11} \\ v_{21} \\ \vdots \\ v_{p1} \end{bmatrix} \tag{3.40}$$

is defined as

$$\begin{aligned} f_1 &= v_{11}x_1 + v_{21}x_2 + \cdots + v_{p1}x_p \\ &= \sum_{l=1}^{p} v_{l1}x_l \\ &= \mathbf{v}_1^T \mathbf{x}. \end{aligned} \tag{3.41}$$

[18]In a more general case, when we do not make assumptions of (i) equal covariance matrices $\mathbf{S}_j$, and (ii) equal prior probabilities q_j for each class, the sample would be classified to the class with the minimum value of the following classification statistic: $\ln|\mathbf{S}_j| + D_j^2 - 2\ln q_j$, where $D_j = \sqrt{(\mathbf{x} - \bar{\mathbf{x}}_j)^T \mathbf{S}_j^{-1} (\mathbf{x} - \bar{\mathbf{x}}_j)}$. When the covariance matrices are assumed to be equal but the prior probabilities are not, we would minimize $D_j^{*2} - 2\ln q_j$ (Huberty and Olejnik 2006). The decision whether to assume equal or not equal prior probabilities for class membership should be made on a case by case basis.

It transforms any p-dimensional point $\mathbf{x}$ into the single dimension defined by the f_1 discriminant function (this dimension represents a *feature*—a new variable, which is a linear combination of the original p variables). Elements of the vector $\mathbf{v}_1$ are the weights of this linear transformation associated with the original variables $x_1, \ldots, x_p$. Let us define a $p \times t$ matrix $\mathbf{V}$ whose columns are the t eigenvectors $\mathbf{v}_1, \mathbf{v}_2, \ldots, \mathbf{v}_t$ associated with the t eigenvalues,

$$\mathbf{V} = \begin{bmatrix} v_{11} & v_{12} & \cdots & v_{1t} \\ v_{21} & v_{22} & \cdots & v_{2t} \\ \vdots & \vdots & \ddots & \vdots \\ v_{p1} & v_{p2} & \cdots & v_{pt} \end{bmatrix}. \tag{3.42}$$

The matrix $\mathbf{V}$ includes weights for all t linear discriminant functions $f_1, \ldots, f_t$. A sample to be classified (e.g., a patient to be diagnosed), represented by a vector $\mathbf{x} \in \Re^p$,

$$\mathbf{x} = \begin{bmatrix} x_1 \\ x_2 \\ \vdots \\ x_p \end{bmatrix} \tag{3.43}$$

in the p-dimensional space of p biomarker variables, can be now represented by a vector $\mathbf{w} \in \Re^t$,

$$\mathbf{w} = \begin{bmatrix} w_1 \\ w_2 \\ \vdots \\ w_t \end{bmatrix} \tag{3.44}$$

in the space of t features defined by the t discriminant functions where

$$\mathbf{w} = \mathbf{V}^T\mathbf{x}. \tag{3.45}$$

As indicated, the t features of the resultant t-dimensional space have the same discriminatory power as the p original variables of our biomarker,

$$T^2(w_1, \ldots, w_t) = T^2(x_1, \ldots, x_p). \tag{3.46}$$

A classification model is built in this t-dimensional space. Classes are represented by t-dimensional hyperspheres, and classified samples by t-dimensional vectors, or points. Since the number of differentiated classes is usually $J \leq 4$, the entire discriminatory information for most classification models can be presented graphically in three or fewer dimensions. For rare models with more than four classes ($J > 4$), this transformation orders the t dimensions according to their decreasing eigenvalues, and the first two or three features often represent most of the discriminatory information. For binary classification ($J = 2$), the discriminatory space can be represented by a single axis.

Besides the dimensionality reduction, there are other advantages of using the t features:

- The features have unitary covariance matrix, which means they are uncorrelated and their variance within each class is equal to 1 (which also means the equivalence of the Mahalanobis and Euclidean distances in the space of these features). Classes are represented by spherical (hyperspherical) areas rather than by ellipsoidal (hyperellipsoidal) areas in the p-dimensional space of original variables.
- The discriminatory power of each feature is equal to the eigenvalue associated with the feature,

$$T^2(w_k) = \lambda_k \quad \text{for } k = 1, \ldots, t. \tag{3.47}$$

This leads to the following property of the T^2 metric of class separation:

$$T^2(w_1, \ldots, w_t) = \sum_{k=1}^{t} \lambda_k. \tag{3.48}$$

- As the eigenvalues are sorted in decreasing order, the discriminatory information content of the features is also sorted in this order. In case we need further reduction in dimensionality, the best t^*-dimensional subspace, $t^* < t$, of the t-dimensional discriminatory space is the one including t^* features corresponding to the first t^* eigenvalues.

To classify a new sample we may estimate the probability of its membership in each of the J classes. Assuming that the vector $\mathbf{w} = [w_1, \ldots, w_t]^T$ representing the unknown sample is the center $\boldsymbol{\mu}_0$ of a hypothetical class $j = 0$, we can perform J significance tests for $j = 1, \ldots, J$,

$$\begin{aligned} H_0 &: \quad \boldsymbol{\mu}_0 = \boldsymbol{\mu}_j \\ H_a &: \quad \boldsymbol{\mu}_0 \neq \boldsymbol{\mu}_j \end{aligned} \tag{3.49}$$

We may use the following test statistic (Ahrens and Läuter 1974; Srivastava 2002; Anderson 2003; Huberty and Olejnik 2006):

$$F_j = \frac{n_j}{n_j + 1} \cdot \frac{N - J - t + 1}{t(N - J)} (\mathbf{w} - \overline{\mathbf{w}}_j)^T (\mathbf{w} - \overline{\mathbf{w}}_j), \tag{3.50}$$

where $\overline{\mathbf{w}}_j$ is the centroid of class j in t-dimensional discriminatory space,

$$\overline{\mathbf{w}}_j = \mathbf{V}^T \overline{\mathbf{x}}_j. \tag{3.51}$$

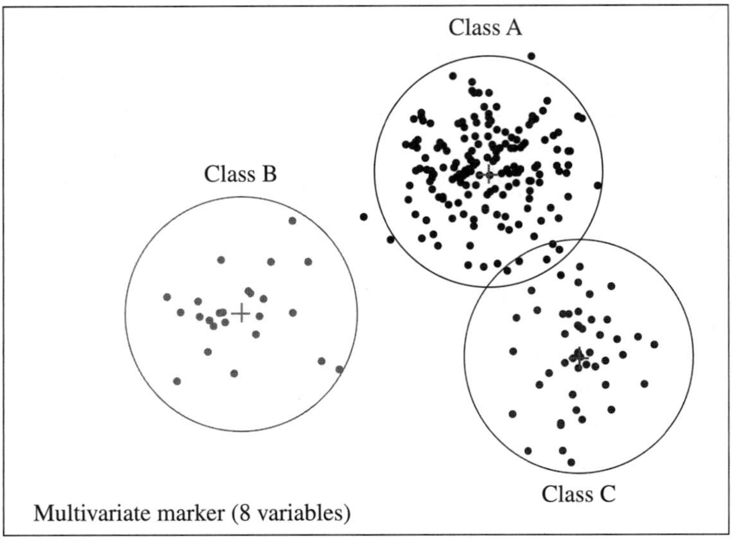

Figure 3.9: Discriminatory space of a classification model built on an eight-gene multivariate biomarker ($p = 8$). For this three-class model ($J = 3$), the discriminatory space is two-dimensional, $t = \min(p, J - 1) = 2$. The circles represent constant density boundaries enclosing 95 percent of the probability for each class. Points represent samples from the training data set. (Graphics from the *MbMD* data mining software.)[20] (See color insert.)

The F_j statistic has an F distribution with t and $N - J - t + 1$ degrees of freedom. The result of each of the J tests can be interpreted in terms of the probability of sample membership in the class j, $j = 1, \ldots, J$. When the tests are performed with a specific significance level α (which may be adjusted for multiple testing), the sample belongs to the class j if $F_j \leq F_\alpha$. Direct interpretation of this approach may result in the sample belonging to one class, to more than one class, or to none of the discriminated classes. Alternatively, we may assign the sample to the class with the smallest F_j value.[19]

For a graphical presentation of the classification results, we may represent each class as a t-dimensional hypersphere (more precisely, it would be a true hypersphere only when the number of classes J is greater than four, for $J = 4$ we would have a three-dimensional sphere, for $J = 3$ a circle, and for $J = 2$ a line segment, see Fig. 3.9). For a given significance level α, the hypersphere with the radius R_j, where

$$R_j = \sqrt{F_\alpha \cdot \frac{n_j + 1}{n_j} \cdot \frac{t(N - J)}{N - J - t + 1}}, \tag{3.52}$$

contains $(1 - \alpha) \cdot 100\%$ samples belonging to the class j (Dziuda 1990).

[19]To account for unequal class prior probabilities $q_j, j = 1, \ldots, J$, we would assign the classified sample to the class with the minimum value of the following classification statistic: $(\mathbf{w} - \overline{\mathbf{w}}_j)^T(\mathbf{w} - \overline{\mathbf{w}}_j) - 2\ln q_j$ (Duda et al. 2001).

[20]www.MultivariateBiomarkers.com

3.3.3 A Stepwise Hybrid Feature Selection with T^2

The goal of feature selection is to find a small subset of variables that can significantly separate differentiated classes and that can be used to build a classifier generalizable to populations represented by the training data. We look for a set with as few variables as possible and with as high discriminatory power as possible. We take into account all p' variables of the training data set (p' is used here to denote the starting number of variables, only to distinguish it from p, which is used more generally—for any number of variables in a set).

A heuristic feature selection process can be designed as a stepwise search for a parsimonious multivariate biomarker that maximizes the Lawley–Hotelling T^2 metric of discriminatory power. The selection of variables into the marker may start with the single most discriminatory variable or with a randomly selected one.[21] At consecutive steps, variables are being added or removed to maximize the discriminatory power of a set of p variables ($p = 1, 2, 3, \ldots$). The discriminatory power of each evaluated set is calculated with the use of the previously defined T^2 metric of class separation. At each step, forward selection is performed first. From the pool of remaining $p' - p + 1$ variables the variable that best complements the current set of $p - 1$ variables is selected, that is, the variable whose addition will maximize the T^2 discriminatory power of a set of p variables. Then, for steps with $p > 2$, an attempt is made to find a $p - 1$ variable subset of the p variables that is more optimal than the previous set of $p - 1$ variables. That is, if the elimination of any one of the p variables from the current set results in a set of $p - 1$ variables with T^2 greater than that for the previously selected best set of $p - 1$ variables, then the variable associated with the minimal T^2 decrease is eliminated. The resulting set of $p - 1$ variables becomes the current selection. At each step, this hybrid (forward and backward) stepwise search is iterated until there is no further gain in T^2. Each eliminated variable is sent back to the pool of available variables, so it can still be reconsidered in subsequent steps. The algorithm stopping criteria are defined by two adjustable parameters: $stop\_T^2$ and $stop\_p$. When $T^2 \geq stop\_T^2$ or $p = stop\_p$, the selection process ends (Table 3.4).

Although the hybrid search through the space of subsets is driven by the T^2 criterion of class separation, each T^2 value is associated with its statistical significance. The significance is calculated in two ways. One method implements the permutation-based generation of the experimental distribution of the T^2 metric. The other uses an approximation of the T^2 distribution with the F distribution, and then applies a conservative Bonferroni correction to the calculated p-value.

[21] A practical note on the random selection of the first variable: The random selection of any of the p' original variables may sound plausible. However, if the selected variable's individual T^2 discriminatory power is particularly small, the best univariate variable is often selected as the second. Then, the starting-point variable may be quickly eliminated and the resulting set may be identical (or very similar) to one starting with the best univariate variable. To avoid this phenomenon, we may restrict our random selection of the first variable to variables that have their individual T^2 above some level. For example, the stepwise hybrid feature selection algorithm implemented in the *MbMD* data mining software uses a *randMax* parameter limiting the random selection to the first *randMax* variables sorted in descending order of their individual T^2 measures (the default value of *randMax* is equal to 1000).

TABLE 3.4: C-Style Pseudocode of the Stepwise Hybrid Feature Selection Algorithm

```
// Stepwise hybrid feature selection
// input:   trainingData(N, p, J)   training data set: p variables, N samples, J classes
//          stop_p, stop_T2         stopping criteria
//          randomFlag              starting point flag
// output:  currentSet              an optimal subset
pool ← trainingData(N, p, J);
currentSet ← NULL;
poolSize = p; markerSize = 0; currentT2 = 0.0;
for (step = 1;    markerSize < stop_p && currentT2 < stop_T2  &&
                  (markerSize < (N-J-2) || (N-J-2) == 0) && markerSize < p;
                  step++ ) {
 maxGain = 0.0;
 markerSize++;
 if (markerSize == 1 && randomFlag) {
    selectedVar = selectRandomVar(pool); // random first variable
    maxGain = calculateT2(selectedVar);
 } else {
       // Forward selection: add the variable that maximizes T^2 of this step.
       for (i=1; i<=poolSize; i++) {
           deltaT2 = calculateT2( currentSet + poolVar (i)) - currentT2;
           if (deltaT2 >= maxGain) {
               maxGain = deltaT2;
                selectedVar = poolVar(i);
           }
       }
 }
 currentT2 += maxGain;
 poolToMarker(selectedVar);      // move selected variable from pool to currentSet
 poolSize−−;
 pEmpirical = bootstrap(currentT2); // Bootstrap estimate of the T^2 p-value
 pF = pValueF(currentT2);           // p-value from F distribution (w/Bonferroni)
 // Calculate multivariate significance of the added variable
 member_F = memberSignificance(currentSet, selectedVar);
 // Backward optimization: if elimination of any variable results in T^2 greater
 // than that of the previous step, eliminate one that minimally decreases T^2.
 minLoss = currentT2;
 if (markerSize > 2 ) {
    for (i=1; i<markerSize; i++) {
        deltaT2 = currentT2 – calculateT2( currentSet – setVar(i));
        if (deltaT2 <= minLoss) {
           minLoss = deltaT2;
           removedVar = setVar(i);
        }
    }
```

(Continued)

TABLE 3.4: *Continued*

```
        if (minLoss < maxGain) {
            markerToPool(removedVar); // move variable from currentSet to pool
            poolSize++; markerSize--;
            currentT2 -= minLoss;
        }
        saveSet(currentSet);     // save set optimal for the current cardinality
    }
}
return currentSet;
```

After the algorithm finishes, we are presented with the marker of size *stop_ p* or with a marker with fewer variables but satisfactory discriminatory power of $T^2 \geq stop\_T^2$. The optimal sets for all considered cardinalities are also available for eventual selection of one of them as our optimal marker.

This heuristic search may be described as the search for an optimal biomarker, where optimization means satisfying a combined criterion of minimal size p and maximal discriminatory power T^2. The search can be performed in a very reasonable amount of time,[22] even for data sets with a very large number of variables p'. This search is, however, a local search and there is no guarantee that a better result cannot be found. To decrease a chance of being trapped in a particularly inefficient local optimum, the search may be performed more than once, each time with a different starting point (a different variable selected as the first one). Additional optimization could include partial backward elimination of the identified biomarker and its stepwise rebuilding to its original size.[23] However, a better optimization method will be introduced in Chapter 4. This method will optimize feature selection by utilizing the *Informative Set of Genes* (identified on the basis of a sequence of alternative multivariate biomarkers) and information acquired from ensembles of classifiers built with the use of the *modified bagging schema* (described later in this chapter).

3.4 SUPPORT VECTOR MACHINES

Support Vector Machines (SVMs) were introduced in the 1990s (Boser et al. 1992; Vapnik 1998) as a supervised method for finding the optimal hyperplane separating two classes (binary classification problem) in a possibly high-dimensional space. Since then, they have become popular in the area of biomedical research.

[22]The cost of the stepwise selection for a marker of p variables out of p' variables is proportional to $p(2p' - p + 1)$. Even under the pessimistic assumption of 10^{-3} seconds per subset this translates into an easily manageable amount of time for the marker selection process.

[23]Optimization of a biomarker should focus on limiting its size and increasing chances for its good generalization. Over-optimization should be avoided as it usually results in overfitting.

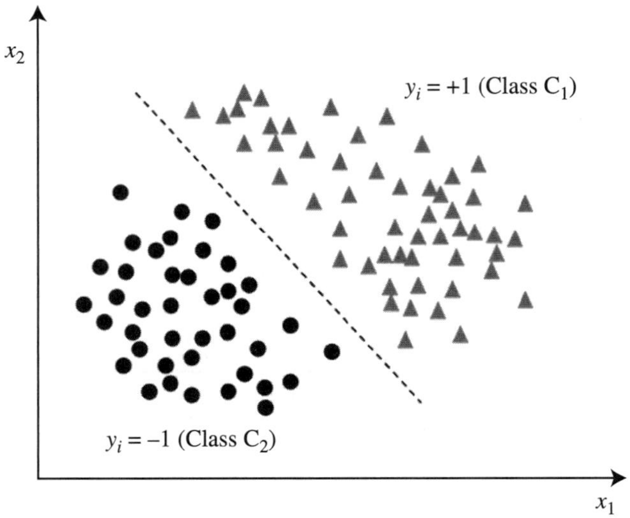

Figure 3.10: A training data set with two input variables (x_1, x_2) and two classes; $p = 2, J = 2$. The classes are linearly separable.

Let the training data set consist of N data points (biological samples) and p variables (for instance, p probe sets or p m/z variables). Each training data point can be represented by a p-dimensional vector $\mathbf{x}_i \in \Re^p$, where $i = 1, \ldots, N$,

$$\mathbf{x}_i = \begin{bmatrix} x_{1i} \\ x_{2i} \\ \vdots \\ x_{pi} \end{bmatrix}. \tag{3.53}$$

The assignment of each training point $\mathbf{x}_i$ to one of the J differentiated classes C_j, where $j = 1, \ldots, J$, can be represented by a pair $(\mathbf{x}_i, y_i)$, where $y_i \in \{C_1, \ldots, C_J\}$. In the context of a two-class SVM ($J = 2$), it is convenient to represent class labels as $y_i \in \{+1, -1\}$. A linear boundary between the two classes is represented by a $(p - 1)$-dimensional hyperplane. The optimal hyperplane is defined as a boundary that maximizes the margin of separation between classes (i.e., maximizes the distance between the boundary and the points that are closest to the boundary).

To illustrate this, let us look at a simple case where we only have two variables ($p = 2$); training points of the two classes are represented in Figure 3.10 by circles and triangles.

3.4.1 Hard-Margin Support Vector Machines

Assume for now that the training data points are linearly separable. This means that all the points on one side of a *separating hyperplane* have the same class label. However, this can be true for an infinite number of hyperplanes. Which one of them is optimal? The SVM learning algorithm looks for the one with the largest margin of class separation.

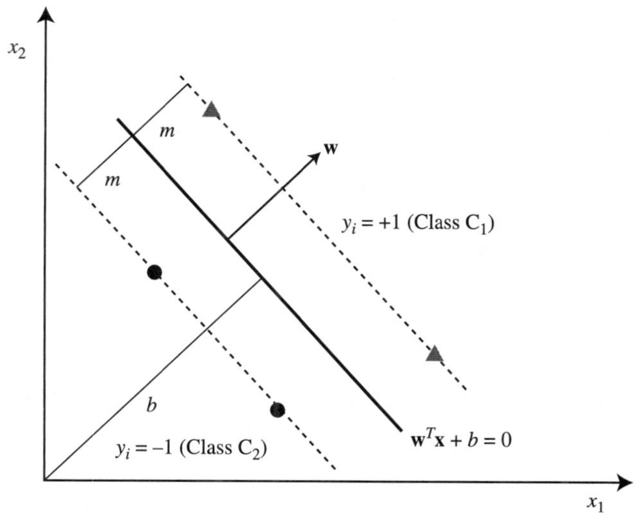

Figure 3.11: Separating hyperplane $\mathbf{w}^T\mathbf{x} + b = 0$. The margin m is the functional distance between the separating hyperplane and the training data point(s) nearest to the hyperplane. Elements of the vector **w** are weights associated with the p variables. The scalar b defines the offset of the hyperplane from the origin. Euclidean distances corresponding to m and b are $m/\|\mathbf{w}\|$ and $|b|/\|\mathbf{w}\|$.

A separating hyperplane (**w**, b) can be defined by the equation

$$\mathbf{w}^T\mathbf{x} + b = 0, \tag{3.54}$$

where the p-dimensional vector **w** is orthogonal to the hyperplane (thus determines its orientation) and the bias term b defines the offset of the hyperplane from the origin[24] (see Fig. 3.11).

For the two-dimensional space defined by variables x_1 and x_2, the equation can be written as

$$[w_1 w_2]\begin{bmatrix} x_1 \\ x_2 \end{bmatrix} + b = 0 \tag{3.55}$$

or

$$w_1x_1 + w_2x_2 + b = 0. \tag{3.56}$$

For linearly separable classes where no training point $\mathbf{x}_i$, $i = 1, \ldots, N$ lies on the separating hyperplane, we have

$\mathbf{w}^T\mathbf{x}_i + b > 0$ for the training points with the class label $y_i = +1$, and
$\mathbf{w}^T\mathbf{x}_i + b < 0$ for the training points with the class label $y_i = -1$.

[24] $\mathbf{w}^T\mathbf{x}$ is a similarity measure between the vectors **w** and **x**. It is called the *inner product* (other names are *scalar product* or *dot product*) and for p-dimensional vectors is defined as $\mathbf{w}^T\mathbf{x} = \sum_{k=1}^{p} w_k x_k$. The superscript T denotes vector transposition.

These two inequalities can be combined into

$$y_i(\mathbf{w}^T\mathbf{x}_i + b) > 0. \tag{3.57}$$

Let us now define two types of margins: the functional margin and the geometric one (Cristianini and Shawe-Taylor 2000). The functional margin of a training data point $\mathbf{x}_i$ with respect to a separating hyperplane ($\mathbf{w}$, b) is equal to the value of the function $y_i(\mathbf{w}^T\mathbf{x}_i + b)$ at the point $\mathbf{x}_i$. The *functional margin* m *of the separating hyperplane* is the minimum functional margin for the training data set,

$$m = \min_i y_i(\mathbf{w}^T\mathbf{x}_i + b). \tag{3.58}$$

The geometric margin of a point $\mathbf{x}_i$ is defined as the Euclidean distance of the point from the separating hyperplane ($\mathbf{w}$, b), and equals

$$\frac{1}{\|\mathbf{w}\|} y_i(\mathbf{w}^T\mathbf{x}_i + b), \tag{3.59}$$

where $\|\mathbf{w}\|$ is the Euclidean norm[25] of the vector $\mathbf{w}$. Therefore, the *geometric margin* γ *of the separating hyperplane* equals

$$\begin{aligned} \gamma &= \min_i \frac{1}{\|\mathbf{w}\|} y_i(\mathbf{w}^T\mathbf{x}_i + b) \\ &= \frac{m}{\|\mathbf{w}\|}. \end{aligned} \tag{3.60}$$

If we move the separating hyperplane parallel to itself in the direction of one class and then in the direction of another until it hits the nearest training point(s) of the classes, we will define two *support hyperplanes* such that the functional margin of their points is m.

As no training point has its functional margin (with respect to the separating hyperplane) less than m, we may rewrite the inequality (3.57) as

$$y_i(\mathbf{w}^T\mathbf{x}_i + b) \geq m. \tag{3.61}$$

The *optimal separating hyperplane* (also called the *maximal margin hyperplane*) is the solution of the optimization problem that maximizes the geometric margin (see Fig. 3.12):

$$\begin{aligned} &\underset{\mathbf{w},b}{\text{maximize}} \ \frac{m}{\|\mathbf{w}\|} \\ &\text{subject to} \ y_i(\mathbf{w}^T\mathbf{x}_i + b) \geq m, \quad i = 1, \ldots, N. \end{aligned} \tag{3.62}$$

[25]The Euclidean norm, or L_2 norm, of a p-dimensional vector $\mathbf{w}$ is defined as the length of the vector in the p-dimensional space, $\|\mathbf{w}\| = \sqrt{\sum_{k=1}^{p} w_k^2}$. The geometric margin is then equal to the functional margin of a normalized vector $\mathbf{w}$ (Cristianini and Shawe-Taylor 2000).

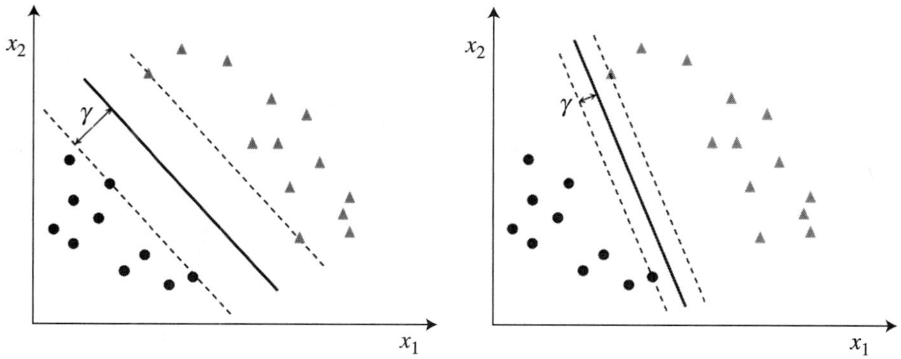

Figure 3.12: Out of the infinite number of (hyper)planes separating the classes, the one that is optimal is the one with the largest geometric margin γ.

It has been shown that instead of maximizing $m/\|\mathbf{w}\|$ we may minimize the inverse $\|\mathbf{w}\|/m$ or the squared inverse $(\|\mathbf{w}\|/m)^2$ subject to the same constraints. Furthermore, we may rescale the hyperplane ($\mathbf{w}$, b)[26] in a way that allows the set up of an arbitrary value of the functional margin m (Cristianini and Shawe-Taylor 2000; Hastie et al. 2001; Shawe-Taylor and Cristianini 2004; Abe 2005). Therefore, we may set the functional margin to be $m = 1$ and state the optimization problem (3.62) as[27]

$$\begin{aligned} &\underset{\mathbf{w},b}{\text{minimize}}\ \|\mathbf{w}\|^2 \\ &\text{subject to}\ \ y_i(\mathbf{w}^T\mathbf{x}_i + b) \geq 1 \quad \text{for all training points } \mathbf{x}_i,\ i = 1, \ldots, N. \end{aligned} \tag{3.63}$$

This optimization problem (known as the *primal optimization problem*) is a quadratic programming problem with a linear constraint that can be solved in the *primal space* of $\mathbf{w}$ and b. The solution is the pair ($\mathbf{w}$, b) that defines the optimal separating hyperplane. The solution vector $\mathbf{w}$ defines the orientation of the hyperplane and is composed of weights assigned to each of the p variables (dimensions). The scalar b defines the offset of the optimal hyperplane from the origin.

The dual representation of this optimization problem has some very interesting properties that can give us a deeper understanding of the solution (Vapnik 1998). By introducing N auxiliary nonnegative variables[28] $\alpha_i \geq 0$, $i = 1, \ldots, N$ (they are the

[26] If the hyperplane ($\mathbf{w}$, b) is a solution to the optimization problem, then a rescaled hyperplane ($a\mathbf{w}$, ab), $a \in \Re^+$ is also a solution. The rescaling will not change the classification function.

[27] A hyperplane with the functional margin $m = 1$ is known as *canonical hyperplane* (Cristianini and Shawe-Taylor 2000). The minimization is performed with respect to both the vector $\mathbf{w}$ and the scalar b (Vapnik 2000).

[28] The dual variable α_i may be interpreted as the importance of a training point $\mathbf{x}_i$ in the solution of the optimization problem (Cristianini and Shawe-Taylor 2000).

Lagrange multipliers α_i associated with the training data points $\mathbf{x}_i$, $i = 1, \ldots, N$), we can represent our optimization problem in the dual space of these Lagrange multipliers.

It has been shown (Vapnik 1998) that our convex optimization problem (3.63) can be solved by finding the saddle point of the following Lagrange function (Lagrangian):

$$L(\mathbf{w}, b, \boldsymbol{\alpha}) = \frac{1}{2}\mathbf{w}^T\mathbf{w} - \sum_{i=1}^{N} \alpha_i[y_i(\mathbf{w}^T\mathbf{x}_i + b) - 1]. \tag{3.64}$$

First, we minimize the Lagrangian $L(\mathbf{w}, b, \boldsymbol{\alpha})$ with respect to $\mathbf{w}$ and b. Solving

$$\frac{\partial L(\mathbf{w}, b, \boldsymbol{\alpha})}{\partial \mathbf{w}} = 0$$

and

$$\frac{\partial L(\mathbf{w}, b, \boldsymbol{\alpha})}{\partial b} = 0$$

results in the following *dual representation* of the Lagrange function:

$$W(\boldsymbol{\alpha}) = \sum_{i=1}^{N} \alpha_i - \frac{1}{2}\sum_{i,j=1}^{N} \alpha_i \alpha_j y_i y_j \mathbf{x}_i^T \mathbf{x}_j. \tag{3.65}$$

Now we need to maximize $W(\boldsymbol{\alpha})$ with respect to $\alpha_i \geq 0$ (i.e., in the nonnegative orthant). This leads to the following *dual representation* of the optimization problem (Cristianini and Shawe-Taylor 2000; Vapnik 2000; Abe 2005):

$$\begin{aligned} &\text{maximize} \quad W(\boldsymbol{\alpha}) = \sum_{i=1}^{N} \alpha_i - \frac{1}{2}\sum_{i,j=1}^{N} \alpha_i \alpha_j y_i y_j \mathbf{x}_i^T \mathbf{x}_j \\ &\text{subject to} \quad \sum_{i=1}^{N} y_i \alpha_i = 0, \\ &\qquad\qquad\quad \alpha_i \geq 0, \quad \text{for } i = 1, \ldots, N. \end{aligned} \tag{3.66}$$

Please note two important properties of this dual representation of the optimization problem:

- The primal optimization problem (3.63) is solved with respect to $p + 1$ variables: the p-dimensional vector $\mathbf{w}$ and the scalar b. The dual problem (3.66) is solved with respect to only N variables: the N-dimensional Lagrange multiplier vector $\boldsymbol{\alpha} = (\alpha_1, \ldots, \alpha_N)^T$, that is, one Lagrange multiplier α_i for each of the N training data points (Abe 2005). For typical microarray gene expression data sets, this is a very significant decrease in the number of optimization problem's variables and in its processing time.

- In the dual representation, training data points appear only in the form of their inner product $\mathbf{x}_i^T\mathbf{x}_j$. This property of support vector machines will facilitate the kernel trick described in Section 3.4.3.

The solution to (3.66) has to satisfy the following conditions (known as the Karush–Kuhn–Tucker, or KKT, complementarity conditions):

$$\alpha_i[y_i(\mathbf{w}^T\mathbf{x}_i + b) - 1] = 0 \quad \text{for } i = 1, \ldots, N. \tag{3.67}$$

It is easy to notice that:

a) α_i can be positive only when $y_i(\mathbf{w}^T\mathbf{x}_i + b) = 1$,

b) all training points, for which $y_i(\mathbf{w}^T\mathbf{x}_i + b) > 1$ must have $\alpha_i = 0$.

The training data points that are on the support hyperplanes $\mathbf{w}^T\mathbf{x}_i + b = \pm 1$ and that have $\alpha_i > 0$ are called *support vectors* (Fig. 3.13).

The solution vector $\mathbf{w}^*$ can be calculated from

$$\mathbf{w}^* = \sum_{i=1}^{N} \alpha_i y_i \mathbf{x}_i, \tag{3.68}$$

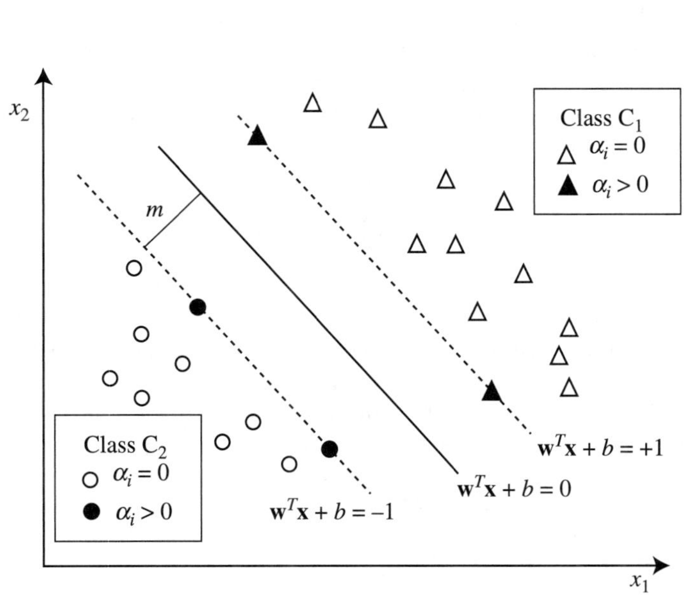

Figure 3.13: Only the training data points that lie on one of the support hyperplanes $\mathbf{w}^T\mathbf{x} + b = \pm 1$ may have nonzero values of the Lagrange multipliers α_i, $\alpha_i > 0$. The training data points with nonzero α_i are called *support vectors*.

which means that the orientation of the optimal separating hyperplane depends only on the support vectors—the training data points with $\alpha_i > 0$. This is another important property of SVMs—the solution is *sparse* in the sense that support vectors are only a small part of the training data.

Once the solution vector $\mathbf{w}^*$ has been found, the offset b^* can be calculated by solving the KKT conditions (3.67) for any of the support vectors,[29] giving

$$b^* = y_i - \mathbf{w}^{*T}\mathbf{x}_i. \tag{3.69}$$

Identification of the optimal separating hyperplane ($\mathbf{w}^*$, b^*) allows us to classify any unknown sample by calculating a signed distance between the vector $\mathbf{x} \in \Re^p$,

$$\mathbf{x} = \begin{bmatrix} x_1 \\ x_2 \\ \vdots \\ x_p \end{bmatrix} \tag{3.70}$$

representing the sample, and the separating hyperplane. The classification function

$$f(\mathbf{x}) = sign(\mathbf{w}^{*T}\mathbf{x} + b^*) \tag{3.71}$$

assigns the unknown sample to class C_1 if $f(\mathbf{x}) > 0$ and to C_2 if $f(\mathbf{x}) < 0$. Points on the separating hyperplane ($f(\mathbf{x}) = 0$) cannot be classified.

Let us look at some implications of the fact that the separating hyperplane is (usually) determined by only a very small number of training points—the support vectors. Once the support vectors are identified, the rest of training data may be discarded. We do not even need to use the vector $\mathbf{w}^*$ explicitly; by rewriting the classification function as

$$f(\mathbf{x}) = sign\left(\sum_{i=1}^{N} \alpha_i y_i \mathbf{x}_i^T \mathbf{x} + b^*\right), \tag{3.72}$$

we may classify unknown samples by using the support vectors directly (again, only these vectors $\mathbf{x}_i$, for which $\alpha_i > 0$). Being dependent only on the support vectors, the optimum hyperplane classifier is called a *support vector machine* (SVM).

According to Cristianini (2001), this duality is the first feature of SVMs, which are linear classifiers that can be represented in the dual form:

$$\begin{aligned} f(\mathbf{x}) &= sign(\mathbf{w}^{*T}\mathbf{x} + b^*) \\ &= sign\left(\sum_{i=1}^{N} \alpha_i y_i \mathbf{x}_i^T \mathbf{x} + b^*\right). \end{aligned}$$

[29]Note that $y_i = \frac{1}{y_i}$. Furthermore, in practice, the optimal value of b^* is usually calculated by averaging b over the set of support vectors.

Maximizing the margin and sparseness of the solution is often considered a way to identify generalizable classifiers. Hastie et al. (2009) note, however, that this depends on the data distribution. If the differentiated classes are normally distributed, then the separating hyperplane will be defined by the noisier data on peripheries of the classes. For such data sets, linear discriminant analysis, which is based on class centroids (depending on all training points), would be a better choice.

3.4.2 Soft-Margin Support Vector Machines

Support vector machines designed under the assumption that the training data points that are linearly separable are called *hard-margin* SVMs. What if the classes cannot be linearly separated in the space of the p input variables? To solve this more common problem, we allow for the optimal separating hyperplane classifier to misclassify some of the training data points. We will then try to find a solution that would be a compromise between maximizing the margin and minimizing the cost of misclassification.

To quantify the classification error, we introduce nonnegative slack variables $\xi_i \geq 0$, $i = 1, \ldots, N$, such that we have $\xi_i < 1$ for the training points $\mathbf{x}_i$ that are correctly classified, $\xi_i = 1$ for the points on the optimal separating hyperplane, and $\xi_i > 1$ for the misclassified training points (Fig. 3.14).

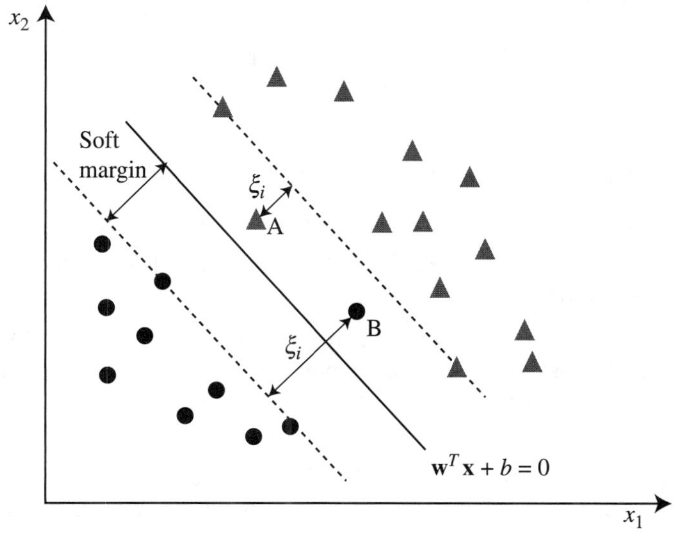

Figure 3.14: The slack variables ξ_i are measures of the margin violations for the training data points. Training points violating the margin may be correctly classified (like the point A). If they are, however, on the wrong side of the separating hyperplane (like the point B), they are misclassified and have $\xi_i > 1$. The Euclidean distance corresponding to a ξ_i equals $\xi_i/\|\mathbf{w}\|$.

The optimization problem will be now formulated as[30]

$$\underset{\mathbf{w},b,\xi}{\text{minimize}} \quad \|\mathbf{w}\|^2 + C\sum_{i=1}^{N}\xi_i$$

$$\text{subject to} \quad y_i(\mathbf{w}^T\mathbf{x}_i + b) \geq 1 - \xi_i \quad \text{for all training points } \mathbf{x}_i,\ i = 1,\ldots,N, \qquad (3.73)$$

where C is a regularization, or trade-off, parameter controlling overlap between the classes (the trade-off between maximizing the margin and minimizing the number of misclassified training points).

Similar to hard-margin SVMs, this optimization problem can be formulated in its corresponding dual form (Cristianini and Shawe-Taylor 2000; Vapnik 2000; Abe 2005; Hastie et al. 2009):

$$\begin{aligned} &\text{maximize} \quad W(\boldsymbol{\alpha}) = \sum_{i=1}^{N}\alpha_i - \frac{1}{2}\sum_{i,j=1}^{N}\alpha_i\alpha_j y_i y_j \mathbf{x}_i^T\mathbf{x}_j \\ &\text{subject to} \quad \sum_{i=1}^{N} y_i\alpha_i = 0, \\ &\qquad\qquad\quad 0 \leq \alpha_i \leq C, \quad \text{for } i = 1,\ldots,N. \end{aligned} \qquad (3.74)$$

Note that the only difference between this optimization problem and the hard-margin SVMs optimization problem (3.66) is the upper bound on the α_i values. However, the Karush-Kuhn-Tucker conditions now include the following constraints:

$$\alpha_i[y_i(\mathbf{w}^T\mathbf{x}_i + b) - 1 + \xi_i] = 0 \text{ for } i = 1,\ldots,N \qquad (3.75)$$

$$\xi_i(\alpha_i - C) = 0 \text{ for } i = 1,\ldots,N. \qquad (3.76)$$

For the solution satisfying these KKT constraints we have the following:

a) The slack variables ξ_i can assume nonzero values only when $\alpha_i = C$.

b) If $\alpha_i = 0$ then $\xi_i = 0$ and the training point $\mathbf{x}_i$ is correctly classified and is outside of the geometric margin $1/\|\mathbf{w}\|$ of the solution.

c) The training points for which $0 < \alpha_i < C$ have $\xi_i = 0$ and $y_i(\mathbf{w}^T\mathbf{x}_i + b) = 1$, thus they lie on one of the two support hyperplanes $\mathbf{w}^T\mathbf{x}_i + b = \pm 1$ and are support vectors of the solution. These support vectors are called *unbounded support vectors* (Fig. 3.15).

d) The training points with $\xi_i > 0$ violate the margin of their class and, according to (a), they have to have $\alpha_i = C$. They are also support vectors because they do

[30]Such a formulated optimization problem is known as L_1 soft-margin SVM, or 1-Norm soft-margin SVM. Another way of defining the optimization problem for soft-margin SVMs is to use squares of the slack variables and minimize $\|\mathbf{w}\|^2 + C\sum_{i=1}^{N}\xi_i^2$. Such SVMs are called L_2 or 2-Norm soft-margin SVMs (Cristianini and Shawe-Taylor 2000; Abe 2005).

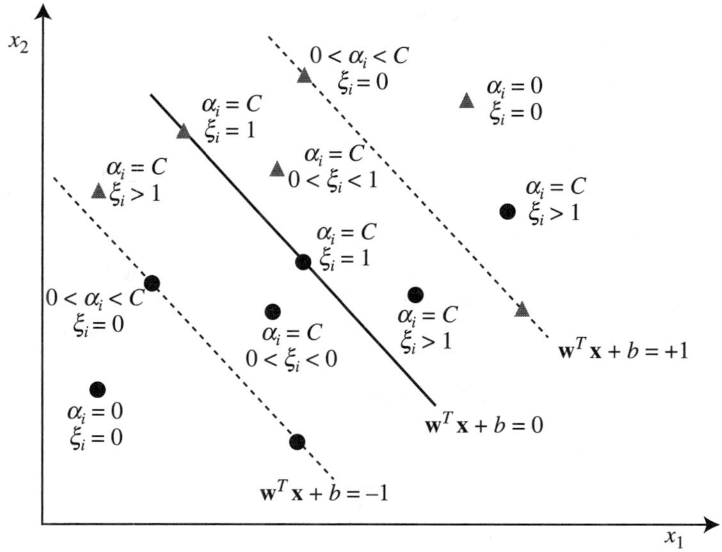

Figure 3.15: This figure represents a soft-margin SVM. The training data points with $\xi_i > 0$ violate their margin. It they have $\xi_i > 1$, they are on the wrong side of the separating hyperplane and are misclassified. They all have $\alpha_i = C$ and are called bounded support vectors (since the α_i multipliers associated with them reach the upper bound value defined by C). The support vectors that are on the support hyperplanes are called unbounded support vectors; they have $0 < \alpha_i < C$ and $\xi_i = 0$.

affect the solution. The training points with $0 < \xi_i < 1$ are correctly classified, and the ones with $\xi_i > 1$ are misclassified. The training points with $\xi_i = 1$ lie on the separating hyperplane and cannot be classified. The support vectors that violate the margin constraint are called *bounded support vectors.*

The solution vector $\mathbf{w}^*$ for a soft-margin SVM can be calculated as

$$\mathbf{w}^* = \sum_{i=1}^{N} \alpha_i y_i \mathbf{x}_i. \tag{3.77}$$

To find the offset b we can use (3.75) and solve it for any[31] of the (unbounded) support vectors with $\xi_i = 0$. We have then:

$$b^* = y_i - \mathbf{w}^{*T} \mathbf{x}_i. \tag{3.78}$$

Comparing (3.68) and (3.69) to (3.77) and (3.78) one may ask whether the solution for the soft-margin SVM is the same as for the hard-margin one. The answer is no.[32] The vector $\mathbf{w}^*$ (the orientation of the separating hyperplane) is defined not

[31] Similarly as for the hard-margin SVM, the optimal value of b^* is usually calculated as the average over all unbounded support vectors.

[32] Assuming the classes are not linearly separable. Otherwise, we would have no bounded support vectors and the solution would be the same as for the hard-margin SVM.

only by the support vectors that lie on the margin, but also by all the bounded support vectors that violate the margin (whether they are classified correctly or misclassified).

What flexibility do we have in selecting the solution? Recall that ξ_i can be interpreted as the amount by which a training point $\mathbf{x}_i$ violates its margin. Therefore, $C\sum \xi_i$ quantifies the total cost of violating the margin constraints. Increasing the value of the regularization parameter C will decrease the margin of the solution whereas decreasing C will result in a larger margin and many more points violating the margin.[33] The cost parameter C is also the upper bound for the α_i Lagrange multipliers, that is, $\alpha_i \leq C$ for $i = 1, \ldots, N$, which limits the impact of outliers.

The classification function for the soft-margin SVM has the same form as for the hard-margin case:

$$f(\mathbf{x}) = sign(\mathbf{w}^{*T}\mathbf{x} + b^*) \tag{3.79}$$

in its primal form and

$$f(\mathbf{x}) = sign\left(\sum_{i=1}^{N} \alpha_i y_i \mathbf{x}_i^T \mathbf{x} + b^*\right) \tag{3.80}$$

in the dual form, which depends only on inner products of the support vectors $\mathbf{x}_i$ and the classified vector $\mathbf{x}$.

Since the inner product $\mathbf{x}_i^T\mathbf{x}$ can be interpreted as a similarity measure between the vectors $\mathbf{x}_i$ and $\mathbf{x}$, the classification is performed by comparing the vector $\mathbf{x}$ representing an unknown sample to all the support vectors (the training points $\mathbf{x}_i$, for which $\alpha_i > 0$).

3.4.3 Kernels

Both hard- and soft-margin SVMs can be used to design classifiers for problems where classes can be efficiently separated (with or without training errors) by a linear boundary. This, however, will not work for classes with intrinsically nonlinear boundaries. Figure 3.16 shows a popular toy example of a training data set (with two input variables x_1 and x_2) for which no linear boundary in the original space (x_1, x_2) can separate the classes (Bennett and Campbell 2000; Jordan 2004; Moulines 2008).

The boundary is nonlinear and such has to be the case for the discriminating function. Or does it? Definitely so, if we limit the solution to the original (x_1, x_2) space. If we, however, map the input space into another space, for example

$$(x_1, x_2) \longrightarrow \left(x_1^2, \sqrt{2}x_1x_2, x_2^2\right), \tag{3.81}$$

then the two classes may be linearly separable in this new three-dimensional space (see Figs 3.17 and 3.18). We could then use the same SVM methods to find the optimal separating hyperplane in this new feature space and define a linear classifier there.

[33]Hence, setting C high may lead to overfitting whereas smaller values of C correspond to more regularized solutions (Hastie et al. 2009).

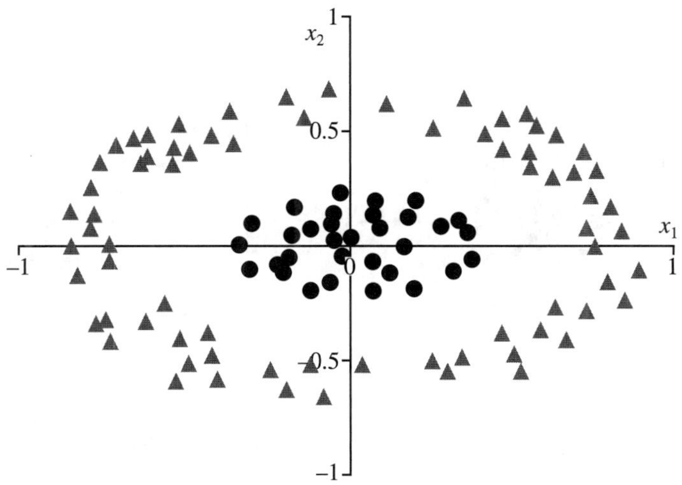

Figure 3.16: A data set with a nonlinear boundary between the two classes.

Generally, solving nonlinear problems with the use of SVMs is based on a mapping Φ from the p-dimensional input space into a usually higher, say s-dimensional *feature space*, in which the classes can be linearly separable:

$$\Phi\colon\ \mathbf{x}_i = \begin{bmatrix} x_{1i} \\ x_{2i} \\ \vdots \\ x_{pi} \end{bmatrix} \in \Re^p \longrightarrow \mathbf{z}_i = \begin{bmatrix} z_{1i} \\ z_{2i} \\ \vdots \\ z_{si} \end{bmatrix} \in \Re^s, \quad i = 1, \ldots, N \tag{3.82}$$

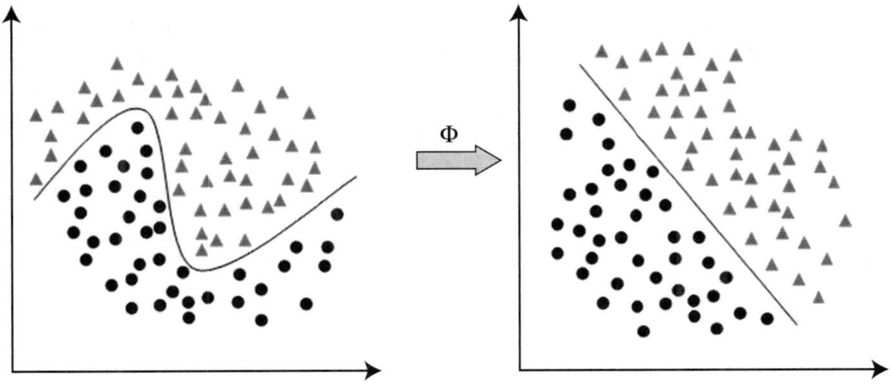

Figure 3.17: The general idea of mapping the space of original variables into a space of features in order to make the classes linearly separable. This figure shows a mapping from a two-dimensional space into a two-dimensional space. Usually, however, the space of original variables is mapped into a higher-dimensional (even infinite-dimensional) space.

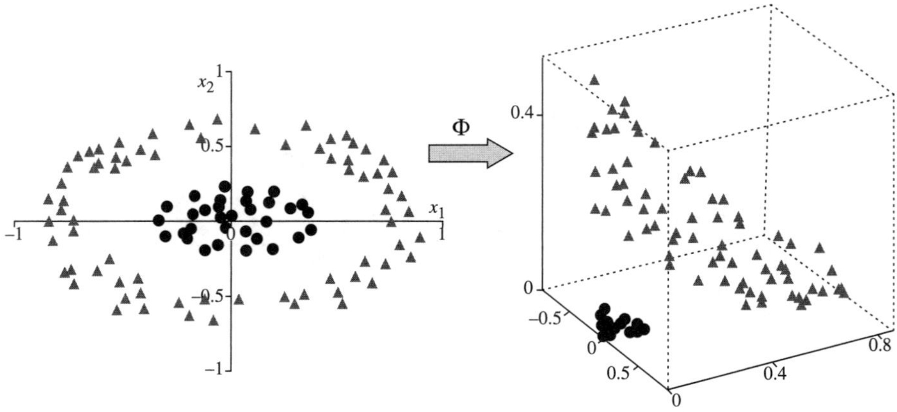

Figure 3.18: Mapping the data into a higher-dimensional space can make the classes linearly separable.

or

$$\Phi\colon (x_1, \ldots, x_p) \longrightarrow (z_1, \ldots, z_s). \tag{3.83}$$

The greatest challenge in practical applications of SVMs is the selection of an appropriate mapping Φ. For the toy example (from Fig. 3.16) with only two variables, it is obvious that the decision boundary can be described by an ellipse. The mapping

$$\Phi\colon (x_1, x_2) \longrightarrow (z_1, z_2, z_3), \tag{3.84}$$

where

$$z_1 = x_1^2; \quad z_2 = \sqrt{2}x_1x_2; \quad z_3 = x_2^2 \tag{3.85}$$

will result in a separating hyperplane in (z_1, z_2, z_3) described by an equation of the form $\mathbf{w}^T\mathbf{z} + b = 0$. By mapping it back into the original space, we will have

$$\begin{aligned}\mathbf{w}^T\mathbf{z} &= w_1z_1 + w_2z_2 + w_3z_3\\ &= w_1x_1^2 + w_2\sqrt{2}x_1x_2 + w_3x_2^2,\end{aligned} \tag{3.86}$$

with

$$w_1x_1^2 + w_2\sqrt{2}x_1x_2 + w_3x_2^2 = 0 \tag{3.87}$$

describing an ellipse in (x_1, x_2).

For genomic and proteomic data sets with thousands of variables, the selection of an appropriate mapping may be far from obvious. For such data, we usually try linear SVMs first and use nonlinear ones only when necessary.

As the feature space $\Re^s$ has higher (and possibly infinite) dimensionality, one may ask about the curse of dimensionality. Are we making the problem computationally much more difficult? To answer this question, let us look first at the classification function in $\Re^s$,

$$\begin{aligned} f(\mathbf{x}) &= sign(\mathbf{w}^T\mathbf{z} + b) \\ &= sign\left(\sum_{i=1}^{N} \alpha_i y_i \mathbf{z}_i^T \mathbf{z} + b\right). \end{aligned} \tag{3.88}$$

Since the point $\mathbf{z}_i$ in the feature space (a support vector in $\Re^s$ when $\alpha_i > 0$) is the image of one of the training points $\mathbf{x}_i$, and $\mathbf{z}$ is the image of the point $\mathbf{x}$ to be classified (unknown sample),

$$\begin{aligned} z_i &= \Phi(x_i), \\ z &= \Phi(x), \end{aligned} \tag{3.89}$$

we have:

$$f(\mathbf{x}) = sign\left(\sum_{i=1}^{N} \alpha_i y_i \Phi(\mathbf{x}_i)^T \Phi(\mathbf{x}) + b\right). \tag{3.90}$$

Not only the classification function but also the dual optimization problem leading to the solution can be presented in terms of the inner product of two points in the feature space $\mathbf{z}_i^T\mathbf{z}$ or the equivalent inner product of the mapping of the two points in the input space $\Phi(\mathbf{x}_i)^T\Phi(\mathbf{x})$. If we now define (for the mapping Φ) a function K, such that

$$K(\mathbf{x}_i, \mathbf{x}) = \Phi(\mathbf{x}_i)^T \Phi(\mathbf{x}), \tag{3.91}$$

then the classification function

$$f(\mathbf{x}) = sign\left(\sum_{i=1}^{N} \alpha_i y_i K(\mathbf{x}_i, \mathbf{x}) + b\right) \tag{3.92}$$

as well as the optimization problem can be formulated in terms of the function $K(\mathbf{x}_i, \mathbf{x})$. Since the only arguments of the function K are vectors in the input space, we can solve the optimization problem and then perform classifications using this function, without explicitly mapping the data.

A function returning the value of the inner product of the images of two vectors in the input space is called a *kernel*. By using kernels, the SVM method can be extended to nonlinear cases. No algorithmic changes are required, just the replacement of the inner product with the kernel function.

Let us illustrate this using our simple example mapping the two-dimensional space into the three-dimensional feature space (Fig. 3.18). To solve the optimization problem and to perform classification, we need the inner product $\mathbf{z}_i^T\mathbf{z}$. For the sake of symmetry of the calculations, consider any two points $\mathbf{z}_i$ and $\mathbf{z}_j$ in the space (z_1, z_2, z_3),

$$\mathbf{z}_i = \begin{bmatrix} z_{1i} \\ z_{2i} \\ z_{3i} \end{bmatrix} \quad \text{and} \quad \mathbf{z}_j = \begin{bmatrix} z_{1j} \\ z_{2j} \\ z_{3j} \end{bmatrix}. \tag{3.93}$$

Their inner product is $\mathbf{z}_i^T \mathbf{z}_j$. Using the mapping defined by (3.84) and (3.85), we have

$$\begin{aligned} \mathbf{z}_i^T \mathbf{z}_j &= z_{1i}z_{1j} + z_{2i}z_{2j} + z_{3i}z_{3j} \\ &= x_{1i}^2 x_{1j}^2 + 2x_{1i}x_{2i}x_{1j}x_{2j} + x_{2i}^2 x_{2j}^2 \\ &= (\mathbf{x}_i^T \mathbf{x}_j)^2, \end{aligned} \tag{3.94}$$

which means that by defining the kernel function as

$$K(\mathbf{x}_i, \mathbf{x}_j) = \Phi(\mathbf{x}_i)^T \Phi(\mathbf{x}_j) = (\mathbf{x}_i^T \mathbf{x}_j)^2, \tag{3.95}$$

we can optimize the problem in the (z_1, z_2, z_3) space and define the classification function without explicitly performing the mapping.

The use of kernel to perform calculations based only on the original data, without explicitly using the mapping Φ is known as *the kernel trick*. Thus, higher or even infinite dimensionality of the feature space causes no computational problems.

The most commonly used kernels are:

- *Polynomial kernel* (of degree d):

$$K(\mathbf{x}_i, \mathbf{x}_j) = (\mathbf{x}_i^T \mathbf{x}_j + 1)^d, \quad d > 0, \tag{3.96}$$

  or its less general version

$$K(\mathbf{x}_i, \mathbf{x}_j) = (\mathbf{x}_i^T \mathbf{x}_j)^d, \quad d > 0, \tag{3.97}$$

  which, with $d = 2$, is the kernel used in our example.

- *Radial basis function kernel*:

$$K(\mathbf{x}_i, \mathbf{x}_j) = \exp\left(-\frac{\|\mathbf{x}_i - \mathbf{x}_j\|^2}{\sigma^2}\right). \tag{3.98}$$

- *Sigmoid kernel* (also known as *hyperbolic tangent kernel*):

$$K(\mathbf{x}_i, \mathbf{x}_j) = \tanh(\beta \mathbf{x}_i^T \mathbf{x}_j + \gamma). \tag{3.99}$$

  Note: This function is a kernel only for some combinations of parameters β and γ (e.g., for $\beta = 2$ and $\gamma = 1$).

3.4.4 SVMs and Multiclass Discrimination

SVMs are binary classifiers. Multiclass problems may be approached either by using a set of binary classifiers, or by trying to extend SVMs to consider all the classes simultaneously (Schèolkopf and Smola 2002; Lee et al. 2004). A set of binary SVMs can be defined either by *pairwise* classifiers or *one-versus-the-rest* classifiers. The latter approach seems to be more popular. It requires fewer classifiers, but its separation power is lower than for the pairwise approach.

3.4.4.1 One-Versus-the-Rest Approach

For J classes, we define a set of J one-versus-the-rest binary classifiers

$$f_j(\mathbf{x}) = \sum_{i=1}^{N} \alpha_{ij} y_i \mathbf{x}_i^T \mathbf{x} + b_j^*, \quad j = 1, \ldots, J. \tag{3.100}$$

They are almost the same as the binary SVMs defined by (3.72) or (3.80); however, they return the value rather than the sign of the expression. Using a discrete classifier that returns just the sign of the expression could result in the assignment of the classified sample into several classes or to none of them. By defining the continuous classifiers in (3.100) we may be able to avoid such unclassifiable regions (Abe 2005). An unknown sample $\mathbf{x}$ is assigned to the class j with the largest value of the classification function $f_j(\mathbf{x})$. Sometimes, however, we may not want this winner-takes-all approach to mask some ambiguity in the classification results. The requirement that the sample $\mathbf{x}$ can be assigned to class j only when the value of $f_j(\mathbf{x})$ exceeds the next largest value by some threshold θ,

$$f_j(\mathbf{x}) \geq f_{j'}(\mathbf{x}) + \theta, \quad j, j' = 1, \ldots, J, \; j \neq j', \tag{3.101}$$

can be added to highlight such ambiguous situations (Schèolkopf and Smola 2002). Please note that the one-versus-the-rest design is intrinsically asymmetric due to the unbalanced size of the classes and possibly the high heterogeneity of the classes in *the-rest* class.

3.4.4.2 Pairwise Approach

In this approach one classifier is designed for each of the $\binom{J}{2} = \frac{J(J-1)}{2}$ possible pairs of classes. Although in practice it usually does not make much sense for a study to differentiate many classes simultaneously, even with a modest number (say 5, 6, or 7) of classes we may need to construct many binary classifiers. For instance, for five classes ($J = 5$) we would need ten classifiers, and for seven classes—twenty one. Although training data sets for these classifiers are smaller than for those in the one-versus-the-rest approach (only a subset of data is used), the times required to train and then to use all the binary classifiers for classification may be larger than for the one-versus-the-rest approach. However, the advantage of the pairwise approach is that the classes are more homogeneous and the design of each comparison is more balanced than for the one-versus-the-rest approach. To classify an unknown sample, we need to use all of the pairwise classifiers. The sample is "voted" into the class to which it was assigned by the largest number of classifiers. Unclassifiable regions may still exist; Abe (2005) describes how such unclassifiable regions may be resolved.

3.4.4.3 All-Classes-Simultaneously Approach

There are also approaches that extend SVMs to directly differentiate between more than two classes (see for example Lee et al. 2004; Abe 2005). Their optimization problems have to deal with all support vectors at the same time (Schèolkopf and Smola 2002), but simultaneous determination of all classification functions allows for avoiding unclassifiable regions (Abe 2005).

3.4.5 SVMs and Feature Selection: Recursive Feature Elimination

Recursive Feature Elimination (Guyon et al. 2002) is an embedded version of the sequential backward feature selection. The algorithm may be described as follows:

- Start with the training data set including all variables.
- Repeat the following steps until the current subset of variables is empty (or a stopping criterion is true):
  - build an optimal classifier for the current subset of variables,
  - for each variable in the current subset estimate its current multivariate importance,
  - remove the feature with the least value of the multivariate importance.

The elimination process ends either when all of the features are eliminated or when a predefined result is achieved (for instance, a combined optimum defined by the size of the subset and its measure of discriminatory power or classification efficiency). In either case we are presented with a sequence of nested feature subsets. The results may have a form that looks like a ranked list of genes (with the last eliminated gene at the top of the list and the one eliminated first at the bottom). Do not confuse, however, the ranked list form of the results with a univariate ranking of the genes. The recursive feature elimination ranks gene subsets. Each of the identified nested feature subsets of size m (where m may assume values from p to 1) can be retrieved from the list by taking the top m genes from the list (assuming the process ended with the elimination of all genes). Support Vector Machine Recursive Feature Elimination (SVM-RFE) utilizes the fact that the optimal separating hyperplane is associated with a vector $\mathbf{w}$ of feature weights w_k,

$$\mathbf{w} = \begin{bmatrix} w_1 \\ w_2 \\ \vdots \\ w_p \end{bmatrix}, \tag{3.102}$$

where $k = 1, \ldots, p$ and p is the number of features (variables, dimensions, genes, ...). A feature that is orthogonal to the vector $\mathbf{w}$ (with the weight w_k equal to zero) does not contribute to classification. The elements w_k of the vector $\mathbf{w}$ may be positive or negative, so we may define the multivariate importance of each feature k as the absolute value of the weight $|w_k|$ (or, for instance, as w_k^2). At each stage of recursive feature elimination, a feature with the minimal multivariate importance $|w_k|$ may be removed. With many thousands of variables in typical gene expression data, training the support vector machine p times may be computationally costly. Some implementations generalize the algorithm by removing more than one feature per step. This can be accomplished either by removing all of the features with a $|w_k|$ value below a threshold level (which may be dynamic and change from step to step) or by removing a predefined proportion of the current number of features. Since the current weight of each feature is determined and meaningful only in the context of all of the features in the current subset, removing more than one feature at a time neglects some interactions between features and may lead to inferior results. There are, however, indications (Guyon et al. 2002) that the results of group elimination are significantly worse only for relatively small subsets of features. Therefore, a reasonable compromise between accuracy and computational costs can be achieved by removing groups of features (even up to 50 percent of them) in the first few iterations and then—once the current subset has only a few hundred variables—switch the mode to eliminating only one variable at a time.

Extensions and variations of SVM-RFE include such approaches as Multiple SVM-RFE (Duan et al. 2005), Recursive SVM, R-SVM (Zhang et al. 2006), and Multiclass SVM-RFE (Zhou and Tuck 2007).

The Multiple SVM-RFE method trains multiple SVMs on different subsets of the training samples at each step of the recursive feature elimination. The multivariate importance of each feature (included in the current feature subset) is based on a statistical analysis of the weight vectors associated with each trained SVM.

The Recursive SVM feature selection (R-SVM) method differs from the SVM-RFE method mainly in the way the multivariate importance of each gene is calculated. Each feature's relative contribution to classification s_k is based not only on its weight w_k, but also on the projection of the distance between class centers onto the feature's dimension,

$$s_k = w_k(c_k^+ - c_k^-), \tag{3.103}$$

where c_k^+ and c_k^- are coordinates of the two class centers in the dimension defined by the feature k.

3.4.6 Summary

Support vector machines (SVMs) are binary classifiers based on an optimal hyperplane linearly separating the classes. For classes with nonlinear boundaries in the input space of original variables, they utilize the kernel function approach mapping the input space into a higher-dimensional feature space in which linear separation may be possible. The kernel trick allows for all calculations to be based on the original training data (in its input space) without explicitly using or even knowing the mapping. Therefore, high, or even infinite, dimensionality of the feature space does not increase the computational cost of training classifiers and performing classification.

The main features of SVMs are:

Maximizing the margin

The optimal hyperplane is designed in a way that maximizes the margin between the classes (whether this is a hard-margin with no training errors, or a soft-margin where the size of the margin is conditioned upon its violations by the training data).

Duality

The dual form of the optimization problem allows training (and then classification) to depend only on the inner product of the vectors in the input space. Furthermore, the dual problem is optimized in the N-dimensional space of Lagrange multipliers, where N is the number of training data points. For typical microarray gene expression data with N biological samples, p genes, and $N \ll p$, this results in a significant decrease in the number of variables as compared to the primal problem whose dimensionality includes all p variables.

Kernels

Kernel functions allow for the extension of linear SVMs to efficiently handle nonlinear cases. Using the original vectors in the input space, kernel functions return the inner product of their mapping into the feature space. This facilitates virtual optimization in the high-dimensional (or even infinite-dimensional) feature space without explicitly mapping training data into the feature space. However, selection of a kernel appropriate for a particular data set may be a time-consuming process.

Sparseness

The solution to the optimization problem is sparse in the sense that it depends only on a small subset of training data—the support vectors.

Convexity

The separating hyperplane is identified by optimizing a convex function, which means there are no local optima and the solution is unique (for a given subset of genes).

3.5 RANDOM FORESTS

Though this be madness, yet there is method in 't.

Shakespeare, *The Tragedy of Hamlet*

3.5.1 Introduction

The *Random Forests* learning algorithm was created by Leo Breiman (Breiman 2001). Though related to decision trees, its utilization of a collection of tree classifiers, as well as randomness in their design, results in better accuracy as well as better generalization and stability of the resulting ensemble classifier. The random forests method (along with other ensemble learning approaches) is gaining popularity in biomedical research. Randomness in the method is represented by bagging and random variable selection. Before describing the random forests learning algorithm, let us take a look at the following methods.

Bootstrap

Generally, bootstrap methods belong to resampling techniques that are used to generate artificial data sets to allow for better estimation of statistical properties of a prediction system or a population. For example, to make some inferences about the distribution of the population from which our statistical sample (training data) was selected, we may investigate the relationship between this sample and a number of its subsamples (bootstrap samples). The bootstrap is most often identified with Efron's *nonparametric bootstrap* in which bootstrap samples are randomly selected from the sample, with replacement, and are of the same size as the original sample (Efron 1979; Efron and Tibshirani 1993). Unknown population parameters may be estimated by averaging estimates from all the bootstrap samples (Duda et al. 2001). The bootstrap approach can be used to estimate the misclassification error rate of a classification system. Bootstrap training sets can be used to build a number of classifiers, which are then used to classify samples from the original training data set. The estimate is more realistic when each bootstrap-based classifier is used to classify only those samples from the original data sets that were not used to train this classifier (Hastie et al. 2001). The random forests method uses Efron's nonparametric bootstrap. Other versions of bootstrap will be mentioned in Section 3.6.

Bagging

Introduced by Breiman (Breiman 1996a), bagging (*bootstrap aggregating*) is a method of combining bootstrap-based classifiers in order to improve the classification accuracy. Multiple classifiers are built from training sets generated as bootstrap samples (with replacement) from the original training data set. Usually, each of the classifiers implements the same classification method (for instance, a decision tree) but being built from different training sets they differ in classification parameters. An unknown sample is classified by all of the classifiers and assigned to the most popular class. By averaging over training sets of the same size selected from the same distribution, bagging is capable of improving the prediction of unstable learning methods[34] by reducing the variance.

Boosting

The general idea of boosting is to combine many weak classifiers to build a powerful ensemble classifier. Weak classifiers are ones whose accuracy is only slightly better than random guessing. Starting with a weak classifier, called a base classifier, the boosting method creates a sequence of classifiers

[34] A method is unstable if a small change in the training set may cause large changes in classifier's parameters and accuracy (Duda et al. 2001).

trained on a modified data set. Modification to the training data may include the assignment of different weights to training samples. At each iteration, emphasis is given to the samples that were misclassified in the previous step. The ensemble classifier is composed from all sequentially created classifiers and classification is based on their weighted vote (Theodoridis and Koutroumbas 2006; Hastie et al. 2009). Although designing classifiers by concentrating on misclassified cases may seem like a recipe for overfitting, Breiman claims that, "*Boosting is a classification algorithm that gives consistently lower error rates than bagging*" (Breiman 2002a). The most popular boosting algorithm, AdaBoost, was proposed in 1997 by Freund and Shapire.

AdaBoost

The AdaBoost (*adaptive boosting*) method (Freund and Shapire 1997) trains a set of classifiers on weighted versions of the original training data. Initially, each of the N training samples is assigned the same weight of $1/N$. At each consecutive step, weights of the samples misclassified by the classifier designed in the previous step are increased. Samples that were correctly classified have their weights decreased. The weights are nonnegative and their sum is kept constant and equal to one (Breiman 2002a). At each step, a new training set is randomly selected, with replacement, according to the current weights assigned to samples. Samples with higher weights have higher probabilities of being selected (or selected multiple times). A sequence of classifiers is built this way and each of them is tested on its own training set to determine the classifier's training error. The process continues for either a specified number of iterations or until a sufficiently low training error of the ensemble classifier is achieved. An unknown sample is classified by the ensemble classifier according to weighted voting of the component classifiers (classifiers with lower training errors are assigned higher vote weights). Usually, the training error of the ensemble classifier decreases exponentially when the number of iterations (and component classifiers) increases. Though the AdaBoost method has been successfully applied to many real-world examples, we should remember that the decrease in training error (based on classification of the training data) does not guarantee better generalization, or the decrease of the classification error measured on an independent test set (Duda et al. 2001). A special caution is due when training data sets include a small number of samples N.

CART

The Classification And Regression Trees (CART) algorithm is a learning algorithm belonging to decision tree methods. A decision tree is a collection of decision nodes connected by branches. It starts with the root node and grows downward by splitting nodes until terminal leaf nodes are reached; more on decision trees may be found in (Duda et al. 2001; Larose 2005; Hastie et al. 2009). The CART model, introduced in Breiman et al. (1984), builds a binary tree with a univariate decision at each node. Generally, decision trees can split a node into more than two branches. However, every multiway split can be represented by a set of binary splits, which enables the decision on how to split each node to be treated as a one-dimensional optimization problem (Duda et al. 2001).

The CART algorithm starts with the entire training data set and splits it at the root node into two nonoverlapping subsets (child nodes). A splitting criterion is a measure of class homogeneity of the samples assigned to each of the child nodes. Decisions are based on the values of a single variable—the variable that offers the best split (highest class purity or lowest class impurity of the two subsets). Each child node is then split and this recursive splitting continues until each of the final nodes (leaves) includes only samples from one class, or when no further splitting is possible. Since a fully grown tree may overfit the training data, the final classification tree is usually the result of a pruning process. Pruning decreases the size of the tree in order to minimize an estimate of the misclassification error rate and improve generalization. After pruning, all nodes that are not split are declared leaves. Each heterogeneous (impure) leaf has to be assigned to one of the classes.

One of the strengths of the CART algorithm is its ability to handle different types of variables—qualitative, quantitative, or their mixture. However, when dealing with numerical variables such as gene expression ones, splitting each node on a single variable results in hyper-rectangular decision regions defined by the cutoff values of the splitting variables (Hand et al. 2001). If the boundaries of such regions are far from alignment with natural boundaries between classes, the CART algorithm may result in inefficient (poorly generalizable) classifiers. To handle such situations, an extension to multivariate trees would be necessary; for example, trees that split nodes on a linear function of a subset of variables. Another disadvantage of decision tree classifiers is their intrinsic instability (high variance) caused by their hierarchical structure—learning errors made at a parent node are propagated to all its child nodes (Hastie et al. 2009).

Gini impurity index

To decide which variable to select to split an internal node, we need to first define a measure of impurity of the child nodes. Assume that we are performing a study differentiating among J classes, and we want to measure the impurity of node α with N_α observations (biological samples). For each class j, where $j = 1, \ldots, J$, define $\hat{p}_j\,(\alpha)$ as the proportion of the N_α training samples that belong to class j. If all N_α samples of node α are in the same class, we want the measure $i(\alpha)$ of the node α impurity to be zero. Increased heterogeneity of the node samples should increase the impurity measure. Commonly used measures are: entropy impurity, variance impurity and its generalization known as the Gini index, and misclassification rate impurity.

- Entropy impurity (also known as information gain impurity) measure:

$$i_{entropy}(\alpha) = -\sum_{j=1}^{J} \hat{p}_j(\alpha)\log_2 \hat{p}_j(\alpha). \tag{3.104}$$

- Misclassification rate impurity:

$$i_{misclassification}(\alpha) = 1 - \max_j \hat{p}_j(\alpha). \qquad (3.105)$$

- Gini impurity index:

$$i_{Gini}(\alpha) = \sum_{j=1}^{J} \sum_{j'=1}^{J} {}_{j' \neq j} \hat{p}_j(\alpha)\hat{p}_{j'}(\alpha). \qquad (3.106)$$

- Variance impurity is a special case of the Gini impurity, when we differentiate only between two classes, thus $J = 2$ and

$$i_{Gini}(\alpha) = \hat{p}_1(\alpha)\hat{p}_2(\alpha). \qquad (3.107)$$

To split a parent node τ into two child nodes α and β, we will select the variable that offers the maximal decrease in the impurity measure. Assuming a binary tree and J differentiated classes, the decrease $\Delta i_{Gini}(\tau)$ in the Gini impurity measure $i_{Gini}(\tau)$ for node τ can be calculated as

$$\Delta i_{Gini}(\tau) = i_{Gini}(\tau) - i_{Gini}(\alpha)\hat{p}_\alpha - i_{Gini}(\beta)\hat{p}_\beta, \qquad (3.108)$$

where $i_{Gini}(\alpha)$ and $i_{Gini}(\beta)$ are the Gini impurity indices of the child nodes and $\hat{p}_\alpha$ and $\hat{p}_\beta$ are the fractions of the training samples in node τ that are assigned to nodes α and β, respectively.

3.5.2 Random Forests Learning Algorithm

The random forests method builds a collection of decision trees—a forest. It utilizes bagging since each tree is grown from a bootstrap sample of the training data (i.e., each tree is trained on a set of biological samples randomly selected, with replacement, from the original training set). The size of each bootstrap training set is the same as the size of the original data set. At each node, a tree is split using the best variable from a small number m of randomly selected variables ($m \ll p$; for instance, $m = \sqrt{p}$ or $m = \log_2(p + 1)$, where p is the number of variables in the training data set). The m variables are selected independently for each node. By doing this, an additional level of randomness is added to the method, on top of bagging. Hundreds or thousands of trees are built to maximum size, without pruning. Classification is based on their unweighted voting for the most popular class. Due to the intrinsic randomness of the design, increasing the number of trees does not lead to overfitting. According to Breiman, this method can handle thousands of variables with no degeneration in accuracy, is more accurate than AdaBoost and is "*relatively robust to outliers and noise*" (Breiman 2001).

Since each bootstrap training set leaves out about one-third of the original samples (they are called *out-of-bag samples*), they may be used as an internal test set to estimate classification error of their corresponding tree. The classification error for each training sample may be estimated by averaging the classification of the sample over all trees for which the sample was out-of-bag. Averaging over all training samples leads to a test-set estimate of the overall classification error, which is called the out-of-bag (OOB) estimate of the misclassification error rate. There are indications that this OOB estimate of the misclassification error rate may be as accurate as estimates based on a test set of the same size as the original training set (Breiman 1996c; 2001). The OOB samples may also be used to select the value of m (the number of randomly selected variables to be considered for splitting a node). Breiman and Cutler's random forests manual (Breiman and Cutler 2004) suggests starting with the default value of $m = \sqrt{p}$, running some 20–30 trees recording the OOB error rate, then decreasing and increasing m until the value of m that minimizes the OOB error rate is found.

The random forests algorithm has only two parameters: the number of trees n_{tree} to build and the number of variables m that are randomly selected to split each node. The steps of the algorithm can be summarized as follows:

1. Randomly select n_{tree} bootstrap training sets, drawing with replacement from the original data set of N samples (and p variables). Each bootstrap set is of size $N \times p$ (the same size as the original data set). On the average, each of the bootstrap training sets includes about 0.632^*N unique samples, which leaves out—on the average—about 0.368^*N samples[35] that are not used to train a particular tree (OOB samples).
2. Loop over the n_{tree} bootstrap training sets and for each of them:
   (a) grow a classification tree to its maximum size (no pruning),
   (b) at each node, randomly select m variables ($m \ll p$) and—using the Gini impurity index—choose the best split based on one of these m variables:
   - find the best split for each of the m variables,
   - select the best split from among the identified m best splits.

   (c) classify each OOB sample (a sample from the original data set that was not selected into the current bootstrap training set) by running it down the current tree; save the results of this OOB classification.
3. Aggregate the results of the OOB classifications to calculate the OOB estimate of error rate.
4. Classify new data by the plurality vote of all n_{tree} classifiers.

The cost to build each of the random forest trees is $O(\sqrt{p}N \log N)$, which makes this method efficient for data sets with large numbers of variables p and not too large numbers of samples N (Tuv et al. 2007). For a comparison of the CART and random forests methods see Table 3.5.

[35]For each selection into the bootstrap set, the probability that a particular sample is selected from the training set of N samples is equal to $1/N$. Thus, the probability that the sample is *not* selected during the N selections is $(1 - 1/N)^N$. For large N, this probability approaches $1/e = 0.368$, which means that on average 36.8 percent of the samples will not be selected into the bootstrap set (Han and Kamber 2006). This approximation is quite good even for relatively small training data sets. For example, when $N = 50$, this probability is 0.364.

TABLE 3.5: Quick Comparison of CART and Random Forests Methods Based on Breiman's Random Forests Manual (2003)

		Random forests	
	CART—tree construction	Construction of a single tree	Notes
Root node	Contains original training data set (p variables and N samples)	Contains a bootstrap training set of the same size as the original training set	Each of the n_{tree} trees is grown from a different bootstrap training set. On average, a bootstrap set leaves out about $1/3$ of the training samples (out-of-bag samples)
Split at each node	All p variables are considered in order to find the best split into two child nodes. The Gini impurity is used as the splitting criterion	m variables ($m \ll p$) are randomly selected. Only these m variables are considered in order to find the best split based on one of them. The Gini impurity is used as the splitting criterion	Breiman suggests setting $m = \sqrt{p}$ and eventually to try using half or twice that number. Larger m values should be used for data sets with many noisy variables
Stop criterion	The tree is grown all the way down and then pruned up to minimize misclassification error for the test set	The tree is grown all the way down and not pruned	
Classification	Single tree classifier	New sample is classified by all n_{tree} trees and assigned to the class with the most votes	OOB samples can be used in lieu of an external test set to estimate the misclassification error rate

3.5.3 Random Forests and Feature Selection

Since a random forest ensemble classifier uses many trees and many variables to classify new samples, it does not provide a biomarker with a small number of variables. Thus, the main goals of biomarker discovery cannot be achieved without additional considerations aimed at feature selection. Feature selection can be embedded within the random forests algorithm and performed by utilizing a measure of variable importance.

Breiman (2001) proposed that changes in the classification of OOB cases, after randomly permuting the data values of a variable, can be used as variable importance metrics. Several metrics have been defined with the following one currently being

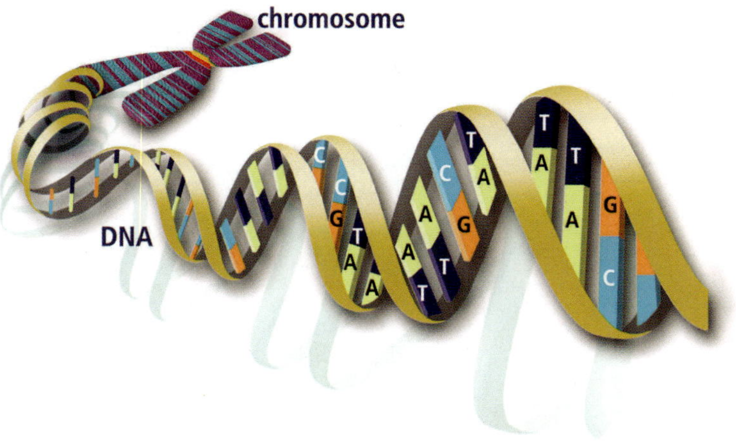

Figure 1.2: The DNA double helix (courtesy: The U.S. Department of Energy Genome Programs, http://genomics.energy.gov). The DNA structure includes two antiparallel deoxyribose-phosphate helical chains, which are connected via hydrogen bonds between complementary bases on the chains (base pairs). The base pairs "rungs of the ladder" are spaced 0.34 nm apart. This double helix structure repeats itself every 10 base pairs, or 3.4 nm (Watson and Crick 1953a; Garrett and Grisham 2007).

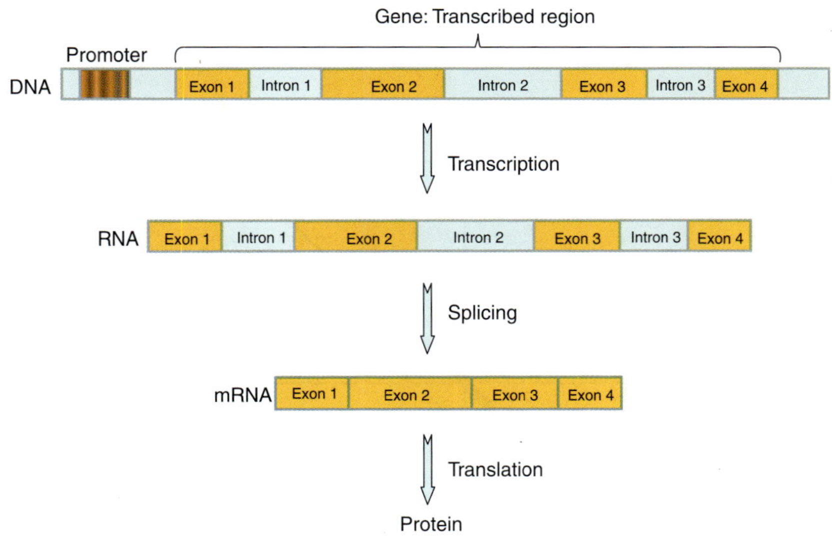

Figure 1.3: A human protein-coding gene: the structure, transcription, splicing, and translation. The top row shows a schematic structure of the gene. The transcribed region consists of exons that are interrupted by introns (on average, introns are about 20 times longer than exons). The promoter associated with the gene is a regulatory sequence that facilitates initiation of gene transcription and controls gene expression. Enhancers and silencers are other gene-associated regulatory sequences that may activate or repress transcription of the gene (they are not shown here as they are often located distantly from the transcribed region). The gene's exons and introns are first transcribed into a complementary RNA (called nuclear RNA, primary RNA transcript or pre-mRNA). Then, the splicing process removes introns, joins exons, and creates mRNA (called also mature mRNA). The mRNA travels from the nucleus to the cytoplasm where, in ribosomes, it is translated into a protein.

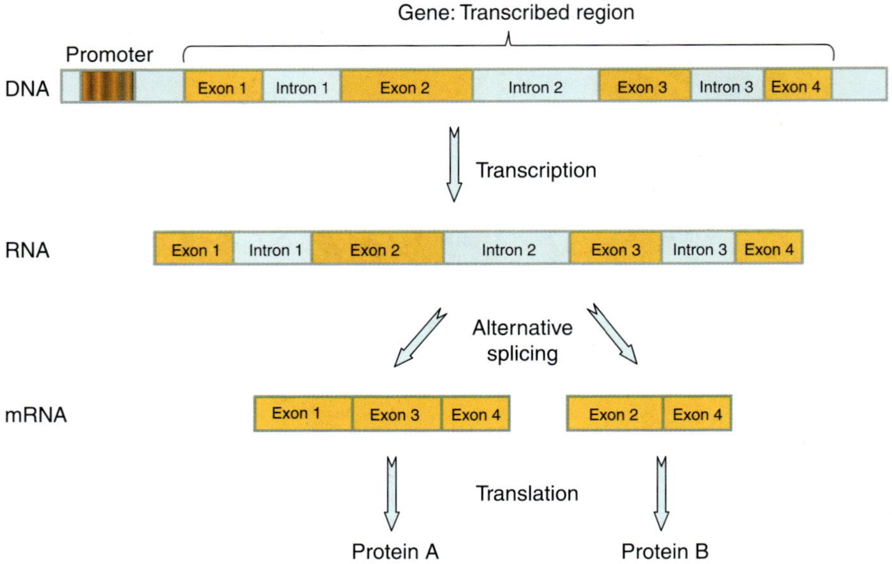

Figure 1.4: A human protein-coding gene: an example of alternative splicing. During alternative splicing of the primary RNA transcript, different subsets of exons are joined to create two (or more) different mRNA isoforms. These mRNA isoforms are then translated into usually distinct proteins. The majority of human protein-coding genes undergo alternative splicing (Stamm 2006).

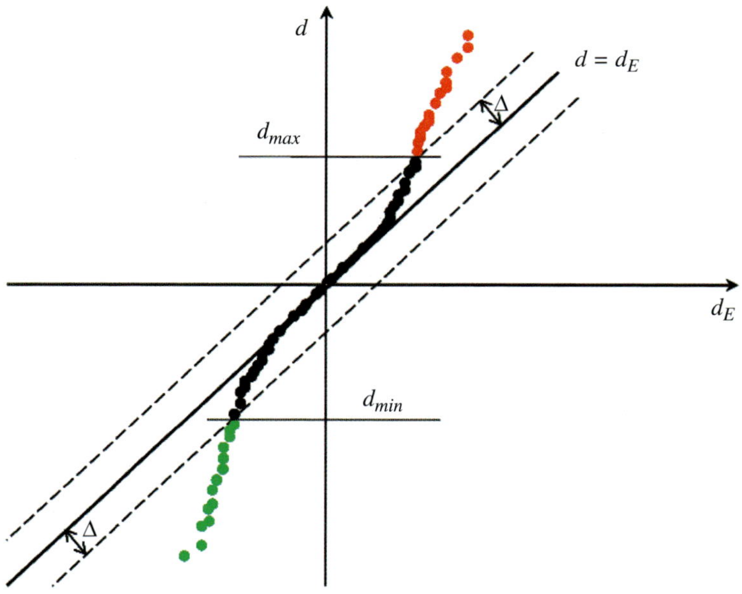

Figure 2.8: A scatterplot of the *observed relative difference d* versus the *expected relative difference* d_E. The dashed lines are at a threshold Δ distance from the $d = d_E$ identity line. The genes represented by the points that are outside of these threshold lines are deemed differentially expressed at the threshold level Δ. Depending on the sign of their relative difference d, SAM calls them either "significant positive genes" or "significant negative genes."

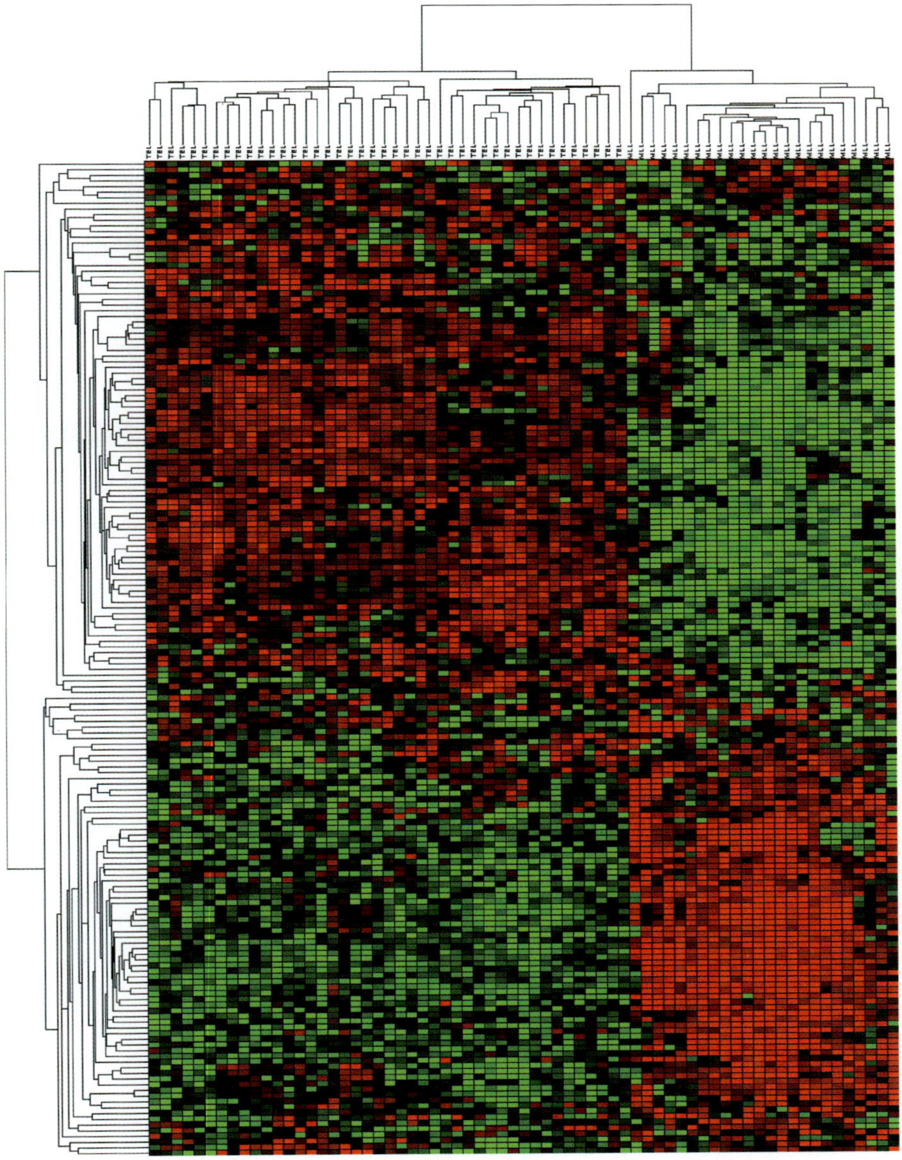

Figure 2.11: An example of a heat map showing the results of independent hierarchical clustering of genes (the dendrogram down the side of the image) and biological samples (the dendrogram across the top of the image). The expression levels are represented as color intensities or as shades of gray. When colors are used, red corresponds to higher expression levels and green to lower ones. When shades of gray are used, brighter spots represent higher expression levels.

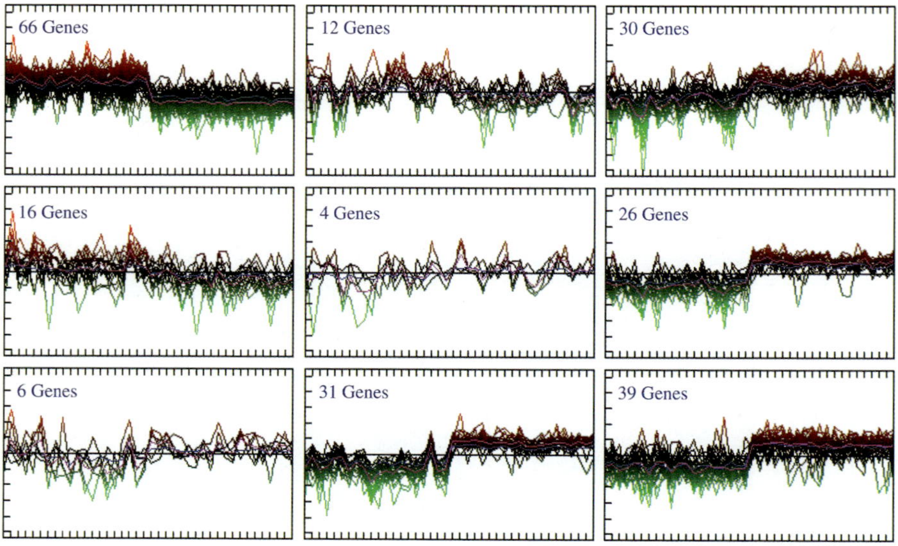

Figure 2.19: An example of graphical presentation of clustering genes by similarity of their expression profiles. Self-organizing map with 3 × 3 rectangular topology and correlation distance have been used. The image was obtained with *MultiExperiment Viewer* software (Saeed et al. 2003).

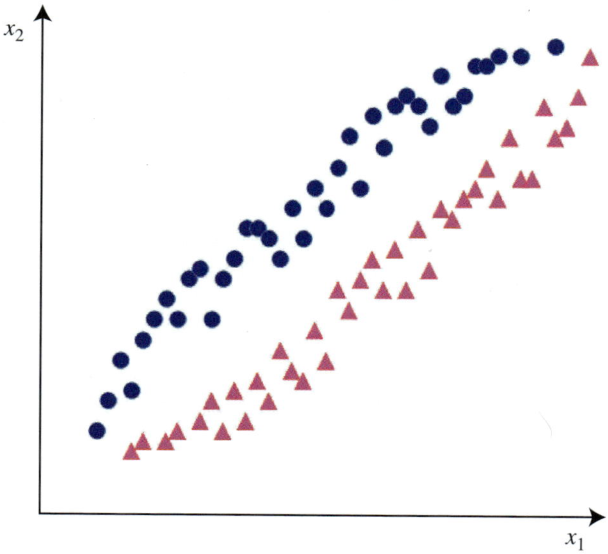

Figure 3.4: A data set with two classes of samples. There are only two variables. Neither of them is univariately significant for the class discrimination. However, as a set of two variables they can perfectly separate the classes.

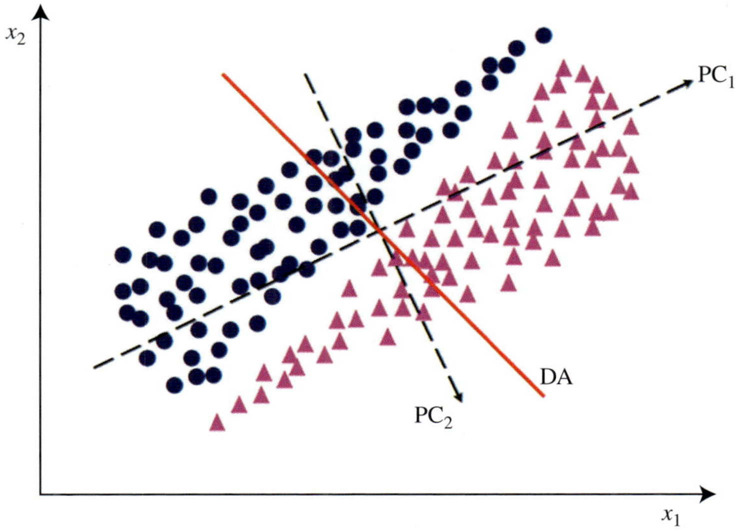

Figure 3.5: A data set with two classes of samples. The direction of the first principal component, PC_1, is aligned with the direction of the most variation in the data. This direction is very different from the direction of DA, which best separates the classes (and can be found by supervised methods such as discriminant analysis). As this is a toy example with only two dimensions, adding the second principal component, PC_2, will preserve the entire variation in the data. However, this would neither decrease the dimensionality nor identify the most discriminatory direction.

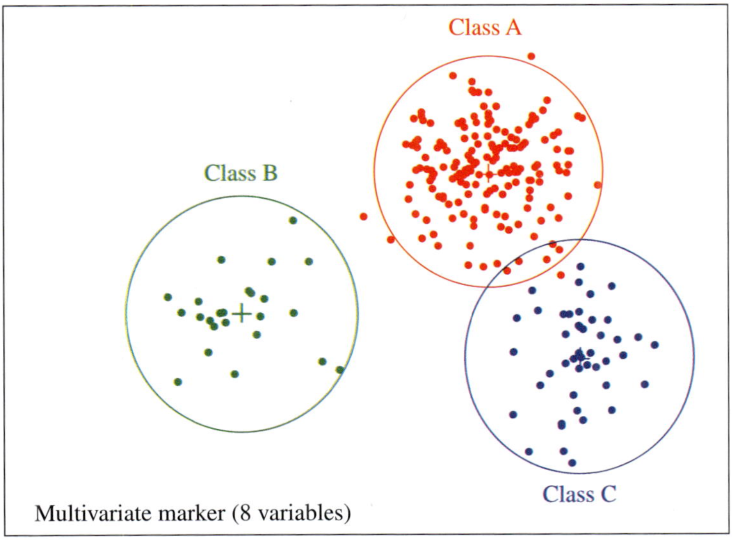

Figure 3.9: Discriminatory space of a classification model built on an eight-gene multivariate biomarker ($p = 8$). For this three-class model ($J = 3$), the discriminatory space is two-dimensional, $t = \min(p, J - 1) = 2$. The circles represent constant density boundaries enclosing 95 percent of the probability for each class. Points represent samples from the training data set. (Graphics from the *MbMD* data mining software.)

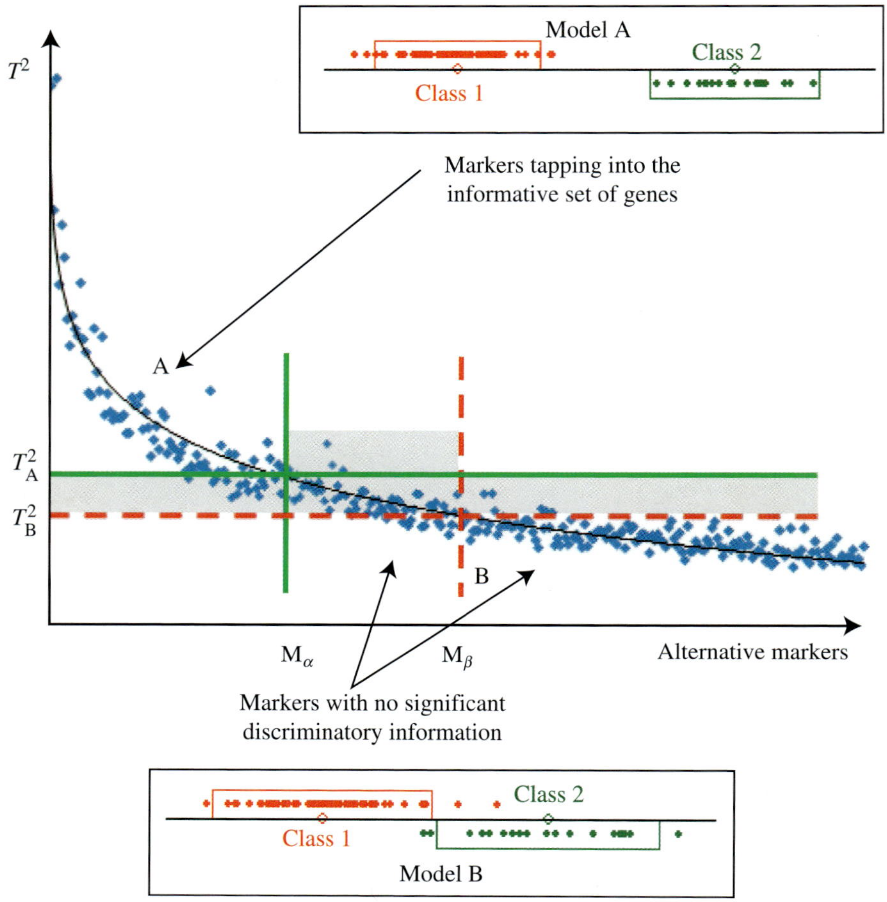

Figure 4.1: Selecting the *Informative Set of Genes*. The scatter plot points represent discriminatory power of the subsequently identified alternative multivariate markers (the plotted data represent the example discussed in Section 6.3 of Chapter 6). Usually, a strong decreasing tendency of this discriminatory power can be well approximated by a logarithmic trend line (a power function may also be tried). Discriminatory spaces and T^2 values of alternative classification models built on the alternative markers represented by the points in area A (above the T^2_A horizontal line and to the left of the vertical line crossing the trend line at T^2_A) indicate good class separation. An example of the discriminatory space and distribution of the training samples for an average model in area A is shown at the top of the figure (Model A). Classification models based on markers from area B (below the T^2_B horizontal line) cannot satisfactorily separate even the training samples. An example of the discriminatory space and distribution of training samples for an average model in area B is shown at the bottom of the figure (Model B). The trend line crosses the T^2_A level of discriminatory power in the vicinity of alternative marker M_α; it crosses T^2_B in the vicinity of alternative marker M_β. Models built on the alternative markers represented by the points in the gray areas (between T^2_A and T^2_B horizontal lines, and above T^2_B and between the vertical lines representing alternative markers M_α and M_β) may have some border line class separation abilities. Since only two classes are differentiated in this example, the discriminatory spaces of Models A and B are one-dimensional. The boxes and vertical offsets of points are used for emphasis only. The boxes represent sections (of the discriminatory dimension) that enclose 95 percent of the probability in each class. The points represent samples from the training data set.

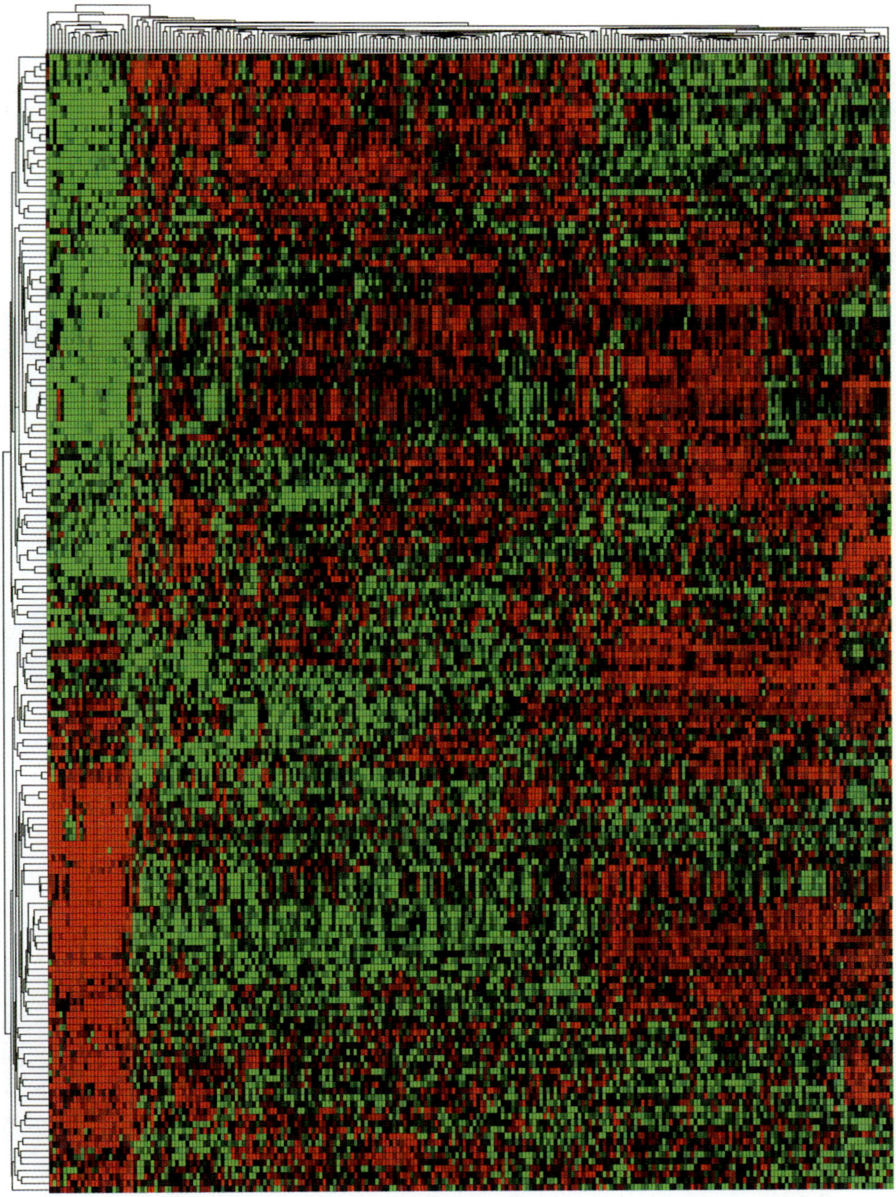

Figure 6.10: A heat map representing the two-way clustering of the INF-176 *Informative Set of Genes*. The dendrogram at the top of the image represents the results of hierarchical clustering of the training samples. The leftmost cluster of this dendrogram includes all twenty four MLL samples. The dendrogram at the left side of the image represents the results of clustering genes. Since the multivariate approach was used to identify the *Informative Set of Genes*, some of them are far from the top of a univariately ordered list of genes. Individually, they do not discriminate the two classes. However, their expression patterns are complementary to the patterns of some other genes in a way that a joint expression pattern of a small subset of such complementary genes provides good separation of the MLL samples from the samples of the other class.

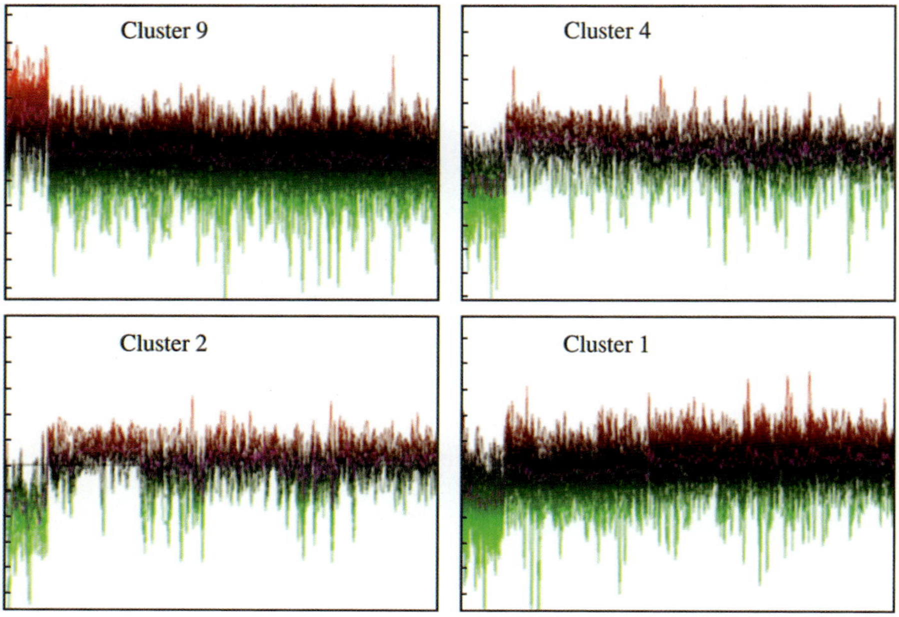

Figure 6.11: The four *primary expression patterns* of the *Informative Set of Genes* identified from the combined training data set (ALL3 + GSE13425).

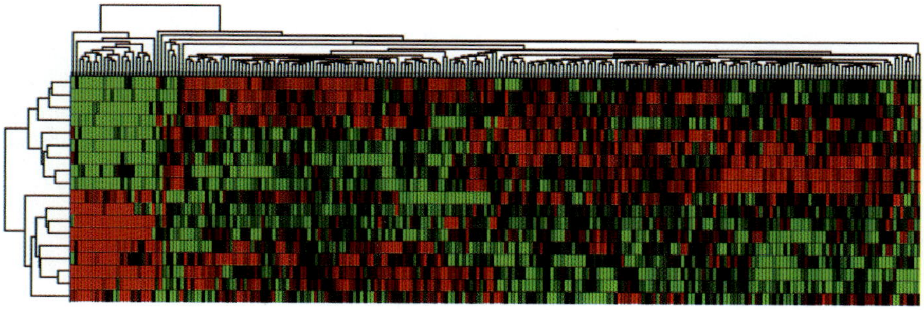

Figure 6.12: A heat map of the expression data of the 18 *most frequent primary* genes of the *Informative Set of Genes*. The leftmost cluster of the top dendrogram includes all 24 MLL samples of the combined training set.

among the most often implemented (Breiman and Cutler 2004; Tuv 2006; Archer and Kimes 2008; Liaw et al. 2008).

To calculate the importance I_k of the variable k, for every tree t in the forest ($t = 1, \ldots, n_{tree}$) take the following steps:

- Classify the samples that are out-of-bag for the tree (not used in the bootstrap training set from which the tree was grown) and count the number of votes for the correct class (i.e., the number of correctly classified OOB samples).
- Randomly permute the data values of the variable k in the OOB samples, run such prepared OOB samples down the tree and count the number of votes for the correct class.
- Subtract the number of votes for the correct class in the OOB samples with the variable k permuted from the number of votes for the correct class in the original OOB data. The resulting difference is the importance $I_k(t)$ of variable k for tree t.

The mean value of this difference over all trees in the forest can be used as the raw importance I_k of variable k,

$$I_k = \frac{1}{n_{tree}} \sum_{t=1}^{n_{tree}} I_k(t). \tag{3.109}$$

By dividing the raw importance I_k by its standard error $\sigma/\sqrt{n_{tree}}$, we can calculate the standardized index

$$z_k = \frac{I_k}{\frac{\sigma}{\sqrt{n_{tree}}}} \tag{3.110}$$

and use it to assign statistical significance to the importance score I_k. Please note, however, that:

(i) the standard error represents the between-tree variance rather than the variance due to sampling from the population (Lunetta et al. 2004),

(ii) we are averaging over all n_{tree} trees, and

(iii) the variable importance is equal to zero for any tree in which the variable is not used to split any node.

Two variable importance measures based on the sample margin have been described by Breiman (Breiman 2002b, 2003). The sample margin is defined as the difference between the proportion of the votes for the correct class and the maximum proportion of votes for each of the other classes (Breiman 2002a). The margin gives us information on the confidence of sample classification. One of the measures defines the importance of the variable k as the mean decrease of the margin over all OOB samples classified after the variable is randomly permuted. The other measure uses the difference between the number of OOB samples with decreased margin and the number of those with increased margin.

The decrease in the Gini impurity index $\Delta i_{Gini}(\tau, k)$ when node τ is split based on variable k can also be used to calculate the variable importance. The variable importance $I_k(t)$ for a single tree t, where $t = 1, \ldots, n_{tree}$ can be calculated as the sum of all impurity index decreases in tree t due to the variable k,

$$I_k(t) = \sum_{\tau \in t} \Delta i_{Gini}(\tau, k). \tag{3.111}$$

Averaging the variable importance $I_k(t)$ over all the trees in the forest gives us the variable importance measure I_k (3.109), which is often consistent with permutation-based measures (Breiman and Cutler 2004).

Using one of the variable importance metrics, feature selection may be performed as an iterative procedure similar to recursive feature elimination. The general idea is to build a sequence of random forests. At each iteration, the least important variables are removed and a new forest is grown on the remaining variables. The criterion to eliminate variables may be either some cut-off value of variable importance or a percentage of the current number of variables. After the sequence of forests is built, a single forest and its set of variables are selected. The selection can be based on the number of variables, on the forest OOB error rate, or on a combination of the two. For example, Diaz-Uriarte and de Andres select the set of variables associated with the forest having the smallest number of genes among all forests with an OOB error rate within a predefined number of standard deviations from the minimum error rate for all forests (Diaz-Uriarte and Alvarez de Andres 2006).

Breiman indicated, however, a weakness in feature selection utilizing random forest based measures of variable importance (Breiman 2001). Variables that are redundant but significantly predictive may simultaneously have a high importance score, even when only one of them would be necessary in a biomarker. This is due to the fact that the variables have similar probabilities of being selected in a random forest and that permutation experiments (and/or importance calculations) are performed separately for each of them. To identify a parsimonious biomarker, one may use the random forests approach to select a subset of relevant variables, and then apply a feature selection method that takes into account all interactions between the variables. The distinction between relevant and irrelevant variables may be based on randomly generated noise variables. Tuv et al. propose generating noise variables by random permutation of the data values of the original p variables across the N samples (Tuv et al. 2007). The values of variable importance calculated for these additional noise variables may be used to determine the ranking cut-off point. To assign statistical significance to the difference in the importance scores, the process of generating noise variables and ranking them is repeated a number of times. Variables whose importance is significantly greater than a selected percentile of importance scores calculated for the noise variables are deemed relevant.

3.5.4 Summary

The random forests learning algorithm utilizes bootstrap sampling from the training data set and the random selection of the variables considered for splitting tree nodes to build an ensemble of decision tree classifiers. The intrinsic randomness of the design results in the ensemble classifier that can be both accurate and relatively robust to noise and outliers. Out-of-bag samples can be used to estimate the misclassification error rate of the ensemble classifier with accuracy comparable to using a test set of the same size as the training data set. One of the internal variable importance metrics (either permutation based or utilizing the Gini impurity index) may be used to perform embedded feature selection. Some external processing may, however, be necessary to identify a truly multivariate biomarker.

3.6 ENSEMBLE CLASSIFIERS, BOOTSTRAP METHODS, AND THE *MODIFIED BAGGING* SCHEMA

3.6.1 Ensemble Classifiers

Ensemble methods build supervised classification systems based on multiple individual classifiers. Ensemble classifiers often outperform single classifiers, especially when the latter are weak or unstable. An unknown sample is classified by all individual classifiers and assigned to one of the classes based on their weighted or unweighted voting. Ensemble classifiers can be built by combining individual classifiers in a parallel or serial fashion (Tuv 2006).

3.6.1.1 Parallel Approach

In the parallel approach, individual classifiers are created independently of each other. Their diversity may be due to different versions of the training data set, due to randomness in the learning process, or both. Ensembles of parallel classifiers are especially useful when they combine individual classifiers that may be accurate but unstable. The random forests method described in Section 3.5 is an example of the parallel approach. Each tree in the forest is grown from a different bootstrap version of the training data. Additional randomness is added at each node by the random selection of a small subset of variables to be considered for splitting the node.

3.6.1.2 Serial Approach

In the serial approach, a series of classifiers is built. The next classifier tries to boost performance by focusing on the samples misclassified by the previous classifier. The resulting ensemble classifier may have high accuracy even when the individual classifiers are quite weak. AdaBoost is the best known example of the serial ensemble approach.

3.6.1.3 Ensemble Classifiers and Biomarker Discovery

One of the main goals of biomarker discovery is to identify a small set of genes whose joint expression pattern can significantly separate the differentiated classes and can be used for efficient classification of new cases. However, the ensemble approach generally does not deliver parsimonious biomarkers. On the contrary, ensemble classifiers usually base their voting on a large number of variables represented in all individual classifiers. These are not the kind of classifiers we would use in clinical practice.[36] Are ensemble approaches useless for gene expression biomarkers? Not

[36]Of course, there could be exceptions to this rule. If a feature selection process embedded in the ensemble approach results in a small (and truly multivariate) set of genes, we could use the ensemble classifier in clinical practice.

necessarily. A single parsimonious biomarker identified from a single training data set may be prone to overfitting and may be unstable. To properly validate the biomarker, we need a large and independent test data set. What if we do not have such a test set? Splitting our original data set into training and test sets is generally not a good idea since many gene expression data sets suffer from small number of samples (often fewer than 100). Though we would not ordinarily use ensemble classifiers directly for the classification task, they may be very useful for finding a more stable biomarker or for providing a reliable estimation of the biomarker generalization.

Whatever feature selection method we use, we may repeat it many times with different versions of our training data set. Furthermore, additional randomness may be added to the feature selection algorithm. We can build an ensemble of classifiers, but not use them for classification. Rather, we can analyze sets of genes selected into the classifiers with the goal of finding a parsimonious but more stable biomarkers than those identified for the individual classifiers. The main idea of this approach is based on the assumption that combinations of the variables that are frequently selected into individual classifiers of an ensemble should lead to more robust biomarkers. The ensemble is therefore used to *vote for variables* rather than for classes (Chan et al. 2007). This may be seen as regularization of potentially unstable feature selection by utilizing a large number of bootstrap-based classifiers as an intermediate step leading to selection of a more stable multivariate biomarker.

If bagging or a similar method of varying training sets that results in out-of-bag samples is used, we can use the OOB samples to estimate the misclassification error rate of our classifier. Let us note that this estimate is for the ensemble classifier. Nevertheless, it can serve as an indirect estimate of the generalization of our parsimonious biomarker derived from feature selection based on the ensemble's individual markers.

3.6.2 Bootstrap Methods

> *"Bootstrap methods are helpful in understanding the variability of all aspects of the prediction problem."*
>
> —(Efron 1983)

As mentioned earlier, Efron's nonparametric bootstrap is not the only bootstrap approach. Here are short descriptions of selected bootstrap methods:

Efron's Nonparametric Bootstrap

Whenever the *bootstrap* term is used without qualification, it most likely refers to Efron's *nonparametric bootstrap* introduced in (Efron 1979). This version of bootstrap makes no assumption about the underlying population, samples with replacement and generates bootstrap samples of the same size as the original sample.

Parametric Bootstrap

Generally, parametric bootstraps are based on some parametric models of the data. Instead of sampling with replacement from the training data, bootstrap samples may be generated from a parametric estimation of the population (Efron and Tibshirani 1993). In regression problems, bootstrap samples may be generated by adding Gaussian noise to the predicted value (Breiman 1996b; Hastie et al. 2009). In classification, parametric noise may be added to the training data vectors (Efron and Intrator 2004).

Randomized Bootstrap

This modification of Efron's nonparametric bootstrap applies only to two-class problems. With some probability, it allows for training observations from one class to be assigned to the other class (Efron 1983).

Double Bootstrap

This method implements bootstrap iteration. Second order bootstrap samples are taken from bootstrap samples. This method of bootstrapping the bootstrap was described by Efron as a way to improve the error rate estimate of linear classifiers. Although a straightforward application of the double bootstrap requires B^2 bootstrap samples, it can be implemented with only *2B* bootstrap samples (Efron 1983).

Without Replacement Bootstrap

This term is usually associated with stratified sampling without replacement, which may be seen as an extension of Efron's nonparametric bootstrap to stratified sampling from finite populations (Gross 1980; Bickel and Freedman 1984). First, disjoint strata of a finite population are independently sampled without replacement. Then, observations in each sample are replicated to result in bootstrap populations of the same size as the population strata. Finally, bootstrap samples of the same size as the original samples are selected without replacement from the bootstrap populations.

m-out-of-n Bootstrap

In the *m-out-of-n* bootstrap, bootstrap samples are of size $m < N$, where N is the size of the original training data.[37] This bootstrap can be implemented with replacement or without replacement (Bickel et al. 1994; Politis and Romano 1994; Bertail 1997; Bühlmann and Yu 2002). There are situations in which the nonparametric bootstrap (with bootstrap samples of size N) fails, but the *m-out-of-n* bootstrap provides consistent estimates (Bickel et al. 1994; Chernick 2008).

3.6.3 Bootstrap and Linear Discriminant Analysis

Efron used the bootstrap approach to estimate the error rate in linear discriminant analysis and presented an example, in which bootstrapping outperformed leave-one-out cross-validation (Efron 1979). Interestingly, Breiman reported that LDA is not improved by bagging (Breiman 1998). However, the four data sets used in his study

[37]In this chapter, we use N to denote the size of the training data set. However, we keep the original *m-out-of-n* name for this bootstrap method. Hence, n in *m-out-of-n* equals N here.

had only a few variables (8–16) and much larger numbers of observations (214 to 1395). We may expect that in such situations (when $p \ll N$) LDA is quite stable and bagging will not produce a significant decrease in the misclassification error rate.

The gene expression data sets we deal with have thousands of variables and much fewer biological samples ($p \gg N$). Therefore, before the bootstrap-based classifiers can be built, we would perform independent feature selection on each bootstrap training set. Each LDA classifier would then be built from a multivariate biomarker identified from its bootstrap training set via a heuristic search. The Lawley–Hotelling T^2 measure of class separation that we would use to drive the feature selection search is based on such LDA assumptions as the independence of training set observations (biological samples) and multivariate normality of its variables. While LDA is quite robust to violations of the normality assumption,[38] we are concerned with the violation of the independence assumption when biological samples are selected into the bootstrap training sets with replacement.

Can we then take advantage of the OOB validation and the more stable biomarkers offered by the bootstrap-based ensemble approach when our LDA-based feature selection is performed in $p \gg N$ situations? Yes, but to avoid shaky ground (Chernick 2008) we would not be sampling with replacement. Instead of using bagging based on the nonparametric bootstrap that samples the training set with replacement, we will modify bagging in a way that bootstrap training sets are generated without violation of the independence assumption. We will use such *modified bagging* to generate a large number of classifiers. By using these classifiers to vote for variables, we expect to find biomarkers that are more robust than a biomarker based on a single feature selection performed on the original training data set (see Chapter 4).

3.6.4 The *Modified Bagging* Schema

Variations of bagging (bootstrap aggregating) may modify the aggregating stage (by utilizing various combinations of information provided by a large number of classifiers),[39] or they may modify the bootstrap approach. Examples of modification to the bootstrap stage may include:

[38]It is worth mentioning that selecting a second order statistical sample (i.e., sampling the training data set, which is already a sample from the underlying population) means selecting from a discrete distribution since the training set includes at most N discrete values of each variable. Thus we may not only violate the normality assumption, but we may depart from the realm of continuously distributed variables (Miller 2008). This is also true for other resampling methods, for example the leave-one-out and K-Fold cross-validations. However, one can argue that such methods are implementing the *what-if* scenario aimed at answering questions like this: "*How would our biomarker or classifier change if they are based on $N - 1$ observations, when one of the biological samples is—for whatever reason—excluded from the training set?*" Stretching this reasoning, one may eventually argue that such methods do not really violate the normality assumption. Nevertheless, even if we agree with this kind of argumentation and extend it to the bootstrap without replacement, it cannot—for small sample sizes—be extended to bootstrapping with replacement.

[39]There are also methods that use bootstrap-generated classifiers but do not aggregate their information. For example, in *bumping* the classifier that best fits the training data is selected from among classifiers built on bootstrap training sets. The bumping approach may be used to avoid models that are finding inefficient local optima (Hastie et al. 2009).

- stratification (independent selection of observations from each class),
- subsampling (*m-out-of-n* bootstrap without replacement),
- selecting all observations from one class and randomly sampling the other class, when the design is heavily unbalanced.

Since the term *bootstrap* may refer to sampling with or without replacement, aggregating classifiers generated with modified bootstrap methods may still be called *bagging*. Nevertheless, other terms have also been proposed. For example, *subagging* may refer to subsample aggregating, and *moon-bagging* to m-out-of-n bootstrap aggregating (Bühlmann and Yu 2002).

We will define a ***modified bagging*** schema as a procedure that generates bootstrap training sets by stratified random sampling of the training data set without replacement. Let the training data set consist of N biological samples assigned to J classes and let each class j, where $j = 1, \ldots, J$, be represented in the training set by n_j samples, so $N = \sum_{j=1}^{J} n_j$. Let us define the γ_{OOB} parameter that represents a desired proportion of the out-of-bag samples, that is, the proportion of the training samples that are not selected into a bootstrap training set. Hence, each bootstrap training set will include about $(1 - \gamma_{OOB})N$ biological samples that will include about $(1 - \gamma_{OOB})n_j$ samples from each class.[40]

Using this *modified bagging* schema, we will be generating a large number, say B, of classifiers that will be based on B bootstrap training sets. Classifying the OOB samples using their respective classifiers will give us an estimate of generalization for the ensemble classifier. However, we may interpret this estimate as a predicted misclassification error rate of an average individual classifier. We may use this estimate during various stages of biomarker discovery and for verification of the selection of the *Informative Set of Genes*[41] (see Chapter 4).

When the standard nonparametric bootstrap is used, the number of possible bootstrap training sets of size N selected with replacement from the original training data set of size N is (Hall 1992; Efron and Tibshirani 1993; Chernick 2008):

$$\binom{2N-1}{N} = \frac{(2N-1)!}{N!(N-1)!}. \tag{3.112}$$

For the *modified bagging* schema, we define the number of OOB samples in each class as

$$nOOB_j = \text{int}(\gamma_{OOB} \cdot n_j + 0.99), \quad j = 1, \ldots, J. \tag{3.113}$$

Hence, the total number of OOB samples for each classifier can be calculated as

$$nOOB = \sum_{j=1}^{J} nOOB_j. \tag{3.114}$$

[40]Note that if $(1 - \gamma_{OOB}) \approx 0.632$, the number of OOB samples will be similar to that of the nonparametric bootstrap with replacement.

[41]The *Informative Set of Genes* will be defined in Chapter 4.

The number of possible bootstrap training sets generated by the *modified bagging* schema is

$$\prod_{j=1}^{J} \binom{n_j}{nOOB_j}. \tag{3.115}$$

This is a large number even for relatively small values of N. The probability that any of the bootstrap training sets is repeated is very low even for large B.

Summing up, to apply the *modified bagging* schema, we will generate hundreds or thousands of classifiers built from bootstrap training sets consisting of a specified proportion of biological samples randomly selected without replacement from the original training data set. For each classifier, a separate feature selection (e.g., heuristic search driven by a multivariate metric of discriminatory power of a subset of variables) will be performed. An additional level of randomness may be added to the schema by introducing randomness into the feature selection process (e.g., by starting a heuristic search from a randomly selected variable). Once the classifiers are generated, we will *not* use them as an ensemble classifier. We will use them to aggregate information pertaining to the misclassification error rate and information about the distribution of variables selected into the classifiers. Each classifier will be tested on its OOB samples. Averaging testing results over a large number of classifiers will provide a reliable estimate of the misclassification error rate. This would be the error rate of the ensemble classifier or its average component. However, it will also give us an estimate of generalization abilities of an optimal biomarker identified by the same feature selection method and of the same size as our bootstrap-based classifiers. In Chapter 4, we will describe the use of the *modified bagging* schema in verification of the *Informative Set of Genes* (by evaluating the amount of discriminatory information remaining in the training set that does not include the *Informative Set of Genes*). Another use of the *modified bagging* schema will be based on aggregating information about variables selected into the bootstrap-based classifiers. By examining the distribution of genes among these classifiers, we will be identifying primary gene expression patterns and frequently used informative genes representing these patterns. This can be seen as using ensembles of classifiers to vote for variables. This approach will also be described in Chapter 4.

3.7 OTHER LEARNING ALGORITHMS

From among other learning algorithms that may be used for classification based on gene expression data, we will briefly describe two nonparametric methods—*k-Nearest Neighbors* and *Artificial Neural Networks*. They may not be our first choice in biomarker discovery based on expression data, but in some situations they may be useful. For example, we would not use the *k*-nearest neighbor approach for classification based on a training set that includes thousands of variables (especially with a small number of training samples). However, after feature selection is performed, a *k*-nearest neighbor classifier applied to data with a small number of variables may be as successful as far more sophisticated approaches. Similarly, we would not use artificial neural networks when the interpretation of the classification model is important (which almost always is the case in biomarker discovery). However, in specific situations where—for whatever reason—it is important to use a nonlinear classification model and where prediction without interpretation is acceptable, neural networks can be among our top choices.

3.7.1 *k*-Nearest Neighbor Classifiers

k-Nearest Neighbor classifiers are conceptually very simple. The training data is used directly to classify a new sample, and no classification model is built. Consider a training data set with p variables, N objects (biological samples) and J classes. Each training sample as well as a new sample to classify can be represented by a point in the p-dimensional space of the p variables. The general idea of k-nearest neighbor classification can be illustrated as follows. Assume we use the Euclidean metric of distance.[42] Imagine a p-dimensional point representing the object to classify. Center a p-dimensional hypersphere at this point. Gradually increase the radius of this hypersphere, from zero until the hypersphere includes exactly k training points—the k nearest neighbors. Each of these k neighbors votes for its class and the new sample is classified into the most popular class. If $J = 2$ and k is odd, the new sample is always voted into one of the classes. In other situations, ties may be resolved by random selection of one of the most popular classes. An extension of the k-nearest neighbor rule can be made to require that at least l of k neighbors, ($l \leq k$), belong to a class in order to declare the class a winner; otherwise the result of classification is considered doubtful (Hellman 1970). When costs of misclassification are different, we may further extend this requirement to allow different l values for different classes (Ripley 1996).

In a straightforward computer implementation of the k-nearest neighbor method, all training points are stored in the memory and used for classification. Advantages of this approach include its simplicity and flexibility—since all training samples are directly used in classification, adding new training samples requires no change to the algorithm (there is no model to retrain). However, every time we classify a new sample, we need to calculate N p-dimensional distances. Hence, the cost of doing this via the straightforward approach is $O(pN)$. To decrease memory requirements and computational costs, various *editing* or *condensing* procedures have been proposed. They try to identify a subset of training points important for k-nearest neighbor classification, and discard the remaining training data. For example, if homogeneous clusters of training samples can be identified, only important exterior points of these clusters may be kept (Hart 1968). However, with such modifications, the k-nearest neighbor approach is no longer flexible in accepting new training data points. Other approaches (such as branch and bound algorithms) keep the entire training data, but focus on speeding up classification by calculating only a subset of distances. They

[42]Euclidean distance is often used in the k-nearest neighbor implementations. If the variables are measured in different units, they should be scaled. Gene (or protein) expression variables are measured in the same units.

utilize the structure of the distances between training data points (which needs to be determined only once) to rule out the training points deduced—without the explicit calculation of their distances to the currently classified sample—to be too far to be included in the set of k nearest neighbors (Hand et al. 2001). Yet another approach computes partial distances in subspaces of the p-dimensional classification space. If a partial distance is greater than the distance between the point to classify and a currently identified kth nearest neighbor, the full distance is not calculated and the training point is ruled out (Duda et al. 2001).[43]

The approach illustrated earlier by the expanding hypersphere works under the assumption that class probabilities are similar in every direction within the neighborhood of the point to classify. If that is not the case, the misclassification error may be large. To adjust for possibly different class probabilities in different directions, we may replace a hypersphere with a hyperellipsoid. The directions and lengths of the axes of the hyperellipsoid can be determined by the ratio of the variation between classes to the variation within classes in a local neighborhood of the point to classify. This modified approach is called the *discriminant adaptive nearest-neighbor* procedure (Hastie et al. 2009). Please note that the parameters of the hyperellipsoid (and adaptation of the distance metric) are determined separately for each sample to classify.

From among the various types of other modifications to the k-nearest neighbor approach, we shall mention those that incorporate weights. The weights may be assigned to variables, to training data points, or they may be incorporated into the voting procedure. In cases where variables have different relative importance, we may assign them different weights and then incorporate the weights into the implemented distance metric (Hand et al. 2001). If—for whatever reason—we want to treat some training data points as more important than others, we may assign weights to the training instances and then multiply each calculated distance by the appropriate weight. Hence, training points with smaller weights are more likely to be among the selected k-nearest neighbors (Morring and Martinez 2004). Finally, voting can be weighted by assigning higher weights to the training points closer to the classified sample (Mitchell 1997; Larose 2005).

Although k-nearest neighbor classification is very simple, it is often successful when class boundaries are highly irregular and classes have many possible prototypes (Hastie et al. 2009). Another advantage of the k-nearest neighbor approach is the easy interpretation of the classification results. However, this approach has the disadvantage of being rendered inappropriate in some situations. For example, k-nearest neighbor methods usually perform poorly when the number of variables p is large. The reason is that the training data is sparse in high-dimensional space and at least some of the k neighbors may be quite far from the sample to classify (Hand et al. 2001). That is one of the reasons why we would not use this approach for a gene expression matrix with thousands of variables. However, when applied *after* feature selection, the k-nearest neighbor methods may perform as well as our preferred learning algorithms (see examples in Chapter 6).

[43] Of course, this approach works only with distances that cannot decrease when the number of dimensions increases. The Euclidean metric belongs to this class.

Even in situations where it is reasonable to use the k-nearest neighbor rule, its performance may highly depend on the choice of k. The best value of k depends on the idiosyncrasies of a particular training data set (and especially on the number of samples N) and should be selected as a compromise between overfitting and generalization, which usually is achieved when k is a small fraction of N. Using $k = 1$ may render unstable classifiers (exceptions include situations where the training data is edited in a way that optimizes the performance of the algorithm for $k = 1$). Setting k too high may define a neighborhood that includes training points quite far from the classified sample (Duda et al. 2001).[44] In practice, cross-validation may be used to select the value of k that minimizes the misclassification error estimate (Hastie et al. 2009).

3.7.2 Artificial Neural Networks

Artificial Neural Networks (ANNs) are learning algorithms inspired by the nonlinearity and parallelism of the human brain learning process. Plasticity, massive parallelism, and fault tolerance are among intrinsic features of this process. Although signal propagation in the human brain is several orders of magnitude slower than in computer chips, humans can efficiently deal with perceptual problems that are far too complex for present-day computer systems. The cerebral cortex of the human brain contains about 10^{11} *neurons*, which are nerve cells consisting of *dendrites* (inputs), the cell body (the processing unit) and the *axon* (output). The high efficiency of the brain's information processing is mostly due to the approximately 10^{14} to 10^{15} interconnections between the neurons (this means that, on average, each neuron is connected to 10^3 to 10^4 other neurons). A neuron receives electrical signals from other neurons via its dendrites, and combines these signals. If the combined input signal exceeds some threshold level, the neuron "fires," that is, transmits its electrical signal output along its single axon. Axons and dendrites of communicating neurons are connected through synapses, which are biochemical units whose neurotransmitters can impose either excitation or inhibition on the receptive neurons. Since the information is stored in the weights associated with synaptic interconnections, the learning process of the human brain may include the creation of new connections between neurons or changes in the weights of existing connections. Due to the distributed nature of processing and storing information, this structure is highly adaptable and fault-tolerant.

Early concepts associated with ANNs include the McCulloch–Pitts model of a neuron (McCulloch and Pitts 1943), the Hebbian learning rule (Hebb 1949), and—inspired by them—Rosenblatt's perceptron (Rosenblatt 1958). The model of a neuron proposed by McCulloch and Pitts consists of binary inputs and a single binary output. Connections between the inputs and the output are either excitatory (with positive weights) or inhibitory (with negative weights). The weights are constant and the neuron calculates the weighted sum of the input signals. If this sum is above a threshold value, the neuron "fires," that is, generates an output signal of 1 (Jain et al.

[44] If we look at the k-nearest neighbor rule as a method to estimate posterior probabilities of class membership in the neighborhood of the classified sample, then large k may result in a neighborhood too large for a reliable estimate.

1996). The Hebbian learning rule states that if two neurons are activated synchronously, their synaptic connection is strengthened.

3.7.2.1 Perceptron

A *perceptron* (Rosenblatt 1958) can be considered a single-neuron network with adjustable real-valued connection weights, bias and an activation function (see Fig. 3.19). Such a simple neural network can be used as a classifier differentiating between two linearly separable classes (Haykin 2008).

Activation signal a is a linear function of all inputs and the bias element,

$$\begin{aligned} a &= w_0 + \sum_{k=1}^{p} w_k x_k \\ &= w_0 + \mathbf{w}^T \mathbf{x}, \end{aligned} \qquad (3.116)$$

where $\mathbf{x}$ is a p-dimensional vector of input signals (corresponding to p input variables) and $\mathbf{w}$ is the vector of synaptic connection weights. Since $\mathbf{w}^T\mathbf{x}$ represents a $(p-1)$-dimensional hyperplane in the p-dimensional space defined by the p input variables, the weight of the bias element w_0 can be interpreted as the offset of the hyperplane from the origin. The output signal y is determined by applying a nonlinear activation function f to the activation signal a,

$$y = f(a). \qquad (3.117)$$

Please note that the bias weight w_0 is directly related to the neuron activation threshold θ, $\theta = -w_0$, which determines the minimum value of the weighted sum of input signals $\sum_{k=1}^{p} w_k x_k$ required for having a non-negative activation signal a.

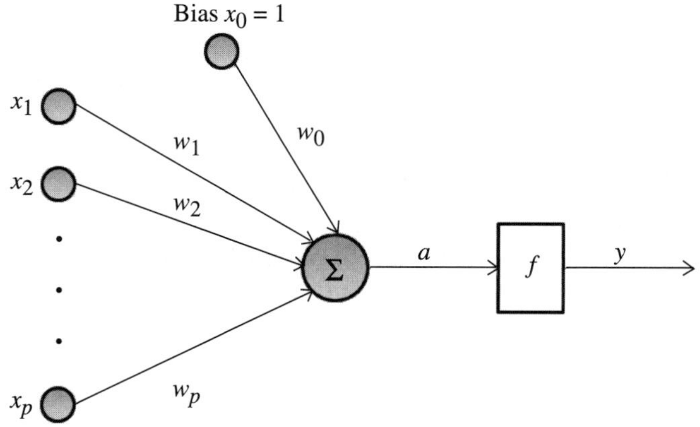

Figure 3.19: A perceptron with p inputs, bias and activation function f. Connection weights between inputs and the summing element are adjustable. The summing node calculates a weighted sum of the p input signals and adds the bias, which is related to the neuron activation threshold.

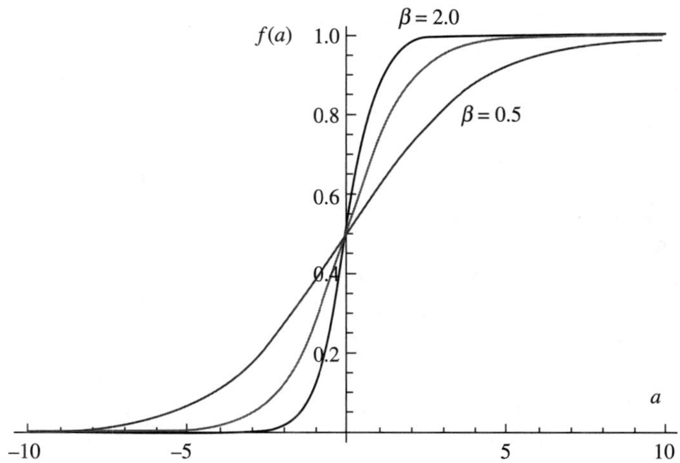

Figure 3.20: The logistic sigmoid function, $f(a) = 1/(1 + e^{-\beta a})$, with three values of the slope parameter, $\beta = 0.5$, $\beta = 1.0$, and $\beta = 2.0$. This is a monotone and differentiable function. Since it returns values in the range (0, 1) for any range of values of the activation signal a, it may be called a *squashing* function.

In a basic version of a perceptron, a hard-limiter activation function is used that returns the sign of the activation signal,

$$\begin{aligned} f(a) &= sign(a) \\ &= sign(w_0 + \mathbf{w}^T\mathbf{x}) \\ &= \left\{ \begin{array}{rl} 1 & \text{if } a > 0 \\ 0 & \text{if } a = 0 \\ -1 & \text{otherwise} \end{array} \right\}. \end{aligned} \tag{3.118}$$

More general versions may use various soft limiting nonlinear activation functions, such as piecewise linear, sigmoidal or Gaussian. The logistic sigmoid function,

$$f(a) = \frac{1}{1 + e^{-\beta a}}, \tag{3.119}$$

where $\beta > 0$ is the slope parameter of the function, is among the most popular ones (Fig. 3.20).

The term *perceptron* can also refer to a single-layer neural network consisting of multiple outputs. Such networks can classify into more than two classes. However, they are still limited to the differentiation of linearly separated classes. To handle nonlinear boundaries between classes, neural networks require a multilayer topology.

3.7.2.2 Multilayer Feedforward Neural Networks

A neural network whose topology contains no feedback loops (i.e., its connection pattern can be represented by a directed acyclic graph) is called a *feedforward* network.

Although such a network does not have to have a layered topology, it usually consists of the *input layer*, some number of *hidden layers* and the *output layer*. Since information in a feedforward network is propagated only in one direction[45] (from input nodes to output nodes), the outputs can be explicitly determined from the inputs and the connection weights; they do not depend on the history of the network state.[46] A connection weight (also known as a synaptic weight) represents the strength of a connection. The weights may be positive (excitatory), negative (inhibitory), or zero (no connection).

For a classification problem with training data in the form of a gene expression matrix with p variables and N biological samples representing J classes, the input layer consists of p nodes, and the output layer contains J neurons.[47] Please note that, similar to the unsupervised version of neural network described in Chapter 2 (the self-organizing map), the input layer is not neuronal (i.e., no processing is performed at this layer's nodes). Here, the input layer nodes correspond to p variables representing a biological sample. The number of hidden layers and the number of neurons at each of them should be adjusted for the complexity of the problem. In most situations, one or two hidden layers can provide satisfactory solutions. For example, a two-layer network whose hidden layer has a sufficiently large number of neurons and sigmoidal activation functions can approximate any continuous input-output mapping function arbitrarily well. However, in some situations, a topology with additional hidden layer(s) may provide a more efficient solution (Bishop 1995).

A network with no hidden layers (i.e., where input nodes are connected directly to output neurons) is called a single-layer neural network (since it contains only one layer of neurons). Neural networks with one or more hidden layers are called multilayer networks. Figure 3.21 depicts a feedforward neutral network with a single hidden layer (hence, a two-layer network). Assume that this is a fully connected network, meaning each neuron is connected to all nodes of the preceding layer. Assume further that there are p input nodes (corresponding to p variables), H nodes (neurons) of the hidden layer, and J neurons of the output layer (corresponding to J differentiated classes). The J network output signals represent the output variables y_j, $j = 1, \ldots, J$. A classified pattern is assigned to class j represented by the output with the highest signal value y_j. If we want to interpret network output signals as class probabilities, the signals should assume values in the range (0, 1) and should sum to 1.

[45]During classification, after the network has already been trained. During supervised learning, a "teacher" system—external to the network—calculates error signals and backpropagates them through the network to adjust the network connection weights. Supervised learning results in a classifier whose parameters are stored in the form of network connection weights.

[46]Hence, feedforward networks are static networks. Networks with feedback connections are called recurrent neural networks or dynamic networks (Dreyfus 2005).

[47]When only two classes are differentiated ($J = 2$), neural networks are usually implemented with only one output neuron.

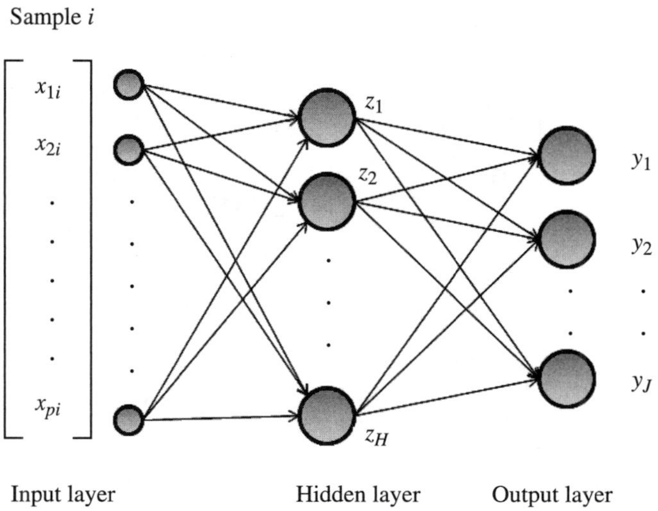

Figure 3.21: Topology of a feedforward neural network with a single hidden layer. The input layer consists of p nodes corresponding to p variables (for instance, expression levels of p genes representing a biological sample). The output layer consists of J neurons corresponding to J differentiated classes. Each neuron of the hidden layer is connected to each input node. Assuming there are H neurons of the hidden layer, each neuron of the output layer will have H input connections. This neural network may be referred to as p-H-J network.

Each training sample can be represented by a p-dimensional vector $\mathbf{x}_i \in \Re^p$ of p gene expression levels,

$$\mathbf{x}_i = \begin{bmatrix} x_{1i} \\ x_{2i} \\ \vdots \\ x_{pi} \end{bmatrix}, \quad i = 1, \ldots, N. \tag{3.120}$$

Each input node is connected to each neuron of the hidden layer, and each of these $p \times H$ connections is characterized by a weight w_{kh}, $k = 1, \ldots, p$, $h = 1, \ldots, H$. Hence, each neuron of the hidden layer is associated with a p-dimensional vector of connection weights (see Fig. 3.22),

$$\mathbf{w}_h = \begin{bmatrix} w_{1h} \\ w_{2h} \\ \vdots \\ w_{ph} \end{bmatrix}, \quad h = 1, \ldots, H. \tag{3.121}$$

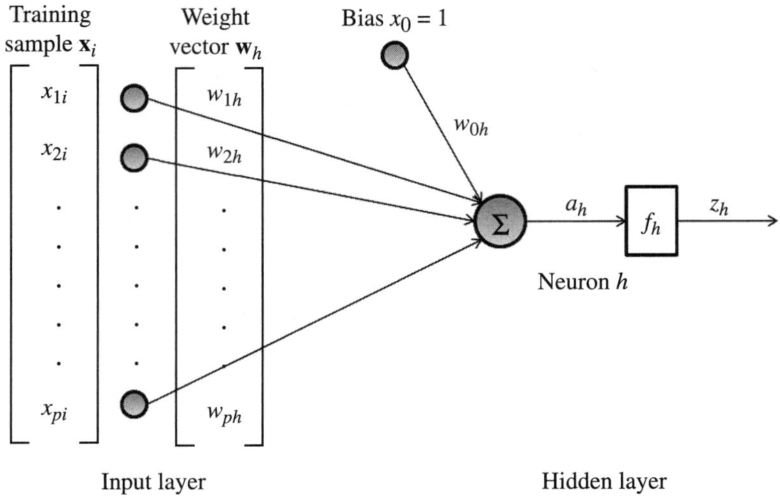

Figure 3.22: Input and output signals for neuron h of the hidden layer of the feedforward neural network with a single hidden layer. There are p connections between input nodes and each neuron h, each with its own weight. Hence, a p-dimensional vector of weights, whose values are adjusted during the learning process, is associated with each neuron h. An additional input $x_0 = 1$ with adjustable synaptic weight w_{0h} represents the bias. The f_h element represents a nonlinear and differentiable activation function.

Similarly, each neuron of the output layer has assigned to it an H-dimensional vector of weights corresponding to its H connections with the hidden layer (see Fig. 3.23),

$$\mathbf{w}_j = \begin{bmatrix} w_{1j} \\ w_{2j} \\ \vdots \\ w_{Hj} \end{bmatrix}, \quad j = 1, \ldots, J. \tag{3.122}$$

Let a_h represent the weighted sum of input signals of neuron h of the hidden layer. If we include the bias w_{0h} of neuron h (see Fig. 3.22), then[48]

$$\begin{aligned} a_h &= w_{0h} + \sum_{k=1}^{p} w_{kh} x_{ki} \\ &= w_{0h} + \mathbf{x}_i^T \mathbf{w}_h, \quad h = 1, \ldots, H. \end{aligned} \tag{3.123}$$

[48]Note that to simplify the notation, we do not include a layer qualification. To add such qualification, the weights could be denoted $w_{kh}^{(1)}$ for the first neuronal layer (the hidden layer) and $w_{hj}^{(2)}$ for the second neuronal layer (the output layer), or more generally $w_{*\dagger}^{(l)}$ for any layer $l = 1, \ldots, L$. Here L denotes the number of layers with adaptive weights.

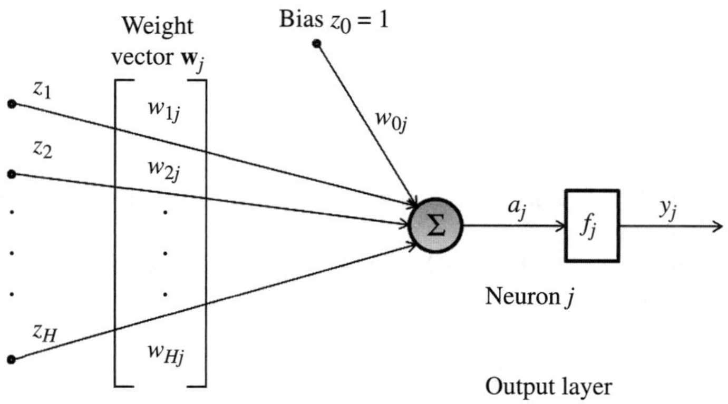

Figure 3.23: Input and output signals for neuron j of the output layer of the feedforward neural network with a single hidden layer. Each neuron j has H input connections and is associated with an H-dimensional vector of weights. f_j represents a differentiable activation function.

If $f_h(\cdot)$ denotes the activation function of neuron h, then the output signal from neuron h of the hidden layer is

$$z_h = f_h(a_h), \tag{3.124}$$

and all output signals of the hidden layer can be represented by an H-dimensional vector $\mathbf{z}$,

$$\mathbf{z} = \begin{bmatrix} z_1 \\ z_2 \\ \vdots \\ z_H \end{bmatrix}. \tag{3.125}$$

Analogically, for neuron j of the output layer, we shall have:

$$\begin{aligned} a_j &= w_{0j} + \sum_{h=1}^{H} w_{hj} z_h \\ &= w_{0j} + \mathbf{z}^T \mathbf{w}_j, \quad j = 1, \ldots, J, \end{aligned} \tag{3.126}$$

and

$$y_j = f_j(a_j), \quad j = 1, \ldots, J, \tag{3.127}$$

where $f_j(\cdot)$ represents the activation function of neuron j. Examples of activation functions that are often used in the output layer include the logistic sigmoid function (see Fig. 3.20) and its generalization, the *softmax* function,

$$f_j(a_j) = \frac{e^{a_j}}{\sum_{g=1}^{J} e^{a_g}}, \quad j = 1, \ldots, J. \tag{3.128}$$

The softmax activation function results in network output signals that lie in the range (0, 1) and sum up to 1.

If we combine (3.123) to (3.127) into a single representation of the network function, we shall have:

$$y_j = f_j\left(w_{0j} + \sum_{h=1}^{H} w_{hj} f_h\left(w_{0h} + \sum_{k=1}^{p} w_{kh} x_{ki}\right)\right), \quad j = 1, \ldots, J. \tag{3.129}$$

3.7.2.3 Training the Network (Supervised Learning)

First, the network is initialized by assigning random values to its weights. Then, the training samples are presented, in a random order, to the network. For each sample $\mathbf{x}_i$, the output values are compared to ones expected for the sample's true class, and error values e_j are calculated for each output y_j. The error values are then used to adjust the connection weights in a way that each adjustment depends on the weight's contribution to the error. The adjustments can be made once per training epoch (an iteration during which all training samples are presented to the network) or after each training sample.

Since the training samples are propagated through the network in a random order, let us use t as the index of learning steps. The goal of training is to minimize some error function E,

$$E = \sum_{i=1}^{N} E_i, \tag{3.130}$$

where $E_i = E_i(t)$ represents the error for training sample $\mathbf{x}_i$ presented to the network at step t. The error E_i may be calculated as[49]

$$\begin{aligned} E_i(t) &= \frac{1}{2}\sum_{j=1}^{J} e_j^2(t) \\ &= \frac{1}{2}\sum_{j=1}^{J} [c_j(t) - y_j(t)]^2, \end{aligned} \tag{3.131}$$

[49]Note that the backpropagation algorithm may use any error function that is differentiable with respect to network connection weights. For example, instead of the sum of squared errors, a cross-entropy error function may be used (Hastie et al. 2009). Furthermore, a *weight decay* regularization approach to reducing the risk of overfitting (by forcing some connection weights to assume values close to zero) may be implemented by adding a complexity penalty term to the error function (Mitchell 1997; Haykin 2008).

where

- $c_j(t)$ is the target value of output j, corresponding to correct classification of sample $\mathbf{x}_i$,
- $y_j(t)$ is the observed value of output j for training sample $\mathbf{x}_i$ presented to the network at step t,
- $e_j(t) = c_j(t) - y_j(t)$ is the error signal at output j at step t.

3.7.2.3.1 Backpropagation

To train the network, we minimize the error E with respect to the network connection weights. This may be accomplished using the *backpropagation* algorithm implementing a *gradient descent* method.[50] Since each derivative of the error term E (with respect to the weights) can be expressed as a sum of the derivatives of E_i over all training samples, we may train the network by minimizing each error E_i separately (Bishop 1995; Izenman 2008). This approach, when training samples are selected in a random order, and weight adjustments are performed every time a training sample has been propagated through the network, is called the *stochastic gradient descent* method.

Recall that t is the index of learning steps; t is increased by one after each training sample is propagated through the network and the network weights adjusted. Using the chain rule for differentiation, the gradient descent adjustment to the weights associated with neuron j of the output layer can be calculated as

$$\begin{aligned} w_{hj}(t+1) &= w_{hj}(t) + \Delta w_{hj}(t) \\ &= w_{hj}(t) - \eta \frac{\partial E_i(t)}{\partial w_{hj}(t)} \\ &= w_{hj}(t) - \eta \frac{\partial E_i(t)}{\partial a_j(t)} \cdot \frac{\partial a_j(t)}{\partial w_{hj}(t)} \\ &= w_{hj}(t) + \eta \delta_j(t) z_h(t), \end{aligned} \tag{3.132}$$

where

- η is the *learning rate* parameter,
- $\delta_j(t)$ is the local gradient at output neuron j (or sensitivity of neuron j),

$$\begin{aligned} \delta_j(t) &= -\frac{\partial E_i(t)}{\partial a_j(t)} \\ &= -\frac{\partial E_i(t)}{\partial e_j(t)} \cdot \frac{\partial e_j(t)}{\partial y_j(t)} \cdot \frac{\partial y_j(t)}{\partial a_j(t)} \\ &= e_j(t) f_j'(a_j(t)), \end{aligned} \tag{3.133}$$

[50] Although backpropagation is not the fastest method for training neural networks, it is among the simplest and most instructive ones (Duda et al. 2001). Conjugate gradient descent is another popular method. The *gradient* of an error function means a *vector of first partial derivatives* of the function with respect to network connection weights.

- $f_j'(\cdot)$ denotes the derivative of the activation function $f_j(\cdot)$,
- $z_h(t)$ is the signal associated with connection h to neuron j, as defined by (3.124).

To use the backpropagation algorithm to adjust weights w_{kh} of the hidden layer, we have to remember that each of these weights is used in calculating signals propagated to all output neurons. Thus the backpropagation algorithm has to take into account the error signals of all the output nodes. Following reasoning similar to that for output neurons, the gradient descent update to the weight associated with a connection between input node k and the hidden layer's neuron h can be calculated as

$$\begin{aligned} w_{kh}(t+1) &= w_{kh}(t) + \Delta w_{kh}(t) \\ &= w_{kh}(t) + \eta \delta_h(t) x_k(t), \end{aligned} \tag{3.134}$$

where

- $\delta_h(t)$ is the local gradient at hidden node h, which depends on the derivative of the activation function of this node and on the weighted sum of local gradients of all output nodes,

$$\delta_h(t) = f_h'(a_h(t)) \sum_{j=1}^{J} \delta_j(t) w_{hj}(t), \tag{3.135}$$

- $f_h'(\cdot)$ is the derivative of the activation function $f_h(\cdot)$,
- $x_k(t)$ is the input signal propagated via connection k to neuron h, i.e., the kth element of the input vector $\mathbf{x}_i$ (3.120), representing the training sample presented to the network at step t.

3.7.2.3.2 The Learning Rate and Momentum Parameters

The magnitude of the weight adjustments depends on the learning rate parameter η, which may assume values $0 < \eta < 1$. Small values of η tend to provide more stable weight estimates but increase the learning time. On the other hand, large values of η may lead to oscillatory behavior, that is, repeated "overshooting" of a minimum due to too large weight adjustments. A compromise can be achieved by adding a momentum term α, $0 \leq \alpha < 1$, which adds inertia to the learning process by making the weight adjustment at step t dependent on the weight adjustment of preceding step $t-1$. For example, to add a momentum term to the weight adjustment $\Delta w_{hj}(t)$ defined within (3.132) for neuron j of the output layer, we shall redefine $\Delta w_{hj}(t)$ in the following way:

$$\Delta w_{hj}(t) = \alpha \Delta w_{hj}(t-1) + \eta \delta_j(t) z_h(t). \tag{3.136}$$

When $\alpha = 0$, backpropagation is carried out without momentum. When $\alpha > 0$ and subsequent adjustments to a connection weight are made in the same direction, the momentum term increases the magnitude of the adjustments and accelerates the gradient descent. When consecutive adjustments have opposite signs, the weight adjustment is decreased (Mitchell 1997; Haykin 2008).

Other modifications to the backpropagation learning process include changing the learning rate η from epoch to epoch. Usually, an initial learning rate is relatively large to speed up early phases of training. Then, it gradually decreases when the network converges (Larose 2005). Furthermore, we may assign different values of the learning rate to different parts of the network (Haykin 2008). Note that for multilayer neural networks, the hypersurface representing the error function[51] may have multiple minima and that the gradient descent method may find a local minimum, not necessarily the global one. However, please recall that a solution associated with a local minimum may be less prone to overfitting the training data than the one corresponding to the global minimum.

3.7.2.3.3 Generalization to a Feedforward Neural Network with Any Number of Layers

Our considerations for a two-layer feedforward network (including only one hidden layer) can be easily extended to a multilayer feedforward network with any number of layers L. Such a network will include $L - 1$ hidden layers and one output layer. Please recall that the depth L of a multilayer neural network is defined by the number of neuronal layers; the input layer—where no processing is performed—is not counted towards the depth. Including the momentum term, we may combine and generalize equations (3.132) and (3.134) as follows:

$$\begin{aligned} w_{*\dagger}^{(l)}(t+1) &= w_{*\dagger}^{(l)}(t) + \Delta w_{*\dagger}^{(l)}(t) \\ &= w_{*\dagger}^{(l)}(t) + \alpha \Delta w_{*\dagger}^{(l)}(t-1) + \eta \delta_{\dagger}^{(l)}(t) f_{*}^{(l-1)}(a_{*}^{(l-1)}(t)), \end{aligned} \tag{3.137}$$

where

- superscript (l) denotes neuronal layer l, $l = 1, \ldots, L$; for hidden layers $l = 1, \ldots, L-1$, for the output layer $l = L$,
- subscript $\dagger$ is a placeholder for the index of neurons of layer l,
- subscript $*$ is a placeholder for the index of neurons (or nodes) of layer $l - 1$; thus $l - 1 = 0$ refers to the input layer,
- $w_{*\dagger}^{(l)}$ denotes the weight of a connection between node $*$ of layer $l - 1$ and neuron $\dagger$ of layer l,
- $f_{*}^{(l-1)}$ denotes the activation function of the previous layer $(l - 1)$; if $l - 1$ refers to the input layer, $f_{*}^{(0)}$ is the identity function and $a_{*}^{(0)}$ corresponds to one of the input variables,
- $\delta_{\dagger}^{(l)}$ is the sensitivity of neuron $\dagger$ at layer l; combining and generalizing (3.133) and (3.135), and using $\diamond$ as a placeholder for the index of neurons of the next

[51]The error hypersurface is defined in a multidimensional space with the number of dimensions equal to the number of connection weights in the network. Since the state of a network can be represented as a point on the error hypersurface, the general idea of the backpropagation algorithm is to update connection weights in a way that this point moves in the direction of the negative gradient of the hypersurface (steepest descent from the current location). The magnitude of each move is determined by the learning rate parameter η.

layer, $l+1$, we shall have:

$$\delta_{\dagger}^{(l)}(t) = \left\{ \begin{array}{ll} e_j(t) f_j'(a_j(t)) & \text{for neuron } j \text{ of the output layer } L \\ f_{\dagger}'^{(l)}(a_{\dagger}^{(l)}(t)) \sum_{\diamond} \delta_{\diamond}^{(l+1)}(t) w_{\dagger\diamond}^{(l+1)}(t) & \text{for neuron } \dagger \text{ of hidden layer } l \end{array} \right\} \tag{3.138}$$

3.7.2.3.4 *Example of Training Algorithm for a Multilayer Feedforward Neural Network*

1. For the implemented topology of the neural network (defined by the number of input variables, the number of layers and numbers of neurons at each layer):
   - initialize the weights of all the network connections to small random values (e.g., between -0.5 and 0.5),
   - set the learning step index $t = 1$ (t will be increased after each training sample is propagated through the network and all weights adjusted),
   - initialize the learning rate η and momentum α,
   - Specify the maximum number of iterations (epochs).
2. Repeat until convergence or until the maximum number of iterations is performed:
   (a) Loop over the data set of the N training samples, selecting samples in a random order. The learning steps are indexed by the consecutive values of t:

   Forward propagation:
   - propagate a single training input pattern $\mathbf{x}_i(t)$ through the network,

   Backpropagation:
   - calculate the error signal $e_j(t)$ for each output j, $j = 1, \ldots, J$,
   - update the network connection weights according to (3.137).

   (b) Decrease the values of the learning rate η and the momentum parameter α.
3. Estimate the misclassification error rate of the resulting neural network classifier, preferably by classifying samples of an independent test data set.

3.7.2.3.5 *Building a Generalizable Neural Network Classifier*

Similar to the situation when we perform only a single feature selection in a search for a multivariate biomarker, training a neural network only once on a training data set with a large number of variables does not necessarily provide an efficient classifier. To look for more generalizable solutions, we may:

- perform cross-validation pretraining runs (for instance, K runs for the K-Fold cross-validation schema) to determine the number of iterations providing the best cross-validation results; then train the network on the entire training set performing that number of iterations (Mitchell 1997),

- train the network multiple times with different starting weights,
- try different network topologies,
- build an ensemble of neural network classifiers based on bootstrap training sets (selected from the original training set with or without replacement); we may either select one of these classifiers (based on some estimate of their generalization abilities) or use all of them as an ensemble classifier,
- combine some of the above.

Additional options include multivariate feature selection to reduce the number of input variables, or regularization by adding a complexity penalty term to the error function.

3.8 EIGHT COMMANDMENTS OF GENE EXPRESSION ANALYSIS (FOR BIOMARKER DISCOVERY)

1. Do not use training data sets with too few (biological) samples. Most likely they do not properly represent the researched populations. Your results may be anecdotal rather than scientific.
2. Do not include in the training data set any sample for which its class membership is doubtful (garbage-in garbage-out).
3. Do not rely on univariate approaches in selecting variables for multivariate biomarkers. By limiting your study to a relatively small number of top univariately ranked variables, most important discriminatory information may be removed from consideration.
4. Do not use unsupervised methods to identify classes as a preprocessing step for classification. With thousands of variables, the identified clusters may have very little to do with the classes you want to discriminate (see #2).
5. Do not use unsupervised methods to decrease dimensionality of the training data set as a preprocessing step for biomarker discovery. The directions of most variation in the data may be very different from the most discriminatory directions (the directions that best separate the differentiated classes).
6. Do not limit evaluation of a classifier to internal cross-validation. Whenever possible, test your classifier on an independent data set. If independent data is not available, use a bagging (or *modified bagging*) approach to estimate the generalization abilities of the classifier.
7. Do not stop at identification of an optimal biomarker and efficient classification system. Extend the biomarker into the *Informative Set of Genes*, which may facilitate biological interpretation of class differences. Furthermore, if no alternative biomarkers can be identified for the same training data set, this may indicate that your optimal biomarker is a result of statistical chance.
8. If possible, do not let your boss override any of the above.

EXERCISES

3.1 Select at least two software packages that can be used (individually or together) for a comprehensive analysis of gene expression data. Consider open source or free for academic use software or trial/test versions of commercial packages. Your selection has to cover the following areas:

- Basic exploratory analysis.
- Unsupervised (taxonomy-related) analysis; for example, hierarchical clustering or self-organizing maps.
- Feature selection; recall that this means multivariate and supervised feature selection (e.g., heuristic stepwise methods or recursive feature elimination).
- Supervised learning algorithms—at least two of the following four: linear discriminant analysis (LDA), support vector machines (SVMs), random forests (RFs), and artificial neural networks (ANNs).
- Ensemble-based approach to estimation of the misclassification error rate.

It is highly recommended to use software that generates low-dimensional visualization of the discriminatory space.

Evaluate considered packages by performing test experiments (for instance, analyzing data sets used in the Chapter 2 exercises). Describe your search, test experiments and results of your evaluation and selection.

3.2 Select a gene expression data set with more than four classes. Design a data-driven multi-stage classification schema.

a) Download the selected gene expression data. Perform quality control of the data and any additional preprocessing deemed necessary. Filter out variables whose expression measurements are not reliable or represent experimental noise, preferably using filters based on detection calls and on the range of expression values.

b) From the software packages evaluated in Exercise 3.1, arrange a working environment that includes feature selection and building ensembles of classifiers (preferably combined into one method).

c) Start with building models that try to separate all classes simultaneously. Prepare the training data (in the form of a gene expression matrix) that consists of as many separate classes as represented in the data set.

d) Build a small ensemble of ten to twenty classifiers based on randomized training sets. Use bagging or similar approach to generate bootstrap training sets (e.g., consider *bagging* for nonparametric learning algorithms, and *modified bagging* for parametric ones). Each classifier should be based on a small multivariate biomarker (no more than ten genes) identified for the classifier by an independent feature selection process.

e) Investigate discriminatory spaces of the generated classifiers looking at their low-dimensional projections representing most of the discriminatory information. A common situation in a multiclass discrimination is that one or two of the classes dominate the class separation, that is, they are clearly separated from the remaining classes that overlap. Identify such a class or classes.

- If only one class is well separated from the overlapping rest of classes, consider a two-class model that differentiates this dominating class against all the remaining training samples combined into a single class.
- If two classes are well separated from the rest, consider a three-class model.

Note:

If your software selection does not offer visualization of the discriminatory space, try to identify the dominating class by interpreting confusion matrices of the classifiers.

f) Estimate the misclassification error rate of the considered model (differentiating either two or three classes). Build a large ensemble consisting of at least 500 classifiers. As before, the classifiers need to be based on bootstrap training sets, and separate feature selection has to be performed for each of them.

g) If the evaluation of the model considered at step (**f**) indicates that this configuration of classes may lead to a classifier with reasonable generalization, end the current stage of the multiclass schema design. Otherwise it is possible that the data set cannot be used for efficient discrimination of the represented classes.

h) At the next stage of the multiclass differentiation, include only the overlapping classes. Remove the dominating class (or classes) from the training data. Using this smaller training data set, repeat steps (**d**) to (**g**) and identify configuration of classes for the next stage of classification. Continue this process until all stages of the multistage classification schema are identified.

3.3 Select one stage of the multistage classification schema designed in Exercise 3.2. Build a preliminary classification model for the groups of samples differentiated at this stage. We call it a preliminary model because we do not utilize here optimization approaches described in Chapter 4.

a) Use the training data set including only groups of samples discriminated at the selected stage. Depending on the class configuration identified for this stage, the samples are grouped either into two or three classes. If necessary, perform additional filtering of noise by criteria specific for the differentiated classes.

b) Perform feature selection utilizing a multivariate supervised approach; for example, a heuristic stepwise search or recursive feature elimination. Identify a parsimonious multivariate biomarker consisting of no more than 10 genes.

c) Build a classifier based on the identified biomarker.

d) Test the classifier using the following methods:

- reclassification,
- internal cross-validation,
- external cross-validation,
- ensemble-based validation,
- if possible, find an independent data set and use it for validation.

e) Compare results of the tests performed at step (d). Based on these results, explain why some of the methods do not provide reliable estimates of the classifier's generalization abilities.

3.4 Repeat all steps of Exercise 3.3 using a different learning algorithm and, preferably, a different feature selection method. It would be best if one of the learning algorithms is parametric and the other nonparametric. Compare results and discuss any significant differences.

3.5 Perform experiments using a training set that combines two data sets.

a) Identify three gene expression data sets differentiating the same set of classes. It is preferable to work with data sets generated in different labs, and maybe in different countries.

b) Set aside one of them as the test set.

c) Use one of the remaining two data sets as a training set. Perform all steps necessary to build a classifier based on a parsimonious multivariate biomarker.

d) Validate the classifier using the test data set.

e) Combine the two data sets into one training set. Use it to build another classifier based on a multivariate biomarker identified from the combined training set.

f) Validate the new classifier using the same test set.

g) Compare and discuss the results of testing both classifiers on the same test set.

CHAPTER 4

THE INFORMATIVE SET OF GENES

4.1 INTRODUCTION

In this chapter, a multivariate method for the identification of the *Informative Set of Genes* will be presented. The term "informative set of genes" is often used without a clear definition and with the implied meaning of a set of genes containing some information relevant for a study. We will define the *Informative Set of Genes* as a set containing all of the information significant for the differentiation of two or more classes represented in a training data set. Various parametric or nonparametric supervised learning algorithms can be used to identify parsimonious multivariate biomarkers. While such biomarkers may contain information sufficient for the efficient classification of new cases, they do not necessarily provide insight into biological processes underlying class differentiation. By our definition, the *Informative Set of Genes* contains *all* significant discriminatory information. As such, it should constitute a much better starting point for the elucidation of biological processes underlying class differences or for the extraction of new knowledge about processes associated with the class differences.

Many studies limit their search for biomarkers and for their biological interpretation to sets of genes identified via univariate or univariately-biased methods. Such methods are missing genes whose discriminatory importance can be identified only by a multivariate approach. In the presented method, the *Informative Set of Genes* can be seen as an expansion of a parsimonious multivariate biomarker. Both of them—the biomarker and the informative set—are identified by a heuristic and multivariate approach to feature selection. Biological processes (or genes) that individually are not crucial for class differences may be very important when considered together with other biological processes (or other genes). Multivariate approaches can discover such relations.

While the *Informative Set of Genes* is identified primarily to facilitate the biological interpretation of class differences, we will also show how using it in combination with ensembles of classifiers generated via the *modified bagging* schema may lead to the discovery of multivariate biomarkers that are more robust than biomarkers resulting from feature selection performed on the entire training data set.

Data Mining for Genomics and Proteomics. By Darius M. Dziuda

4.2 DEFINITIONS

In Chapter 3, we defined a *multivariate biomarker* as a set of genes whose joint expression pattern is predictive of class membership, and an *optimal multivariate biomarker* as a parsimonious multivariate biomarker that provides the best compromise between overfitting and generalization. Now, we will add a definition of the *Informative Set of Genes* and definitions of alternative biomarkers that will be used for its identification.

Informative Set of Genes—a set of genes containing *all* of the information significant for class differentiation.[1] Since such a set is usually considerably larger than the optimal multivariate biomarker, it provides better foundation for the elucidation of the biological processes associated with class differences.

Alternative Multivariate Biomarker—a set of genes that has satisfactory discriminatory power, but does not contain any gene included in the previously identified optimal multivariate biomarker.

Secondary Alternative Biomarker—a set of genes that has satisfactory discriminatory power, but does not contain any gene included in the optimal biomarker and the previously identified alternative marker. Extending this definition, we can define a tertiary and then subsequent alternative markers.

Sequence of Alternative Biomarkers[2]—a series of alternative markers identified via the same multivariate feature selection method when each successively identified alternative marker is based on a training data set that does not include genes that were already selected into the optimal multivariate biomarker or into the previously identified alternative markers.

Alternative Models—classification models based on alternative markers. Each of them is built to estimate the amount of discriminatory information contained in the training set used for the identification of a particular alternative marker.

4.3 THE METHOD

To identify the *Informative Set of Genes*, we build a sequence of alternative multivariate biomarkers (Dziuda 2007; Dziuda and Zhou 2007; Dziuda 2008). Although various supervised methods can be considered for this purpose, we recommend using only such methods that are able to: (i) identify parsimonious multivariate biomarkers; and (ii) provide a measure of discriminatory power of each alternative biomarker. An example of such methods is the stepwise hybrid feature selection driven by the T^2 measure of class separation (described in Chapter 3).

[1]More precisely, all such information contained in the training data set.

[2]All of the markers we discuss are multivariate. Hence, whenever the term *marker* is used without qualification, it means *multivariate marker*. Furthermore, we use the terms *marker* and *biomarker* interchangeably and treat them here as synonymous.

The process leading to the *Informative Set of Genes* consists of the following steps.

- The identification of an optimal multivariate biomarker.
- Estimating generalization abilities of this optimal biomarker.
- Generating a sequence of alternative multivariate biomarkers.
- Selecting the *Informative Set of Genes*.
- The verification of the *Informative Set of Genes*.

We assume that the *Informative Set of Genes* represents gene expression patterns important for the discrimination of phenotypic classes represented in the training data set. To identify these patterns and the genes that best represent them, we perform further analysis of the informative set:

- identification of the primary expression patterns of the *Informative Set of Genes*,
- identification of the most important genes of the primary patterns.

4.3.1 Identification of the *Informative Set of Genes*

Identification of an Optimal Multivariate Biomarker

Since the *Informative Set of Genes* can be seen as an extension of an optimal multivariate biomarker,[3] we start with the identification of the optimal biomarker. We use the entire training data set—the gene expression matrix with all the genes that are represented there after filtering unreliable variables and noise. Using stepwise hybrid feature selection mentioned earlier, we identify the optimal multivariate biomarker[4]—a small set of genes with satisfactory discriminatory power.

Estimating Generalization Abilities of the Optimal Biomarker

Before generating a sequence of alternative markers, we estimate the generalization abilities of a classifier built on our optimal biomarker. For this purpose, we can use either an independent test data set or the *modified bagging* approach. Nevertheless, even if we have the independent test set available, it may be advantageous to set it aside (we will discuss its better use at a later stage) and proceed, at this stage, with the *modified bagging* approach to validation of the optimal biomarker.

As described in Chapter 3, to apply the *modified bagging* schema, we generate hundreds or thousands of classifiers (an ensemble) built on randomized training sets. Each classifier[5] is based on its own training set consisting of a specified proportion of biological samples randomly selected—without replacement—from the original training set. The samples that are not used to train a classifier constitute its out-of-bag (OOB) samples. For each classifier, a separate feature selection is performed and a multivariate

[3]In a sense that the *Informative Set of Genes* is used to facilitate identification of biological processes represented by the optimal biomarker.

[4]This marker represents an optimal compromise between overfitting and generalization—it is small enough to minimize chances of overfitting, but is large enough to allow accurate classification of new samples.

[5]If we use the T^2-based feature selection, then linear discriminant analysis (LDA) will be used to design classifiers (see Chapter 3).

marker of a particular size (the same as the size of the optimal biomarker) is identified. An additional level of randomness may be added to the feature selection process (for example, we may start each heuristic search from a randomly selected variable). Once the classifiers are built, each of them is used to classify its own OOB samples. For example, if each of the randomly generated training sets consists of 80 percent of all training samples, each classifier will be tested on the remaining 20 percent of samples that were not used in training the classifier. By averaging test results over a large number of classifiers, we are provided with a reliable estimate of the misclassification error rate of an average classifier of the ensemble. We also use it as an estimate of generalization abilities of our optimal multivariate biomarker. Our experience indicates that such an estimate is quite stable when based on an ensemble of 1,000 or more classifiers.

When—and only when—this estimate indicates that our optimal biomarker may be able to provide accurate classification of new samples (above some predefined level of sensitivity, specificity, and the overall rate of correct classification), we will proceed with the identification of the *Informative Set of Genes*. Otherwise, it is very likely that our training data set does not contain enough discriminatory information to be used as a base for building efficient classifiers.

Generating a Sequence of Alternative Multivariate Biomarkers

While the optimal multivariate biomarker may contain information sufficient for efficient classification of new cases, it does not necessarily provide insight into the biological processes underlying class differences. The *Informative Set of Genes*, as we defined it, contains *all* of the information significant for class differentiation. Analysis of the *Informative Set of Genes* should allow for the elucidation of biological processes associated with class differences and may result in new biomedical knowledge.

To find the *Informative Set of Genes*, we start by generating a sequence of alternative multivariate markers. After the optimal biomarker is identified, its genes are removed from the training data set and feature selection is performed again to identify the alternative multivariate marker. This process is then repeated—the genes of the alternative marker are also removed from the training set and a subsequent alternative marker is identified. The process continues until one of its stopping criteria is satisfied. For example, when we use the T^2-driven heuristic feature selection, the stopping criterion may be defined by two adjustable parameters:

- *min_*T^2—the minimum required discriminatory power of an alternative multivariate marker. This parameter should be set to a T^2 value clearly indicating that discriminatory information of the training data is exhausted. The process of generating alternative markers ends when the T^2 measure of discriminatory power of subsequently identified alternative markers drops below *min_*T^2.
- *max_ALT*—the maximum number of identified alternative multivariate markers.

Since the removal of variables selected into subsequently identified alternative markers decreases the amount of discriminatory information remaining in the training set, we expect that the discriminatory power of successively identified alternative markers will also have decreasing tendency. To facilitate the comparison of discriminatory power of subsequent alternative markers and to detect the exhaustion of the discriminatory information of the training set, all of the identified alternative markers should be of the same size. This size may be determined by evaluating the discriminatory space of a classifier built on the optimal multivariate biomarker, and by probing the discriminatory power of multivariate markers of different cardinalities. Usually, when differentiated classes seem to be quite homogeneous and not too difficult to separate by a small optimal marker providing relatively large discriminatory power, the size of alternative markers may be the same as the size of the optimal marker. Otherwise, this size may be slightly greater than the size of the optimal marker.[6] In Chapter 3, we stated that the preferable size of a truly multivariate biomarker should not exceed ten variables. The same holds true for alternative markers. We do not recommend alternative markers that include more than 10 variables. Such markers would be likely to overfit their training data and include variables that are less likely to be important for discrimination and more likely to be selected by fitting noise.

> ***Note:***
>
> If no alternative multivariate biomarker of reasonable size and reasonable discriminatory power can be identified, then it is very likely that the training data set does not include discriminatory information sufficient for the efficient discrimination of the represented classes. Even if we have already identified a single (and seemingly promising) optimal multivariate biomarker, the absence of alternative markers suggests that this marker is a result of overfitting and does not represent biological processes associated with class differences.

Selecting the Informative Set of Genes

To decide which alternative markers should be included in the *Informative Set of Genes*, we need to select a cut-off point at which significant discriminatory information of the training set is considered exhausted. Although parametric or nonparametric methods can be used for identification of the *Informative Set of Genes*, we focus on the parametric method using the T^2 measure of discriminatory power.[7] When using this method, the T^2 level is an important but not the single factor in deciding about the cut-off point. Please recall from Chapter 3 that the T^2 measure is based on the parametric assumptions of linear discriminant analysis (such as the multivariate normal distribution of variables and homogeneity of variance–covariance matrices). Although LDA is quite robust to violations of these assumptions, known or suspected departures from the assumptions should be taken into account. Visual examination of the distributions of training samples in relation to the LDA-defined distributions of the differentiated classes should provide us with additional information necessary for making the cut-off decision.[8]

When we plot discriminatory power of subsequent alternative markers, it should exhibit decreasing tendency that is often well approximated by a logarithmic trend line (Fig. 4.1). By examining the discriminatory spaces of alternative classification models

[6]For illustration of both cases, see Chapter 6.

[7]This method is illustrated by the exercises presented in Chapter 6.

[8]Among other factors to consider are:

- information about how well the training data represents the underlying populations (this usually depends on the size of the training set),
- information about potential heterogeneity of classes (for example, when a single class includes distinct subpopulations—this is common in some multiclass discrimination schemas).

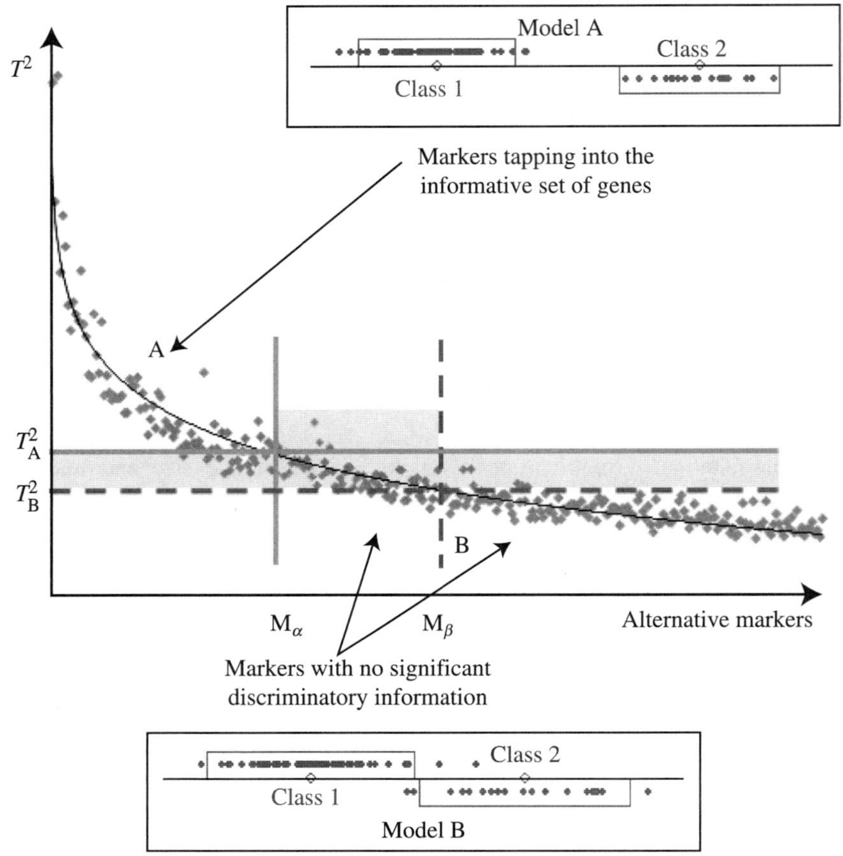

Figure 4.1: Selecting the *Informative Set of Genes*. The scatter plot points represent discriminatory power of the subsequently identified alternative multivariate markers (the plotted data represent the example discussed in Section 6.3 of Chapter 6). Usually, a strong decreasing tendency of this discriminatory power can be well approximated by a logarithmic trend line (a power function may also be tried). Discriminatory spaces and T^2 values of alternative classification models built on the alternative markers represented by the points in area A (above the T^2_A horizontal line and to the left of the vertical line crossing the trend line at T^2_A) indicate good class separation. An example of the discriminatory space and distribution of the training samples for an average model in area A is shown at the top of the figure (Model A). Classification models based on markers from area B (below the T^2_B horizontal line) cannot satisfactorily separate even the training samples. An example of the discriminatory space and distribution of training samples for an average model in area B is shown at the bottom of the figure (Model B). The trend line crosses the T^2_A level of discriminatory power in the vicinity of alternative marker M_α; it crosses T^2_B in the vicinity of alternative marker M_β. Models built on the alternative markers represented by the points in the gray areas (between T^2_A and T^2_B horizontal lines, and above T^2_B and between the vertical lines representing alternative markers M_α and M_β) may have some border line class separation abilities. Since only two classes are differentiated in this example, the discriminatory spaces of Models A and B are one-dimensional. The boxes and vertical offsets of points are used for emphasis only. The boxes represent sections (of the discriminatory dimension) that enclose 95 percent of the probability in each class. The points represent samples from the training data set. (See color insert.)

with different levels of discriminatory power we decide on the T^2 cut-off value below which the alternative models do not provide good separation of the classes.[9]

Figure 4.1 shows an example of a discriminatory power plot with the marked areas representing different amounts of discriminatory information. Examination of alternative models (their discriminatory spaces and distributions of the points representing reclassification of training samples) based on alternative markers with different levels of discriminatory power T^2 allows for the identification of two levels of T^2: T^2_{A} and T^2_{B}.

Alternative markers with discriminatory power $T^2 \geq T^2_{\mathrm{A}}$ are deemed to represent significant discriminatory information. Alternative markers with $T^2 < T^2_{\mathrm{B}}$ (area B) carry no such information. Markers M_0 to M_α, identified before the trend line crosses the T^2_{A} level, seem to tap into significant discriminatory information still present in the subsequently decreasing set of variables of the training data.[10] However, some of them may be trapped in particularly inefficient local optima, especially when the trend line approaches the T^2_{A} level. Conversely, some markers identified after M_α may find local optima with $T^2 > T^2_{\mathrm{A}}$. There may be arguments for and against the inclusion of these two types of markers in the *Informative Set of Genes*. We recommend to start by defining the *Informative Set of Genes* as the set of genes included in the alternative markers represented by the points in area A (i.e., included in these of M_0 to M_α markers whose discriminatory power is not less than T^2_{A}). We can verify our decision by estimating residual discriminatory information left in the training set after all markers of area A are excluded.

Verification of the Informative Set of Genes

To verify our selection of the *Informative Set of Genes*, we may compare estimated sensitivity and specificity of average classifiers of the three ensembles:

1. The ensemble built on the entire training set (the same ensemble that was used for validation of the optimal multivariate biomarker).
2. The ensemble built on the *Informative Set of Genes*.
3. The ensemble built on the training set without variables included in the *Informative Set of Genes*.

In each case, the *modified bagging* schema may be used to build a large number (hundreds or thousands) of classifiers trained on a particular proportion of biological samples randomly selected from the original training set. Independent feature selection is performed on each bootstrap training set and each classifier is built on its own multivariate biomarker of the same size as the size of the previously identified optimal biomarker. For each of the three ensembles, the average sensitivity and specificity of its classifiers is estimated via classification of the OOB samples. The classifiers built on multivariate markers selected from the *Informative Set of Genes* are expected to provide similar results as the classifiers built on markers selected from the entire

[9]Most often, this cut-off level is in the range between $T^2 = 3$ and $T^2 = 5$.

[10]M_0 denotes the optimal multivariate biomarker; the first alternative marker is denoted M_1, the second M_2, etc.

training set (the results may even be better since the multivariate markers selected from the *Informative Set of Genes* should be less likely to fit noise). The sensitivity or specificity of the classifiers built on markers selected from the training set that does not include the *Informative Set of Genes* is expected to be significantly lower and too low for efficient classification. For example, in the exercise described in Section 6.3 of Chapter 6, the average sensitivity drops more than 20 percent (from 98.2 percent for the ensemble using only the *Informative Set of Genes* to 77.1 percent for the ensemble using all variables but the ones included in the *Informative Set of Genes*).

4.3.2 Primary Expression Patterns of the *Informative Set of Genes*

Usually, the optimal multivariate biomarker consists of up to ten genes (see Chapter 3) whereas the informative set—of hundreds of them. Since our *Informative Set of Genes* is identified in a multivariate way, it has a much better chance of pointing to all biological processes significant for differences between classes than sets identified via univariate or univariately-biased approaches. The biological processes, which by themselves—in isolation from other processes—are not crucial for the class differences, cannot be identified by univariate methods. However, such processes may be very important when considered together with other biological processes. Multivariate approaches are capable of extracting such relations.

We assume that the *Informative Set of Genes* includes all gene expression patterns associated with biological processes important for the differentiation of classes represented in the training data. To facilitate biological interpretation of class differences, we will identify these patterns. Various clustering methods may be used to group genes with similar expression patterns. Please note that the unsupervised approach is used here *after* the supervised analysis is performed, when we already know that the clustered genes of the *Informative Set of Genes* are associated with differences between the classes of interest.[11]

Clustering Genes of the Informative Set of Genes

Self-organizing maps (SOM) and hierarchical clustering are methods frequently used for grouping genes by their expression patterns. SOM clustering (see Chapter 2) is usually more informative since it preserves relations between groups of genes—expression patterns of neighboring clusters are more similar to each other than to patterns of clusters that are farther away on the grid. Like for other clustering methods, results of SOM clustering depend on the distance measure and other parameters of the clustering process (such as the grid topology, the learning rate, or the definition of a local neighborhood). To group genes with similar shapes of their expression patterns across samples (rather than grouping genes with similar expression levels), we may use the correlation distance. As for the topology of the self-organizing map and the number of output layer's neurons, we may use a rectangular grid with the number of neurons corresponding to the average cluster size of 20–25 genes.

Sizes of the identified clusters may vary significantly. The size of alternative markers has been selected in a way that provides a good compromise between overfitting and

[11]We do need to guess whether a gene correlated with a gene associated with class differences is also important for class separation—all genes of the *Informative Set of Genes* were deemed as such.

TABLE 4.1: ***Primary Expression Patterns*** **Identified by Calculating the Use of Clusters and Their Genes in Ensembles of Classifiers Built with the** ***Modified Bagging*** **Schema**

Cluster use in ensemble of 1000 classifiers built on training set with 8950 variables				Cluster use in ensemble of 1000 classifiers built on 355 variables of the informative set				Cluster use in 641 perfect OOB classifiers of the ensemble built on training set with 8950 variables				Cluster use in 778 perfect OOB classifiers of the ensemble built on 355 variables of the informative set			
Cluster	Size	Cluster Use	Average Use	Cluster	Size	Cluster Use	Average Use	Cluster	Size	Cluster Use	Average Use	Cluster	Size	Cluster Use	Average Use
12	10	767	76.70	12	10	815	81.50	12	10	569	56.90	12	10	694	69.40
13	25	548	21.92	13	25	608	24.32	13	25	408	16.32	13	25	497	19.88
1	55	997	18.13	1	55	1193	21.69	15	16	181	11.31	1	55	813	14.78
15	16	246	15.38	15	16	255	15.94	1	55	573	10.42	15	16	235	14.69
11	12	109	9.08	11	12	181	15.08	11	12	67	5.58	3	4	54	13.50
14	24	144	6.00	3	4	51	12.75	14	24	116	4.83	11	12	140	11.67
5	38	213	5.61	5	38	447	11.76	5	38	162	4.26	14	24	238	9.92
2	35	169	4.83	14	24	265	11.04	2	35	140	4.00	5	38	346	9.11
6	24	114	4.75	16	40	440	11.00	16	40	142	3.55	6	24	202	8.42
16	40	177	4.43	6	24	244	10.17	6	24	80	3.33	16	40	304	7.60
8	23	74	3.22	2	35	280	8.00	3	4	8	2.00	2	35	245	7.00
3	4	12	3.00	8	23	97	4.22	8	23	33	1.43	7	11	29	2.64
7	11	23	2.09	7	11	45	4.09	9	4	4	1.00	8	23	56	2.43
4	28	28	1.00	9	4	15	3.75	7	11	11	1.00	9	4	8	2.00
10	6	4	0.67	4	28	54	1.93	10	6	4	0.67	4	28	23	0.82
9	4	1	0.25	10	6	9	1.50	4	28	10	0.36	10	6	4	0.67

In this example (see Section 6.3 of Chapter 6), the *Informative Set of Genes* consists of 355 genes. One ensemble of 1,000 classifiers was built on the training set including only these 355 genes. Another ensemble of 1000 classifiers was built on the training set including all 8,950 variables (remaining in the analysis after filtering noise and unreliable probe sets). SOM algorithm with a 4×4 rectangular grid and the correlation distance measure was used to analyze the *Informative Set of Genes* and identify sixteen gene expression clusters. The same four clusters have the highest average use of their genes in both ensembles, whether the calculations are performed for all classifiers or for the perfect OOB classifiers only. These four clusters are called the *primary clusters* and we hypothesize that they represent the *primary expression patterns* associated with the most important biological processes underlying class differences.

generalization. Furthermore, the alternative markers selected into the *Informative Set of Genes* have been identified when their training data sets still included a significant amount of discriminatory information. Nevertheless, it is possible that some genes included in some of these alternative markers were selected due to fitting noise. However, we may assume that it is quite unlikely that the same noise-fitting pattern was selected into many alternative markers. This suggests that clusters with one or very few genes may represent expression patterns selected into alternative markers by chance.

Using Ensembles of Classifiers to Identify Primary Gene Expression Patterns

To identify primary expression patterns (the patterns that are associated with the most important biological processes underlying class differences), we examine the distribution of each cluster's genes among a large number of classifiers built with the *modified bagging* schema. We use two of the three ensembles of classifiers generated and used for the verification of the *Informative Set of Genes*—one built on the training set including all variables, the other built on the variables of the *Informative Set of Genes* only.

For each cluster and ensemble combination, we calculate two parameters:

Cluster Use—a number of times the cluster genes are selected into the ensemble's classifiers.

Average Use of Cluster Genes—*Cluster Use* divided by the number of genes in the cluster. This represents the average number of times a gene from the cluster is selected into the ensemble's classifiers.

We can calculate these parameters either using all classifiers of an ensemble or only those that provide accurate or perfect classification of their respective OOB samples. Let us define the latter.

Perfect OOB Classifier—a classifier that correctly classifies all of its OOB samples.

Accurate OOB Classifier—a classifier that correctly classifies at least a particular proportion of its OOB samples (e.g., at least 90 percent of its OOB samples).

When classes can be efficiently separated and the OOB-based estimates indicate high sensitivity and high specificity of average classifiers of the two ensembles, we may expect that either of these approaches would identify the same set of clusters that are most often used by the classifiers of the two ensembles.[12] These primary clusters represent the primary expression patterns.

Primary Clusters—clusters (of the *Informative Set of Genes*) with the highest average use of their genes by classifiers of the two ensembles—one ensemble built on the entire training data set, the other built on the training set including only genes from the *Informative Set of Genes*.

[12] If there are discrepancies, priority is given to the classifiers built on the *Informative Set of Genes* since the ones built on the training set including all variables may be more prone to fitting noise.

Primary Expression Patterns—gene expression patterns represented by the primary clusters.

For example, in the exercise sketched in Section 6.3 of Chapter 6, the same four clusters are at the top of all four lists of clusters sorted in descending order of the average use of cluster genes. The four lists represent the four combinations of the two ensembles and two ways of calculating cluster use—one including all classifiers, the other including only the perfect OOB classifiers (see Table 4.1).

Although all clusters (with the possible exception of ones that include only very few genes that are rarely selected into ensemble classifiers) of the *Informative Set of Genes* may be used for biological interpretation, priority should be given to the *primary clusters* since they are most likely associated with biological processes that are most important for class differentiation.

4.3.3 The Most Frequently Used Genes of the Primary Expression Patterns

The genes included in the *primary clusters* (that represent the *primary expression patterns* of the *Informative Set of Genes*) are not necessarily equally important for classification and for elucidation of biological processes associated with class differences. We make the assumption that the genes that are most often selected into the perfect OOB classifiers (which are built from training sets consisting of randomly selected subsets of all training samples) are most important for class discrimination. To identify such genes, we analyze the distributions of the primary cluster genes among the perfect OOB classifiers of the same two ensembles built on the *Informative Set of Genes* and on all variables. We introduce two additional definitions:

Frequent Primary Genes—the genes of the *primary clusters* that are selected into at least a particular proportion (for example, one percent) of perfect OOB classifiers of the ensemble built on the training set including only genes of the *Informative Set of Genes*.

Most Frequent Primary Genes—the genes of the *primary clusters* that are selected into at least a particular proportion of the perfect OOB classifiers of each of the two ensembles (one built on all variables, the other on the *Informative Set of Genes*).

The *frequent primary* genes or the *most frequent primary* genes may be used to represent the primary expression patterns if we prefer to focus our biological interpretation on a smaller number of genes than all the genes included in the primary clusters. These genes may also be important for identification of robust multivariate biomarkers.

4.4 USING THE *INFORMATIVE SET OF GENES* TO IDENTIFY ROBUST MULTIVARIATE BIOMARKERS

Although our initial reason for identification of the *Informative Set of Genes* was to facilitate elucidation of biological processes associated with class differences, we can also use the *Informative Set of Genes*, its primary expression patterns and its frequent primary genes for regularization of feature selection that may lead to identification of a more robust biomarker.

One of prerequisites of building the sequence of alternative biomarkers was the identification of a parsimonious multivariate biomarker[13] and its positive validation (for example, by estimating sensitivity and specificity of an average multivariate biomarker of the same size by classification of OOB samples of a large number of classifiers built from randomized training sets). Now, we can limit our heuristic search to the *frequent primary* genes. Since they represent the *primary expression patterns* associated with class separation and are frequently used by the perfect OOB classifiers, we may expect that their optimal combination will provide not only a more robust multivariate biomarker, but also one with a plausible biological interpretation.[14] If we have an independent test data set and have not used it yet, we should use it now to validate our final multivariate biomarker.

4.5 SUMMARY

We described a method for the identification of the *Informative Set of Genes*, which we defined as a set of genes containing all of the discriminatory information significant for differentiation of phenotypic classes represented in the training data set. We also demonstrated the utilization of the ensemble paradigm in validation of multivariate biomarkers, in verification of the *Informative Set of Genes*, and in identification of more robust biomarkers.

The search for the *Informative Set of Genes* is performed after we have evidence that the training data support identification of parsimonious and generalizable multivariate biomarkers. This is done by the identification and validation of a multivariate biomarker that represents optimal compromise between overfitting and generalization (i.e., is small enough to minimize the danger of overfitting, but large enough to allow for the accurate classification of new samples). This optimal multivariate biomarker is a small set of genes whose joint expression pattern provides satisfactory class separation. We estimate the generalization abilities of the optimal biomarker by classifying the OOB samples of a large number of classifiers built using the *modified bagging* schema.

While it is possible that this optimal biomarker can correctly classify new cases,[15] a small set of genes included in this marker does not necessarily provide

[13]Recall that we called this marker an "optimal" multivariate biomarker since it was the best compromise between overfitting and generalization we were able to achieve before the identification of the *Informative Set of Genes*. Now, using a set of the *frequent primary* genes, we may be able to find a more robust biomarker.

[14]Different algorithms may result in different sets of genes deemed as an optimal multivariate biomarker. However, those seemingly different results may represent a stable solution to the classification problem, as long as these different sets of genes represent the same set of biological processes.

[15]Our OOB-based validation gives us information about the classification efficiency of an average biomarker of the size of the optimal biomarker. Although it can be used as an indication of accuracy of the optimal biomarker, we will not be sure about the actual performance of this particular biomarker until we test it on an independent test set.

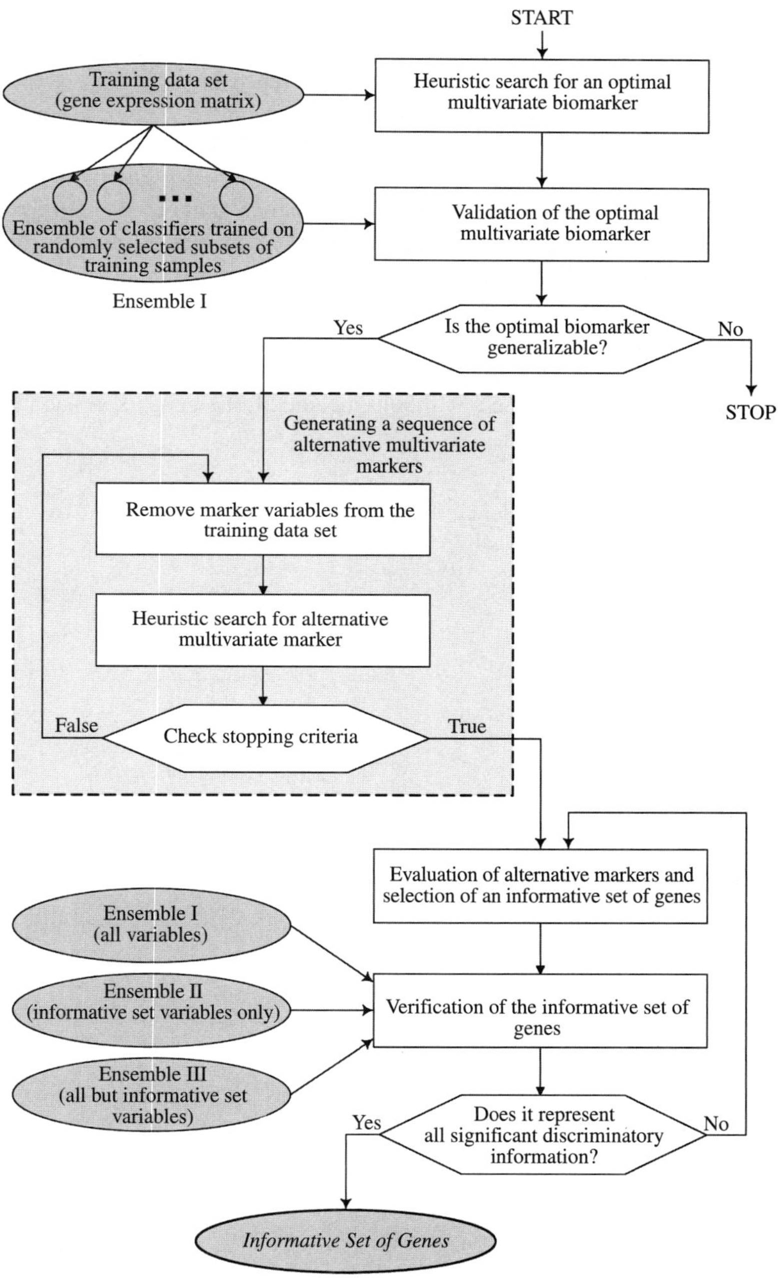

Figure 4.2: The key steps in the identification of the *Informative Set of Genes*. First, the optimal multivariate biomarker is identified and validated. If it can be well generalized, a sequence of alternative multivariate markers is generated. The discriminatory power of these alternative markers and the discriminatory spaces of (based on them) alternative classification models are analyzed and a candidate informative set of genes is selected. Then, using ensembles of classifiers built on randomized training sets, we verify whether the informative set of genes includes all significant discriminatory information.

insight into biological processes underlying class differences. The primary reason for identifying a larger *Informative Set of Genes* is to facilitate biological interpretation of gene expression patterns associated with class differences.

To identify the *Informative Set of Genes* we generate a sequence of alternative multivariate markers. Each of them is a result of the heuristic feature selection process performed on training data from which all the variables selected into the optimal biomarker and into all previously identified alternative markers are removed (see Fig. 4.2). The *Informative Set of Genes* is defined as the set of genes included in the alternative markers whose discriminatory power is above a particular level and which are identified before the significant discriminatory information of the training data is deemed exhausted.

By its definition, the *Informative Set of Genes* contains all of the significant discriminatory information. As such, it should include all the gene expression patterns

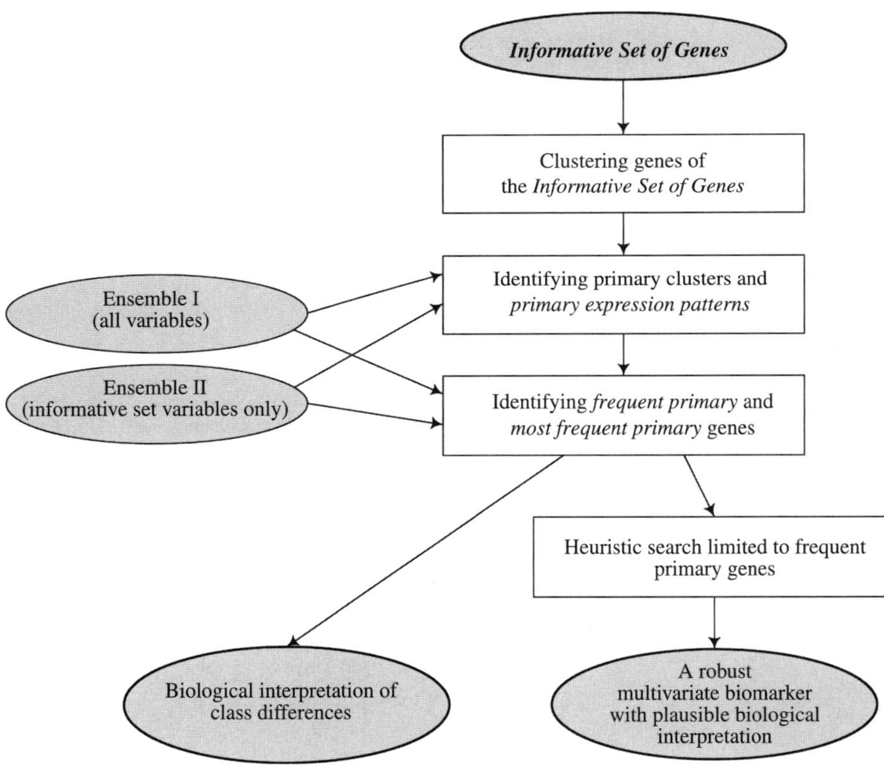

Figure 4.3: The analysis of the *Informative Set of Genes*. First, the genes of the *Informative Set of Genes* are clustered by their expression patterns. Then two ensembles of classifiers built via the *modified bagging* schema are used to identify the primary clusters. These are the clusters whose genes are most frequently used by the ensembles' classifiers. By analyzing the distribution of individual genes of the primary clusters among the perfect OOB classifiers of both ensembles, we identify the *frequent primary* and the *most frequent primary* genes. These are the genes that are most important for biological interpretation of class differences as well as for identification of robust multivariate biomarkers.

associated with biological processes whose combinations may lead to class distinction. To identify these patterns, we perform clustering of the *Informative Set of Genes*. Then, using the ensemble approach and the *modified bagging* schema, we evaluate the importance of each of these patterns. Two ensembles of classifiers are used for this purpose—one built on the training set including all variables, the other built on the training set including only genes of the *Informative Set of Genes*. The more important the pattern is for classification, the more classifiers will include genes from the pattern. By examining the distribution of each cluster's genes among classifiers of the two ensembles, we identify the primary clusters and the *primary expression patterns* (see Fig. 4.3). Although all clusters of the *Informative Set of Genes* may be used for biological interpretation, we hypothesize that the primary expression patterns represent biological processes whose combinations are most important for class differentiation.

After finding the primary expression patterns of the *Informative Set of Genes*, we are looking at the individual genes included in the *primary clusters* representing these patterns. By analyzing the distribution of these genes among the perfect OOB classifiers (of the same two ensembles we used for finding the primary clusters), we identify a set of *frequent primary* genes and its subset of the *most frequent primary* genes. These are the genes that are most often selected into the classifiers that perfectly classify their OOB samples. For biological interpretation of class differences, we should focus first on the frequent primary genes. Whereas a single multivariate biomarker includes only a few genes and while the *Informative Set of Genes* usually consists of a few hundred genes, the set of frequent primary genes may represent the optimal amount of information allowing elucidation of biological processes underlying phenotypic differences between classes. If necessary, biological interpretation may be stratified into four layers that will be based on: the *most frequent primary* genes, the *frequent primary* genes, all genes of the *primary expression patterns*, and eventually on the entire *Informative Set of Genes*.

Finally, we described another important utility of the *Informative Set of Genes*. In addition to its primary function of facilitating biological interpretation of class differences, we can use the *Informative Set of Genes*, its primary expression patterns and its frequent primary genes for identification of potentially more robust biomarkers. For example, if we limit heuristic feature selection to the frequent primary genes, we may expect that their optimal combination will provide a robust multivariate biomarker with a plausible biological interpretation.

EXERCISES

4.1 Select a publicly available gene expression microarray-based data set with two or three classes (diseases, therapy outcomes, phenotypes, biological states, etc.) and at least 50 (preferably more) biological samples. Select a biomedical area that you are interested in. If you do not have preferences, consider cancer- or CNS-related research. You may reuse one of the data sets identified for the Chapter 2 or Chapter 3 exercises, if it fits this description. Perform all steps necessary for preparation of the training data set.

a) If only raw data (CEL files) are available (or if you are not sure how the available probe set level data was preprocessed), perform low-level preprocessing. Use the *Expression Console* software or other low-level preprocessing software of your choice.

b) Having probe set level expression data available, perform quality control of the data and any necessary additional scaling, normalization, or transformation.

c) Filter out variables that are unreliable or represent experimental noise.

d) Prepare the training set in the form of a gene expression matrix.

4.2 Using the training data set prepared in Exercise 4.1, identify the *Informative Set of Genes*.

a) Perform feature selection experiments resulting in multivariate biomarkers of different cardinality (number of variables included in the set). Decide on the optimal size of the multivariate marker—your optimal biomarker should be a small set of genes with satisfactory discriminatory power.

b) Estimate the generalization abilities of the optimal biomarker using the ensemble-based validation. Use the bagging (or *modified bagging*) approach to generate bootstrap training sets, and build an ensemble of at least 500 classifiers. If the results of OOB classification are satisfactory, proceed to step (c). Otherwise consider different cardinality of a biomarker or use a different data set.

c) Generate a sequence of alternative multivariate biomarkers.

d) Investigate discriminatory power of subsequently identified alternative markers and discriminatory spaces of alternative models. Decide on the cut-off defining the *Informative Set of Genes*.

e) Prepare a second training set including only the genes from the *Informative Set of Genes*.

4.3 Verify your selection of the *Informative Set of Genes* using three ensembles of at least 500 classifiers each:

a) The ensemble built on the entire training set.

b) The ensemble built on the *Informative Set of Genes*.

c) The ensemble built on the training set without variables included in the *Informative Set of Genes*.

4.4 Using hierarchical clustering or self-organizing maps, group genes of the *Informative Set of Genes* into clusters of similar gene expression patterns.

4.5 Identify *primary expression patterns* by examining the distribution of each cluster's genes among classifiers of the ensembles (**a**) and (**b**) from Exercise 4.3.

4.6 Using the same two ensembles (as in Exercise 4.5), identify the set of *frequent primary* genes and its subset of the *most frequent primary* genes. To achieve this, examine the distribution of individual genes of the primary clusters among classifiers of the two ensembles.

4.7 If you have biology background, try to provide a biological interpretation of the *most frequent primary* genes. You may extend your interpretation to all *frequent primary* genes, and then to biological interpretation of the primary gene expression patterns.

4.8 In a search for robust multivariate biomarkers with a plausible biological interpretation, perform feature selection limited either to *frequent primary* genes or to the *Informative Set of Genes*.

a) Use more than one method of feature selection to identify more than one parsimonious multivariate biomarker.

b) Use more than one learning algorithm to build classifiers based on the biomarkers identified in step (**a**).

c) Test the classifiers on an independent test set and compare their performance.

CHAPTER 5

ANALYSIS OF PROTEIN EXPRESSION DATA

5.1 INTRODUCTION

The Human Genome Project, rapidly developing technology for more efficient genome-level sequencing, and mature gene expression microarrays are the main factors in the unprecedented change in the way biomedical research is conducted. Statistics, data mining, multivariate analysis, and computer science met to change the face of computational biology and create bioinformatics. Steady advances in sequencing and microarray technologies have been driving quantitative evolution of genomic research. The next, qualitative leap is expected when protein chip technology matures. Large-scale proteomic studies utilizing direct and simultaneous measurement of protein expression at a whole-proteome level will follow.

Proteomics can be defined as the study of proteins, their structures and functions, their post-translational modifications, interactions with other proteins, and their role in the metabolic pathways. The ultimate goal of proteomics is to study all proteins within a cell or tissue. As proteins are direct players in cell physiology, analysis of their expression should yield better understanding of biological processes. High throughput proteomic techniques coupled with multivariate analysis may enable the investigation of disease-related processes at the individual patient level (Gulmann et al. 2006). Multi-protein expression patterns may be used as biomarkers for early diagnosis, personalized therapy selection, evaluation of treatment results, drug discovery, and in other situations where we want to assign a new sample to one of the differentiated classes. To support personalized medicine, a disease may be redefined in a way that will partition the disease into segments according to characteristic molecular profiles. In genomics, the profile is a gene expression pattern. Proteomics, however, is capable of deeper partitioning since a disease with the same clinical characteristics and the same gene expression pattern may still be associated with different protein expression patterns (DePalma 2003).

Due to alternative splicing, post-translational modifications, and interactions between proteins, the human proteome is much larger than the human genome. Current estimates of the size of the human proteome range from several hundred thousand proteins to one million proteins (Pollard et al. 2008). Gene expression analysis, even when performed at the whole-genome level, cannot elucidate processes such as

Data Mining for Genomics and Proteomics. By Darius M. Dziuda

post-translational modification of proteins, or protein–protein interactions. Many cellular functions cannot be appropriately analyzed at the level of gene expression (Maercker 2005).

The main difficulty in protein chip technology is related to the three-dimensional structure of a protein molecule, which is the crucial factor since protein functions are determined by their three-dimensional folding. The genomic microarray technology of printing DNA on a two-dimensional surface cannot be applied to proteins. Proteomic material immobilized on the microarray has to be fully functional and in such a position that its biding sites are accessible to target molecules (Tanaka et al. 2006). Furthermore, whereas in DNA microarrays, immobilized probes and their targets represent complementary strands of DNA that can hybridize, protein microarrays require high-affinity capture reagents (such as antibodies) that will bind target proteins. Yet another challenge in proteomics and proteomic arrays is detection of low-abundance proteins—there are no protein amplification techniques such as polymerase chain reaction (PCR)[1] widely used in genomics.

Many companies are developing protein arrays, but we are yet to see a high-throughput technology that would allow simultaneous and *direct* measurement of expression of hundreds of thousands of proteins, and eventually the entire human proteome.

For researchers involved in genomics since its beginning, the current technologies of proteomics recall those of genomics in the 1990s. Then, before the completion of the draft version of the human genome and before the maturity of genomic microarray technologies, electrophoresis-based studies were the main stream of gene expression analysis. The main advantage of electrophoresis-based experiments was the ability to identify (and quantify) any sequence, whether known or unknown. In contrast to this *unbiased* technology, microarray experiments are *biased*—gene expression can be measured only for the known sequences (corresponding to genes or tentative genes) that are represented by probe sets on the array. The main advantage of microarray technologies is, however, the *direct* measurement of gene expression whereas electrophoretic experiments resulted in intensities of bands representing the abundance of DNA fragments with approximately known lengths and known sequences of few nucleotides on both ends.[2] A many-to-many relationship between the bands and genes (or unknown sequences) required specialized algorithms for 'translation' of the band data into gene expression data; see for example (Bader et al. 2006). As evolving microarrays were covering more and more known sequences, electrophoresis-based genomic studies were vanishing from the gene expression research.

[1]Polymerase chain reaction (PCR) is a widely used method for amplification of the amount of a specific DNA fragment. The reaction is performed in cycles and the amount of target DNA is amplified exponentially (theoretically, it is doubled at each cycle) (Garrett and Grisham 2007).

[2]Restriction enzymes are proteins that cut DNA in specific locations corresponding to known recognition sequences. The resulting DNA fragments are then separated by their length during gel electrophoresis. Because the enzymes cut DNA in known spots, the experimental length of the fragments (bands) could be compared with the results of 'theoretical' digestion of DNA sequences.

Although biased proteomic technologies (such as antibody or protein microarrays) are constantly evolving,[3] they have yet to achieve a high-throughput whole-proteome level of *direct* measurement of relative protein expression. Unbiased and *indirect* proteomic technologies, such as mass spectrometry and two-dimensional gel electrophoresis, are still the *flavor du jour*.

In this chapter we will discuss both approaches. Our main focus—as in Chapter 3—will be multivariate biomarker discovery, which for proteomics means the identification of a small set of proteins whose joint expression pattern can separate the differentiated classes and can be used for efficient classification of new cases.

Multivariate Proteomic Biomarker

A small set of proteins, or variables representing proteins (for instance, a small set of m/z mass spectrometry peaks) whose joint expression pattern can significantly separate the differentiated classes. Such a set can be used to build a classification system capable of accurate prediction of class membership for new samples. Analogically as for multivariate genomic biomarkers, the stress here is on the predictive power of the *set of proteins* rather than on the discriminatory power of *individual* proteins. Similarly as in genomics, a common misconception in early proteomic biomarker discovery studies was the assumption that each protein included in a multi-protein biomarker has to be *individually* correlated with the disease or phenotype.

Before moving on to the first topic (protein chip technologies), we start with the introduction of a few basic terms (Ganten and Ruckpaul 2006; Eidhammer et al. 2007):

Antibody

Antibodies (or immunoglobulins) are proteins used by the immune system to bind with specific antigens of foreign cells (to neutralize bacteria or viruses).

Antigen

Antigens are foreign molecules that trigger *anti*body *gen*eration in the organism.

Peptide

Peptides are short chains of amino acids (not longer than several dozen amino acids). Longer chains of amino acids are called polypeptides.

Polypeptide

A macromolecular chain of amino acids. A polypeptide may be a protein or one of its components.

Protein

A protein is a macromolecule composed of one or more polypeptides. In some texts, each of these single polypeptide chains is called a protein, and assemblies of polypeptides are called protein complexes.

[3]The concept of multi-analyte assays for simultaneous measurement of concentration of hundreds of proteins was introduced about 20 years ago (Ekins 1989).

Enzyme

Enzymes are proteins that act as catalysts of biochemical reactions.

Molecular mass (m)

The mass of a single molecule of a substance. The molecular mass is commonly expressed in Daltons. Dalton (Da) is defined as $1/12$ of the mass of the carbon isotope ^{12}C, and equals approximately $1.661 * 10^{-24}$ g.

Isoelectric point (pI)

The isoelectric point of a protein is defined as the pH value of a solution at which the protein is electrically neutral. pH, or *potential of Hydrogen*, measures the acidity of a solution and is defined as the negative common logarithm of the concentration of H^+ (hydrogen ions) in the solution. Neutral solutions have $pH = 7$ $(-\log_{10} 10^{-7})$. A protein has a net positive charge at pH lower than its isoelectric point, and a net negative charge at pH higher than its isoelectric point.

Peptide-mass fingerprint

A set of molecular masses of the peptides that are generated by enzymatic digestion of a protein. The peptide masses are determined by mass spectrometry. The resulting *peptide-mass fingerprint* is used for protein identification by comparing the fingerprint to the peptide masses calculated for a theoretical (*in silico*) digestion of protein sequences in a database.

5.2 PROTEIN CHIP TECHNOLOGY

The ultimate goal of protein chip technology supporting the analysis of protein expression for biomarker discovery is the development of *whole-proteome* microarrays. Such arrays would enable direct and simultaneous measurement of expression levels of all of the proteins of a proteome as large and complex as the human proteome. Another major goal of protein chip technology is the identification of protein functions on a whole-proteome scale. The *whole-genome* DNA microarrays—although technologically much more mature than proteomic microarrays, and very important for biomedical research—cannot elucidate processes such as post-translational modification of proteins, or protein–protein interactions. Although our focus is analysis of protein expression data, in this section we will describe the currently most popular types of protein chips, whether their main application is protein expression profiling or protein functional analysis.

Protein microarrays can be divided into two categories: *forward-phase* microarrays and *reverse-phase* microarrays. The forward-phase protein microarrays consist of molecules (such as antibodies, proteins, protein fragments, enzymes, or peptides) immobilized in a matrix pattern on small surfaces (coated glass slides, microplates or membranes). These immobilized *capture* molecules, or *detector* molecules, are to react with proteins of the analyzed samples, or *analytes*, in order to determine the presence, and to measure abundance, of target proteins. The most popular among the forward-phase arrays are the antibody microarrays that use protein-specific antibodies as the immobilized capture molecules. The protein microarrays with immobilized peptides or whole proteins that are used as the capture molecules are among other popular forward-phase arrays. In reverse-phase protein microarrays, the analytes (the samples to be analyzed) are immobilized on the array surface. This allows for the simultaneous analysis of small amounts of tissue from many samples.

Depending on the design of protein microarrays, either the analytes or the detection molecules are labeled in a way that allows for their quantification. The common methods of labeling include fluorescence, chemiluminescence, and radioactivity. Fluorescent labeling is currently the most popular one (Korf 2006).

The analysis of protein expression data generated by biased protein chip technologies is similar to the analysis of gene expression data. Once the signal associated with the targeted proteins is measured and the raw data is available, the data has to be preprocessed to represent the concentration of each of the targeted proteins in each of the analyzed biological samples. Although details of the preprocessing depend on a particular microarray technology and the labeling method used in the experiment, the general approach to preprocessing is similar to that previously described for the gene expression microarrays. The measured intensities have to be adjusted for background, quality controlled, normalized across all arrays of the experiment and log transformed (if necessary). After this low-level preprocessing, we have the experimental data in the form of a protein expression matrix, akin to the gene expression matrix described in Chapter 2. The rows of the protein expression matrix represent the targeted proteins (either directly or via the capture or detection molecules associated with the proteins), and the columns represent the analyzed samples. Each data point (a cell in the protein expression matrix) represents the concentration of a protein (row) in a sample (column). Once the data is represented in the form of the protein expression matrix, the same methods of the basic exploratory analysis and the more advanced multivariate supervised or unsupervised analysis can be applied as described in Chapters 2 and 3 in relation to data represented by the gene expression matrix.

5.2.1 Antibody Microarrays

Antibody microarrays are currently most popular among the protein chips. They are protein-detection and profiling arrays. Protein-specific antibodies are immobilized on the solid surface of a chip in such a way that they are biologically active and may interact with proteins or antigens (Tanaka et al. 2006; Astle and Kodadek 2008). Due to similarities in the structure of antibodies against different proteins, these arrays can be manufactured with standardized spotting conditions. Similarly as DNA microarrays are used for gene expression profiling, the antibody microarrays can be used to quantify the concentration of specific known proteins (Maercker 2005). In particular, they may be used for biomarker discovery, that is, to identify a subset of proteins whose joint expression pattern can differentiate between disease states. As the concept of antibody-based protein detection is well known and commonly used in ELISA (enzyme-linked immunosorbent assays), the antibody microarrays can be seen as the multiplexed form of the ELISA technique. The microarrays, however, allow for simultaneous detection of hundreds of protein targets, are more sensitive and require a much smaller amount of biological material (Pang et al. 2005; Korf 2006).

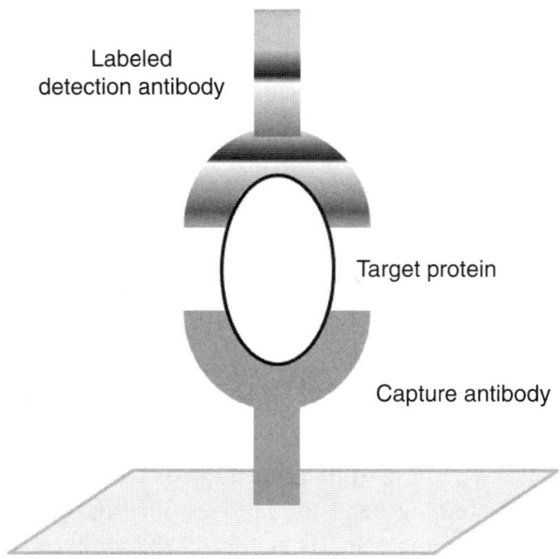

Figure 5.1: Sandwich type of antibody microarray technology.

There are two types of antibody microarrays—*sandwich* technology arrays and *single-antibody* technology arrays (Figs 5.1 and 5.2). In the sandwich type arrays, two protein-specific antibodies are used for each target protein. Capture antibodies are immobilized on the array to bind first with the sample proteins. Then fluorescently labeled protein-specific detection antibodies are added to bind with the captured proteins and allow for their quantification. The requirement that two independent protein-specific antibodies bind to each targeted protein increases the specificity of protein identification.

In the single-antibody arrays, the analyzed samples are themselves labeled allowing for direct detection of proteins bound to the immobilized capture antibodies. This approach allows for identification of proteins for which paired antibodies are not known. However, as all proteins in the analyzed sample are labeled, background noise increases.

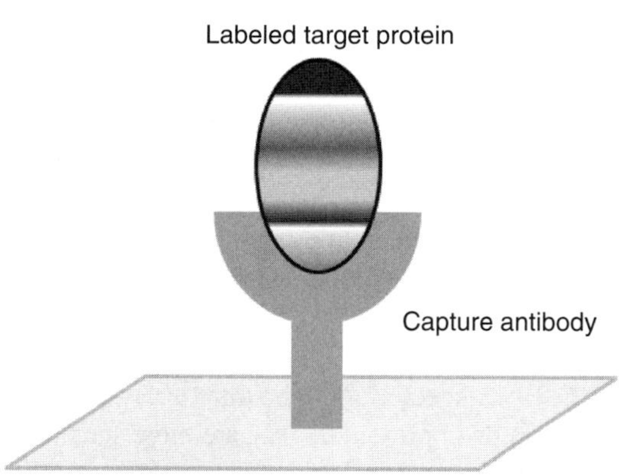

Figure 5.2: Single-antibody type of antibody microarray technology.

5.2.2 Peptide Microarrays

Peptide chips are microarrays consisting of thousands of immobilized peptides playing the role of capture molecules. Since many protein activities are directed towards peptides, the peptide microarrays can be used to study diverse biological or biochemical activities of proteins (Min and Mrksich 2004; Volkmer-Engert 2006). Fluorescent labeling of samples is a typical method of detecting proteins bound to the array peptides. Two main approaches are used to generate peptide arrays. One is the *in situ* synthesis of peptides, and the other is the immobilization, on the array surface, already synthesized peptides.

The *in situ* approach is usually implemented either by utilizing photolithographic synthesis (Fodor et al. 1991) or the SPOT-synthesis technique (Frank 1992). The mask-based photolithographic technology is similar to one used in manufacturing Affymetrix gene expression microarrays. Peptide sequences are synthesized in parallel directly on the microarray surface—amino acid by amino acid—by adding at each step a single amino acid to each of the currently unmasked spots. Since preparation of photomasks is both expensive and time consuming, other photolithographic approaches have been proposed. One of them is the light-activated parallel synthesis of peptides that utilizes digital photolithography and photogenerated reagent chemistry (Pellois et al. 2002; Gao et al. 2004). The SPOT *in situ* technique allows for manual or automated synthesis of peptides by delivering small volumes of activated amino acids to addressable spots of a membrane support (such as pure cellulose chromatography paper). Although the SPOT-synthesis technique does not result in array densities as high as achieved by photolithographic approaches, it is very flexible and does not require any expensive equipment (Frank 2002). Another main approach to preparation of peptide microarrays is to immobilize, on the array, already presynthesized peptides. This method is preferable in situations when we want to use the same set of peptides to prepare a large number of identical microarrays.

5.2.3 Protein Microarrays

Protein microarrays consisting of thousands of immobilized complete proteins can be used as *functional assays* enabling high-throughput detection of protein biochemical activities or protein interactions with proteins and other biomolecules (Ge 2000; Maercker 2005). "*The ultimate form of a functional protein array would consist of all of the proteins encoded by the genome of an organism*" (Schweitzer et al. 2003). To build a whole-proteome array, we need to know the sequence of an organism genome and all of its *open reading frames* (ORFs).[4] Each ORF needs to be cloned and then used in protein production and purification. Such prepared proteins are spotted onto the microarray surface (Kung and Snyder 2006). Although whole-proteome microarrays can already be built for such organisms as yeast (Zhu et al. 2001), due to the much higher complexity of the human proteome we are yet to see functional arrays for the complete human proteome (Korf 2006).

[4] *Open reading frame* is a sequence of nucleotides that could potentially encode a protein or an RNA product. The sequence begins with the start codon (ATG) and ends with one of the three stop codons, TAA, TAG, or TGA (Singleton 2008).

5.2.4 Reverse Phase Microarrays

Protein chips with immobilized capture molecules, such as those described in previous sections—antibody, peptide, and protein arrays—are called *forward-phase* protein microarrays. Another type of protein chips are *reverse-phase* microarrays (Paweletz et al. 2001), in which the samples to be analyzed (rather than the capture molecules) are spotted on the microarray. Then an antibody for a specific protein can be used to assess the abundance of this target protein in the samples. The reverse-phase protein microarrays consist of immobilized tissue or cell *lysates*,[5] thus each microarray spot may represent the whole proteome of a particular cell or tissue.[6] These arrays can be used to study protein modifications and the differences in their expression levels among groups of cells or tissues. This may lead to the identification of proteomic biomarkers specific to diseases or their states. Since samples from hundreds, or even thousands, of patients can be spotted on a single array, this technology allows for screening such groups of samples for the presence, or absence, of specific target proteins. Coupling reverse-phase microarray technology with laser capture microdissection allows for the analysis of samples represented by specific subpopulations of tissue cells (Charboneau et al. 2002; Hamdan and Righetti 2005; Astle and Kodadek 2008). Various applications of the reverse-phase protein microarrays have been recently reported. These applications include the detection of changes in protein expression in response to different treatments, studying dynamics of protein signaling networks, analyzing sensitivity of cell lines to a compound, preclinical biomarker discovery, and the validation of potential drug candidates (Balboni et al. 2006; Glaser 2007; VanMeter et al. 2007; Spurrier et al. 2008).

5.3 TWO-DIMENSIONAL GEL ELECTROPHORESIS

Two-dimensional gel electrophoresis (2D gel electrophoresis) is unbiased technology used in proteomics for the identification and quantification of proteins by separating them in two dimensions. Known since the 1970s (O'Farrell 1975), 2D gel electrophoresis is still the core technology of proteomic studies (Donoghue et al. 2008).

In the first dimension, the proteins are separated by their isoelectric point (charge).[7] A pH gradient (usually, strips of chemical compounds arranged in increasing order of their pH values) is embedded into a polyacrylamide gel[8] and then an electric field is generated by applying an electric potential to the gel across the pH gradient. The sample proteins that have a positive charge will move towards the negative electrode, and those negatively charged towards the positive one. The migration of

[5]A *lysate* is a solution that contains cells whose cellular membrane has been broken. This allows for the analysis of the cell proteins.

[6]Sometimes the cell or tissue lysates are separated into fractions to decrease the number of proteins at each spot.

[7]The separation of analytes by their isoelectric point is called *isoelectric focusing* (Ganten and Ruckpaul 2006).

[8]A polyacrylamide gel is commonly used in 2D gel electrophoresis experiments. Therefore, the technique is often called *two-dimensional polyacrylamide gel electrophoresis*, or 2D-PAGE (Jungblut 2006).

each protein will stop once the protein become electrically neutral. This will happen when the protein enters the pH zone corresponding to its isoelectric point.

In the second dimension, orthogonal to the first one, the sample proteins are separated by their molecular mass. To make the protein progress through the gel dependent on their masses, their natural three-dimensional forms have to be unfolded. This is achieved by denaturing them with sodium dodecyl sulfate (SDS). SDS is not only unfolding proteins into rodlike-shaped particles, but it also makes them negatively charged, with the charge proportional to the protein mass (which means that the net charge per unit mass is approximately the same for most of the proteins). An electric field is generated again, this time in the direction orthogonal to the direction of the electric field used for the separation by charge. The sample proteins will move towards the current position of the positive electrode. Their velocities will depend on the lengths of their unfolded rodlike forms (Hamdan and Righetti 2005; Eidhammer et al. 2007). Since these lengths are proportional to proteins' molecular masses, the velocity of the protein migration will be approximately in reverse proportion to their masses. Smaller proteins will move faster. When the smallest of them reaches the end of the gel, the electric field is turned off and all sample proteins are immobilized at spots whose spatial coordinates depend on their isoelectric points and molecular masses. Since the isoelectric point and the molecular mass are independent protein properties, separation by both of them may result in a quite uniform distribution of proteins across the two-dimensional area of the gel (O'Farrell 1975).

To visualize and detect protein spots in the gel, staining with fluorescent or nonfluorescent dyes or metals (usually silver), or radioactive labeling can be used. However, none of these detection techniques can simultaneously achieve all of the sensitivity, reproducibility, and linearity requirements that would guarantee accurate quantification of all types of proteins. Due to this and other limitations of 2D gel electrophoresis, low abundant proteins, proteins with very small or very large masses, and proteins with very low or very high values of their isoelectric point are often not detected. Furthermore, some spots may contain more than one protein posing an additional challenge for protein quantification. (Beranova-Giorgianni 2003; Jungblut 2006; Eidhammer et al. 2007; Donoghue et al. 2008).

Once the gel is scanned, its digital image is used as the raw data for further analysis. Spots (representing proteins) are identified and their intensities are quantified, adjusted for noise and for the background, and normalized across the gel. Data for each spot include its spatial coordinates, spot size, and its normalized intensity. Since each gel represents a single biological sample in the experiment, the spots have to be matched across all of the gels (usually, standard reference spots are used to align the gels). After preprocessing, the format of the data used for statistical analysis can be similar to the gene expression matrix, with the variables (rows) being here the spots representing different proteins, columns corresponding to samples, and data points representing the spot intensities. The same univariate and multivariate methods (as used for the gene expression data) can be applied to identify differentially expressed variables or multivariate biomarkers.

However, once such spot variables are selected, they have to be traced back to proteins. The spatial characteristics of the spots, which are translated into the approximate mass and charge values, may be used to associate the spots with the proteins they represent. Unfortunately, these characteristics are often insufficient for unique identification of the proteins. Commonly, mass spectrometry of each trypsin-digested protein of interest is subsequently performed to determine the protein *peptide-mass fingerprint*. The fingerprint is then used to search protein sequence databases in order to identify the protein (Henzel et al. 1993; Fenyo and Beavis 2008).

Fluorescent labeling of samples allows for processing of more than one sample on a single gel. Up to three samples can be labeled with different cyanine dyes (Cy2, Cy3, and Cy5), mixed, and have their proteins separated on one gel. When three samples are multiplexed, they usually include two experimental biological samples and one reference sample, which is the internal standard that is run on all of the gels of the experiment. The internal standard sample may be used to normalize spot intensities within and between gels and to improve mapping of spots between gels. This technique, called *two-dimensional difference gel electrophoresis* (DIGE), allows for minimizing experimental variation in experiments aimed at the identification of proteins that are differentially expressed between samples representing two populations (Unlü et al. 1997; Kreil et al. 2004; Ganten and Ruckpaul 2006; Kultima et al. 2006; Donoghue et al. 2008).

Although two-dimensional gel electrophoresis can separate thousands of proteins in a specific tissue, this technique has relatively low throughput, low sensitivity and poor reproducibility. This technology is constantly improved, but eventually it will be replaced by *biased* proteomic technologies, once they achieve the whole-proteome level of direct quantification of protein expression.

5.4 MALDI-TOF AND SELDI-TOF MASS SPECTROMETRY

Mass spectrometry can be used in proteomics as high-throughput technology for profiling and quantification of proteins by measuring the ratio of mass to charge either for whole proteins or for their fragments (peptides). Mass spectrometry is performed in three key steps. First, the analyte (proteins or peptides) is ionized into gas phase ions. Then, the ions are separated according to their mass-to-charge ratio (m/z). At the third step, a detector counts the ions, separately for each m/z value (Vlahou and Fountoulakis 2005; Dubitzky et al. 2007). Due to the very large dynamic range of protein expression (the expression levels may differ by ten orders of magnitude), direct application of mass spectrometry to protein expression analysis of complex biological samples (such as plasma or serum) is limited. Usually, mass spectrometry is performed after the sample proteins are separated by 2D gel electrophoresis or after they are prefractionated with the use of chromatography (Poon 2007; Bertucci and Goncalves 2008).

Recently, two mass spectrometry techniques are among the most commonly used:

- MALDI-TOF—*Matrix-Assisted Laser Desorption and Ionization Time-Of-Flight* mass spectrometry.
- SELDI-TOF—*Surface-Enhanced Laser Desorption and Ionization Time-Of-Flight* mass spectrometry, which can be considered a modified version of MALDI-TOF.

The most significant disadvantage of these mass spectrometry techniques is that the m/z ratios do not provide direct identification of proteins. 'Translation' of spectral peak data into protein data is necessary—again, similarly as in the 1990s for the unbiased genomic methods, specialized algorithms for identification of significant associations between selected spectral peaks (for instance, the m/z peaks selected into a multivariate biomarker) and proteins are necessary.

5.4.1 MALDI-TOF Mass Spectrometry

The MALDI-TOF technique is often used to identify proteins after they are separated by 2D gel electrophoresis. The gel spots of interest (for instance, the spots selected in the result of the statistical analysis of the gel image data) are excised from the gel and digested with an enzyme, usually trypsin.[9] The digestion of the spot protein results in peptides that are subsequently analyzed by MALDI-TOF mass spectrometry. The MALDI-TOF analysis determines the masses of these peptides, that is, the peptide-mass fingerprint of the unknown protein represented by the analyzed spot. The identification of the protein is based on a database search for the match between this experimental peptide-mass fingerprint and the theoretical peptide masses calculated for the proteins included in a sequence database (Beranova-Giorgianni 2003; Hamdan and Righetti 2005; Vlahou and Fountoulakis 2005). One may ask, "Why do we cleave the spot protein into peptides (and then use the peptide-mass fingerprint to identify the protein) rather than determining the mass of the whole protein and match it to the masses calculated from the sequences of known proteins?" There are several reasons. The protein of interest may be a modified version of the protein in the database and its measured mass may not match the mass calculated from the sequence stored in the database. Furthermore, the sequence in the database may have errors preventing the correct matching at the protein level. Searching by the peptide-mass fingerprint is more robust. Only peptides affected by modifications or errors will not match. Matching all other peptides may be sufficient for the positive identification of the protein. Furthermore, the accuracy of mass measurement and the sensitivity of detection are better in the peptide mass range than for larger masses of whole proteins (Gobom 2006).

To prepare an analyte for MALDI-TOF, its molecules are embedded in a *matrix* of usually small organic molecules that can absorb the laser energy (of a specific frequency).[10] The mixture is crystallized and then irradiated by very short pulses of the laser light. The matrix absorbs the laser energy (thus protecting the analyte from being damaged) and facilitates *desorption* and *ionization* of the analyte.[11] This results in predominantly singly charged ions of the analyte molecules. Therefore, the m/z ratio can be interpreted as the molecular mass[12] (Gobom 2006; Dubitzky et al. 2007; Eidhammer et al. 2007). An electric field accelerates the ions to velocities that are

[9]Enzymes cut proteins in known (enzyme-specific) spots. For instance, trypsin cuts proteins at the sites of two amino acids—lysine and arginine.

[10]Commonly used are nitrogen lasers emitting ultraviolet light.

[11]Current knowledge of this process is mostly empirical rather than based on full understanding of its details (Gobom 2006; Eidhammer et al. 2007).

[12]This is true for both MALDI-TOF and SELDI-TOF mass spectrometry experiments.

in reverse proportion to their masses. When an ion reaches the detector, its *time-of-flight* (TOF) can be used to calculate its mass (or, more generally, its mass-to-charge ratio). A signal generated by ions created by each of the laser pulses is a spectrum of peaks defined by m/z ratios and the peak intensities. Averaging spectra from different pulses allows for decreasing the signal-to-noise-ratio.

Generally, MALDI-TOF mass spectrometry may be used to determine molecular masses ranging from hundreds of Daltons to several hundred thousand Daltons. However, the range for a single experiment is not that large and depends on whether we are measuring proteins or peptides.[13] For example, an average trypsin-cut peptide consists of about ten amino acids, which gives an average molecular mass of about 1000 Da. Although very short fragments (say, less than 500 Da) can be measured with high accuracy, they are not informative (for they can be associated with a very large number of proteins). On the other hand, long protein fragments (say, with a mass above 6000 Da) are often the result of miscleavage, in which case their mass should not be used for the protein identification.[14] For these reasons, the analyzed spectra of peptide masses are usually limited to the range 500–6000 Da (Eidhammer et al. 2007).

5.4.2 SELDI-TOF Mass Spectrometry

SELDI-TOF mass spectrometry,[15] also known as SELDI ProteinChip technology[16] (Wright 2002; Reddy and Dalmasso 2003; Hamdan and Righetti 2005; Reddy et al. 2006), is often used in the discovery of protein expression biomarkers. ProteinChip arrays, used by the SELDI-TOF technology, consist of chemically or biochemically active surfaces that can capture either specific proteins or specific classes of proteins. The chemically active surfaces (hydrophobic, hydrophilic, cation exchange, anion exchange, metal affinity, or other solid chromatographic surfaces) are used to capture subsets of proteins with specific physical or chemical properties. ProteinChip arrays with this type of surfaces represent an *unbiased* technology and are used in protein expression and biomarker discovery studies. SELDI-TOF ProteinChip technology can also be used as a *biased* technology targeting specific known proteins. In such arrays, the biochemically active surfaces contain immobilized capture molecules (such as antibodies) that will interact with specific target proteins. Since the capture molecules target particular known proteins, the arrays with this type of surfaces are

[13] Although MALDI-TOF mass spectrometry can be performed either on peptides or whole proteins, the problem with whole proteins is that the ionization techniques available for them work well only when the sample contains about equal amounts of different proteins. Biological samples are dominated, however, by a small number of proteins with the high abundance (such as albumins). Such proteins would overshadow other proteins.

[14] This mass is different from the theoretical mass of a properly cut trypsin-digested peptide, so it should not be compared—during a database search—to the masses calculated from theoretical trypsin-digestion of known protein sequences.

[15] What is currently known as SELDI-TOF mass spectrometry was introduced by Hutchens and Yip as "*laser-assisted surface-enhanced affinity capture (SEAC) time-of-flight mass spectrometry*" and "*surface-enhanced neat desorption (SEND) time-of-flight mass spectrometry*" (Hutchens and Yip 1993).

[16] SELDI ProteinChip technology was commercialized by Ciphergen Biosystems, Inc. (currently Vermillion, Inc.). In 2006, the technology was acquired by Bio-Rad Laboratories, Inc. (www.bio-rad.com).

used in the studies interested in protein interactions or in the quantification of the specific target proteins.

Each ProteinChip array has eight spots to process eight samples in parallel. Studies focusing on protein expression profiling and biomarker discovery use arrays with chemically active surfaces. SELDI-TOF mass spectrometry utilizing arrays with chromatographic surfaces is a combination of chromatographic prefractionation of sample proteins and MALDI-TOF mass spectrometry. Since arrays with diverse types of substrates have different affinity to different classes of proteins, this can be seen as separation of the classes of proteins. Separate mass spectra are generated for different classes of proteins. This way complex biological samples can be analyzed with improved resolution and enrichment of at least some low-abundant proteins. Combining results from various arrays allows for the detection of up to 2000 proteins (Poon 2007; Bertucci and Goncalves 2008).

Once unbound proteins are washed off, processing of the ProteinChip array is similar to the MALDI-TOF technique. Usually, a solution with energy absorbing molecules (matrix) is added to the array. After the crystallization of the mixture of this solution and the captured proteins, the proteins are subject to laser-initiated desorption and ionization. Then, the gas-phase protein ions are detected by time-of-flight mass spectrometry (Bio-Rad Laboratories 2008). The generated spectral data usually consist of 15,000–50,000 variables corresponding to the detectable m/z ratio values. The height (the second dimension) of the spectrum represents the relative signal intensity for each m/z variable (the relative abundance of spot proteins of a given molecular mass).

SELDI-TOF mass spectrometry is a high-throughput technology allowing for the processing of a large number of samples in a relatively short time. Analyzing the generated (and appropriately preprocessed) mass spectra with the use of data mining methods—such as those described in Chapter 3—may lead to identification of multivariate proteomic biomarkers. However, these biomarkers are sets of m/z variables that do not provide direct identification of proteins. Although one can imagine using the m/z variables for diagnostic or prognostic purposes, such sets of SELDI peaks—by themselves—identify neither the proteins nor biological processes associated with class differentiation. Additional processing is necessary to identify proteins corresponding to the selected SELDI peaks. Since the measured m/z value represents only the approximate molecular mass of the protein, searching protein databases may result in many proteins with the molecular masses within the error range of the m/z measurement. To identify the protein generating a SELDI peak, the sample proteins may be purified and then separated by either one- or two-dimensional gel electrophoresis (this, however, is not a high-throughput process). The gel spot corresponding to the selected SELDI peak is excised and digested by trypsin. Molecular masses of the resulting peptides are identified by mass spectrometry, and then protein sequence databases are searched for this protein's peptide-mass fingerprint (Mazzatti et al. 2007; Poon 2007).

Although SELDI-TOF mass spectrometry can be used to identify markers with m/z values up to 250 kDa, the current resolution of this technology is satisfactory only for proteins with molecular masses up to about 20 kDa. Other limitations of this technology include questionable stability and reliability of its results, and difficulty in the identification of low-abundant proteins in complex biological samples. Nevertheless, in part due to its high throughput, relatively low cost and ease of use, SELDI-TOF mass spectrometry is already a popular and important tool in the discovery of proteomic biomarkers. It is expected that further development of this technology (and its eventual coupling with other technologies) will lead to significant improvements in its sensitivity and reliability, and to direct identification of the biomarker proteins (Poon 2007; Bertucci and Goncalves 2008).

We have to remember that SELDI-TOF, when used for biomarker discovery, is an unbiased technology. Although currently this is to its advantage, this advantage will be disappearing when our knowledge of proteomes becomes more complete. Just as the unbiased genomic methods of the 1990s were replaced by the biased methods capable of addressing entire genomes, we may expect future shifts in proteomic methods towards the biased whole-proteome approach.

5.5 PREPROCESSING OF MASS SPECTROMETRY DATA

5.5.1 Introduction

Although we will focus here on SELDI-TOF m/z data, preprocessing of any m/z mass spectrometry data resulting from experiments generating predominantly single-charged ions can be performed in a similar way.

Raw mass spectrometry data can be represented in the form of a matrix similar to the gene expression matrix (see Table 5.1). The rows of the matrix (variables) correspond now to the recorded mass-to-charge, or m/z, ratios and the columns represent biological samples (for instance, tissues from different patients). The m/z ratios are measured in Daltons (Da) per fundamental unit of charge (the charge of an electron).[17] The data points represent intensities recorded for the m/z values. For the supervised analysis, we have also meta data that assigns samples to classes (phenotypes, diseases, etc.) we want to differentiate.

Although the format of the data is similar to that of gene expression matrix, the data itself is more similar to the band intensities resulting from early genomic experiments based on gel electrophoresis of the restriction enzyme-digested DNA fragments (bands).

The intensities associated with neighboring variables (m/z ratios, or basically molecular masses) are not independent variables; in fact, they are usually very highly correlated. In a biomarker discovery study, our ultimate goal is to identify a small subset of proteins whose multivariate expression pattern can be used for efficient classification of new samples. The first step of this higher-level analysis focuses on peaks and our biomarker would be composed of, preferably, just a few m/z peaks. Then, the peaks have to be associated with proteins. Typically, however, this higher-level analysis is not performed on the raw data. The raw data represented by the m/z intensity matrix is first preprocessed in order to remove multicollinearities, identify and quantify true peaks, and align and normalize them across all spectra of

[17]For MALDI-TOF and SELDI-TOF mass spectrometry, we may assume single-charged ions, thus the m/z ratios could be seen as measured in Daltons.

TABLE 5.1: ***m/z* Intensity Matrix**

	Class 1						Class 2		***	Class J		
m/z	Sample 1	Sample 2	Sample 3	Sample 4	Sample 5	***	***	***	***	***	Sample $N-1$	Sample N
500.0223	25.9588	26.6526	24.4485	29.7677	26.1579						24.4203	25.9005
500.4307	28.6677	29.4781	26.9261	33.0075	29.8984						27.7164	29.0940
500.8573	33.8643	35.1272	31.9417	38.9422	34.3449						31.4630	33.0437
501.2751	38.0995	40.2232	35.2459	43.2378	36.7521						33.3736	35.1533
501.6931	39.6762	42.3087	35.7506	44.4585	38.0613						34.1921	36.1327
502.1113	39.7707	42.6244	35.3323	44.2172	38.0231						34.1357	36.2976
***												
***												
***												
15985.13	5.1260	4.9562	5.1565	4.4309	5.6946						4.5777	4.5680
15988.53	5.1059	4.9439	5.1494	4.4263	5.6690						4.5746	4.5636
15992.61	5.0613	4.9013	5.1072	4.4050	5.5761						4.5602	4.5554
15995.77	5.0525	4.8931	5.0944	4.4004	5.5560						4.5534	4.5514
15998.93	5.0297	4.8816	5.0752	4.3936	5.5237						4.5499	4.5437

Columns represent samples (which, for supervised learning, are assigned to classes); rows represent values of the m/z variables.

the study. Typical elements of this low-level preprocessing are: baseline correction, smoothing and noise reduction, peak detection, alignment, and normalization. So far, there is no standard method for preprocessing of m/z mass spectrometry data (Cruz-Marcelo et al. 2008). There are many approaches that differ not only in the algorithms used but also in the set and order of performed preprocessing steps.

5.5.2 Elements of Preprocessing of SELDI-TOF Mass Spectrometry Data

Low-level preprocessing of mass spectrometry data can be seen as extracting the true signal from the observed raw signal. Theoretically, we could represent the observed signal as a composition of the true signal, baseline, and noise (Coombes et al. 2007),

$$O(x) = n \cdot S(x) + B(x) + N(x) \tag{5.1}$$

where

- $O(x)$ is the observed signal,
- $S(x)$ represents the true signal,
- n is a normalization factor,
- $B(x)$ is the baseline,
- $N(x)$ is the noise,
- x represents either the recorded time of flight t, or the mass-to-charge ratio m/z calculated from the time of flight.

Whereas we can model the true signal as a sum of independent (and eventually overlapping) peaks representing proteins and we can use white noise as a model for $N(x)$, we do not have a plausible model for the baseline component (Coombes et al. 2005, 2007).

5.5.2.1 Quality Assessment

A biomarker discovery study may use hundreds or even thousands of SELDI-TOF spectra. Before the spectra are preprocessed and then analyzed, their quality should be evaluated. Low quality spectra (for instance, ones with low signal-to-noise ratios) indicating failed experiments should be removed from further processing. Both visualization techniques and computational methods may be used for quality assessment. Heat maps of all spectra (or the spectra from one class of samples) may be used to visually inspect peaks that are observed in many spectra; their alignment may give us information about the quality of mass calibration (Hilario et al. 2006). Coombes and colleagues used principal component analysis (PCA) to detect outliers among SELDI chips. This approach may be extended to assessing the quality of individual spectra (spots) of SELDI chips (Coombes et al. 2003). Similarly, as for microarray data, clustering samples by their intensity data may reveal serious quality problems when the samples are grouped by factors other than their biological characteristics (for instance, by the date of their processing).

5.5.2.2 Calibration

Each ion detected by time-of-flight mass spectrometry is associated with the intensity of its signal and the recorded time t of its flight to the detector. Calibration is the process of mapping the time t scale into the mass-to-charge ratio m/z scale (Coombes et al. 2007). Typically, calibration is performed at the beginning of the low-level preprocessing of mass spectra. However, the preprocessing steps may as well be carried out on the time-scale data, with calibration performed at the end of preprocessing. Coombes and colleagues point out that using the time scale for peak alignment yields more reproducible results since we are avoiding errors introduced to the data by the calibration step (Coombes et al. 2007).

5.5.2.3 Baseline Correction

The baseline offset of the spectrum, $B(x)$, is attributable mainly to chemical noise generated by the molecules of the energy absorbing matrix. For each spectrum, this offset line can be approximated and subtracted from the raw spectrum intensities. Usually, the baseline is highest at the low range of m/z values and exponentially decreases with the increase in m/z values. Popular methods of the baseline approximation fit polynomial or exponential functions to the local minima of the spectrum. Other approaches may be based on fast Fourier transform or wavelets (Hilario et al. 2006; Eidhammer et al. 2007). Coombes et al. proposed a method that combines baseline correction with the peak detection step (Coombes et al. 2003). Instead of explicit fitting of the baseline for the entire spectrum, they defined the baseline locally, for each identified peak, as the local minimum in the fixed-width window containing the peak. The baseline-adjusted height of the peak is calculated simply as the difference between the local maximum and local minimum. However, in situations when peaks overlap, the local minimum may be significantly higher than the real baseline and the height of the peak may be underestimated (Coombes et al. 2007).

5.5.2.4 Noise Reduction and Smoothing

The random noise component of the observed signal, $N(x)$, is mainly of electronic origin. A simple way of reducing the noise is to perform smoothing of the spectrum by using a sliding window and replacing the intensity values in the window by a single value based on all of the values in the window, for example their weighted average. Fourier transform, smoothing splines, and wavelets are among more sophisticated approaches to noise reduction (Hastie et al. 2001; Hilario et al. 2006; Eidhammer et al. 2007). For example, Coombes et al. use the undecimated discrete wavelet transform (UDWT) for denoising SELDI spectra (Coombes et al. 2005).

5.5.2.5 Peak Detection

A discretized representation of a mass spectrum may contain tens of thousands of m/z values. The intensities of neighboring m/z values are often highly correlated when signals representing proteins (or peptides) are wider than the sampling m/z intervals.

Although it is possible to perform a higher-level analysis (such as feature selection and biomarker discovery) using the raw data, peak detection is recommended for at least two important reasons. First, it significantly increases chances for plausible biological interpretation of an identified biomarker. Second, it decreases the possibility that the biomarker is the result of fitting noise. Furthermore, it also reduces the dimensionality of the data used for the higher-level analysis. Many peak detection algorithms identify peaks as local maxima with intensities above some threshold of the signal-to-noise ratio. Some methods consider also the area under the peak or even the shape of the peak. Additional approaches (such as wavelet transforms) may be necessary to separate overlapping peaks (Beyer et al. 2006; Hilario et al. 2006; Coombes et al. 2007; Eidhammer et al. 2007).

Although many preprocessing algorithms perform peak detection individually for each analyzed spectrum, it is preferable to do it simultaneously for all spectra in the experiment. This way such issues like mass-shift of corresponding peaks from different spectra may be more properly addressed. One of the proposed approaches is to perform peak detection in the mean spectrum (Coombes et al. 2007). First, all spectra have to be aligned, the mean spectrum calculated and denoised. Peak detection is then based on finding local maxima and minima in the mean spectrum. Another approach combines peak detection with peak alignment and supervised peak filtering[18] (Bader et al. 2006). A general idea of this method can be described as follows.

- Identify all local maxima from all spectra (all samples); a local maximum may be defined as being required to monotonically decrease over some m/z range (parameter) on each side of the peak.
- Treat the local maxima corresponding to m/z values differing less than some maximum shift (parameter) as the same maximum. This means that any calculations for such maxima (especially their comparison) should use the intensities corresponding to the maximum point rather than the intensities for the same m/z values.
- Sort all of the identified local maxima according to some measure(s) of their relevance. For instance, sort them first in ascending order of their p-value of an ANOVA F-test (based on the ratio of the between-class variance to the within-class variance), and then in descending order of some measure of their N-fold change between the classes.
- Select the top peak (local maximum) from the above list and move it to another list—the list of selected peaks; then remove from the original list all local maxima within the mass-window distance (parameter) of the selected peak. Repeat these two steps until the original list of sorted peaks is empty.

5.5.2.6 *Intensity Normalization*

Due to experimental variations, peaks from different spectra should be normalized before comparing their intensities. One popular method of normalization is to divide each intensity of a spectrum by the sum of all intensities of the spectrum

[18]The goal of this supervised filtering is to alleviate the problem of inherent multicollinearity of raw m/z data and to select the most prominent peak (if any) in each m/z window. This is not feature selection for biomarker discovery.

(the total ion count). Sometimes, more robust results may be achieved when the intensities are divided by the median intensity of the spectrum. If internal standards (known proteins added to the sample) are available, peak intensities may be normalized in relation to such standards (Coombes et al. 2007; Eidhammer et al. 2007).

5.5.2.7 Peak Alignment Across Spectra

Due to measurement errors, peaks corresponding to the same protein may, in different spectra, be associated with different m/z values. The m/z errors are usually estimated as not greater than 0.3 percent of the m/z values. Peaks with their m/z values within such m/z error intervals should be aligned across spectra and treated as the same peak. For example, the identified peaks may be first sorted by their intensity values or their signal-to-noise ratios. Then, starting from the most prominent peaks, we may match peaks from different spectra if their m/z values differ less than an appropriate m/z error interval (Coombes et al. 2005; Bader et al. 2006). Peak alignment based on hierarchical clustering of peaks from all considered spectra has also been proposed (Prados et al. 2004). Peaks are clustered by their m/z values, with constraints based on the m/z measurement error rate, m_{err}. The distance between two m/z values (or two clusters of m/z values) is calculated in relation to their mean, so it can be directly compared to the relative measure of the m/z error. Although the centroid linkage distance is used by Prados and colleagues to identify clusters that are candidates for merging, two clusters may be merged only if their complete linkage distance[19] is below $2 \times m_{err}$, the doubled mass measurement error.

5.6 ANALYSIS OF PROTEIN EXPRESSION DATA

Whether we use biased or unbiased technologies to generate raw protein expression data, after preprocessing such data can be represented in the form of a *protein expression matrix* (see Table 5.2). This matrix form can be used as a common form for a wide range of protein expression studies. Like the gene expression matrix introduced in Chapter 2, the protein expression matrix consists of N columns representing biological samples and p rows representing variables. The variables presumably represent proteins, either directly or indirectly. For biased proteomic technologies, such as protein microarrays, the variables directly represent proteins. When these technologies mature to the current maturity level of genomic technologies, the protein expression matrix may represent whole-proteome level expression data.

For unbiased technologies, such as SELDI-TOF or MALDI-TOF mass spectrometry, the variables represent the identified spectra peaks, which hypothetically represent proteins or peptides. While it would be more precise to call such data the *peak expression matrix*, it has the same form as the more general protein expression matrix and can be analyzed in exactly the same way. Only after an optimal biomarker is identified, we have to remember the necessity of matching its peaks to proteins, before looking for biological interpretation of the biomarker.

[19]Centroid linkage and complete linkage are described in Chapter 2.

TABLE 5.2: Protein Expression Matrix: *N* Samples (columns), *p* Variables (rows), and *J* Classes

	Class 1						Class 2		***	Class *J*		
	Sample 1	Sample 2	Sample 3	Sample 4	Sample 5	***	***	***	***	***	Sample $N-1$	Sample N
Variable 1	50.6083	45.8562	56.6637	66.0191	48.7441						20.3157	22.0191
Variable 2	81.3635	80.8688	86.9904	99.9598	80.8768						35.7969	48.9452
Variable 3	30.7451	21.9715	28.7226	25.6417	25.2696						16.3164	15.8974
Variable 4	25.2859	24.6415	32.1971	30.4539	23.9356						32.3733	65.2203
Variable 5	75.9739	99.6320	95.6099	73.0236	50.5500						86.1237	94.3348
Variable 6	31.3223	61.3896	40.8125	30.7833	25.0216						43.7728	47.9578
***												
***												
***												
Variable *p*-3	26.2785	36.7642	30.6385	30.8155	24.0845						40.8044	43.6642
Variable *p*-2	17.9427	1 9.6682	20.0322	20.0845	16.7964						34.6466	40.4123
Variable *p*-1	19.5918	20.5732	22.3328	22.2665	18.7692						19.7486	25.0679
Variable *p*	75.2334	90.1458	89.2247	75.6991	75.5829						87.5451	74.6747

The first column identifies variables. For biased technologies, the variables are proteins, which may be represented by their names or identifiers. For unbiased technologies (such as SELDI-TOF or MALDI-TOF mass spectrometry), the variables are peaks represented by their m/z values. Generally, only one row at the top of the table is required to identify samples. For supervised learning, it is convenient to add one more row defining the assignment of samples to classes. Alternatively, this meta-data information may be supplied in a separate file.

The protein expression matrix has the same form as the gene expression matrix introduced in Chapter 2. Furthermore, the goals of protein expression studies are basically the same as the goals of gene expression studies. For example, biomarker discovery aims at the identification of small sets of proteins (or m/z peaks) whose joint expression pattern can significantly separate differentiated classes. Therefore, the methods of analysis and mining proteomic data represented by the protein expression matrix are the same as those described in Chapters 2 and 3, which were dealing with gene expression data.

Similarly as for gene expression data, there is no single workflow common for all protein expression studies (see Figs 5.3 and 5.4). Low-level preprocessing depends on the type of raw protein expression data and on the technology that generated the data. For example, it will be different for mass spectrometry data, and different for protein microarray data. It will also differ for various technologies of proteomic microarrays.

Whatever are the idiosyncrasies of low-level preprocessing, this step of protein expression analysis should result in a protein expression matrix. The higher level analysis and data mining of protein expression data include additional preprocessing, basic exploratory analysis, unsupervised learning, and supervised learning—feature selection, biomarker discovery, and classification.

5.6.1 Additional Preprocessing

Depending on the technology generating raw expression data and the steps of low-level preprocessing, some additional preprocessing of protein expression matrix data may be necessary. It may include filtering, transformation, and additional quality assessment. The approaches may be identical or similar to those described for gene expression data (refer to Section 2.6). For example, a logarithmic transformation[20] of the intensities may be necessary to stabilize the variance (i.e., to transform multiplicative noise into additive noise).

5.6.2 Basic Exploratory Data Analysis

The univariate exploratory analysis is the common first step in analyzing protein expression data. Usually, a t test (when there are two classes to differentiate) or an ANOVA F test (in cases with three or more classes) are used to identify differentially expressed variables (see Sections 2.7.1 and 2.7.2). The variables may be ordered by p-values representing the significance of their differential expression. Due to a large number of simultaneous univariate tests (equal to the number of variables), the p-values have to be corrected for multiple testing (see Section 2.7.5). Since the univariate approach neglects relations between variables, it has a very limited use for biomarker discovery. Nevertheless, it may give us some feeling about the data at hand and maybe even some information that can be used during the multivariate supervised analysis (for instance, during stepwise feature selection).

[20]Depending on a data set, other transformations (e.g., the *arsinh* or the *cube root* transformation) may result in better stabilization of the variance (Huber et al. 2002; Coombes et al. 2003; Beyer et al. 2006).

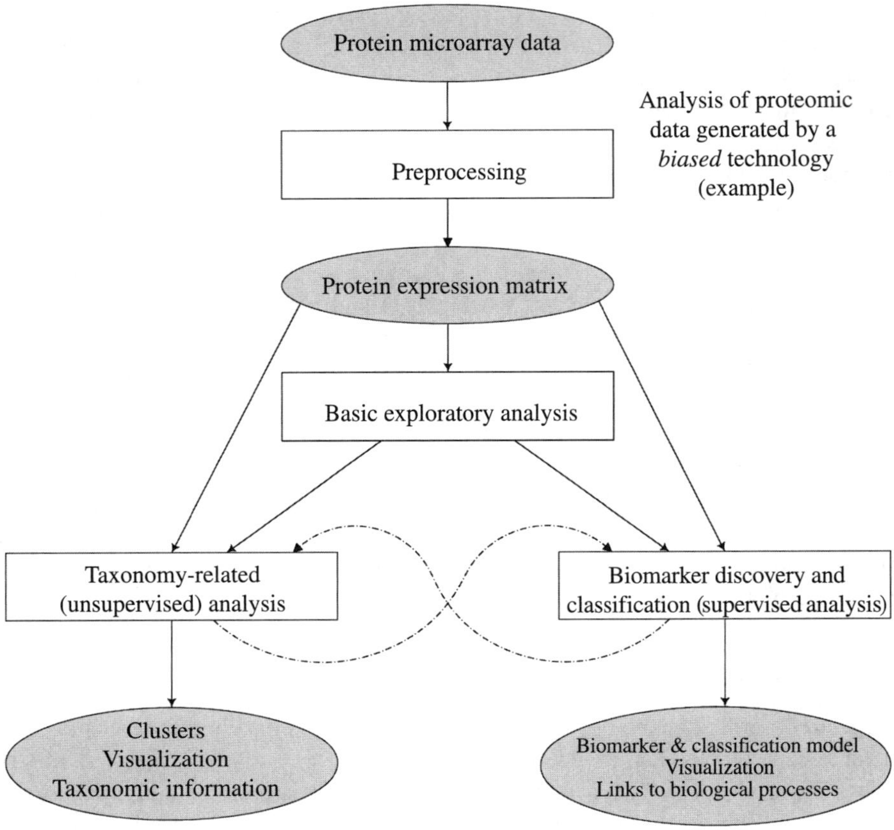

Figure 5.3: Analysis of protein expression microarray data—an example. The raw data is generated by a biased proteomic technology. Preprocessing of the raw data depends on the idiosyncrasies of specific microarray technology. The protein expression matrix is the basic data form used for higher-level analyses. The basic exploratory analysis is commonly performed as the initial step of the higher-level analyses. Unsupervised learning is the main approach for studies looking for new taxonomic information. For supervised learning studies, unsupervised analysis is sometimes performed as a part of the exploratory analysis. However, unsupervised analysis should not be used for variable selection or dimensionality reduction performed as a preprocessing step for supervised analysis because it will most likely result in discarding important discriminatory information. The dashed line from *Biomarker discovery* to *Taxonomy-related analysis* represents the use of unsupervised analysis to find associations between proteins identified during the supervised analysis (for instance, to cluster proteins selected into the informative set of proteins—see Chapter 4).

5.6.3 Unsupervised Learning

For studies seeking new taxonomic information, unsupervised methods may be used to group samples or variables, or samples and variables simultaneously. Clustering methods are used to identify groups of similar objects. Clustering biological samples may lead to the identification of new subtypes of diseases. By clustering variables representing proteins (either directly or indirectly) we may gain information about

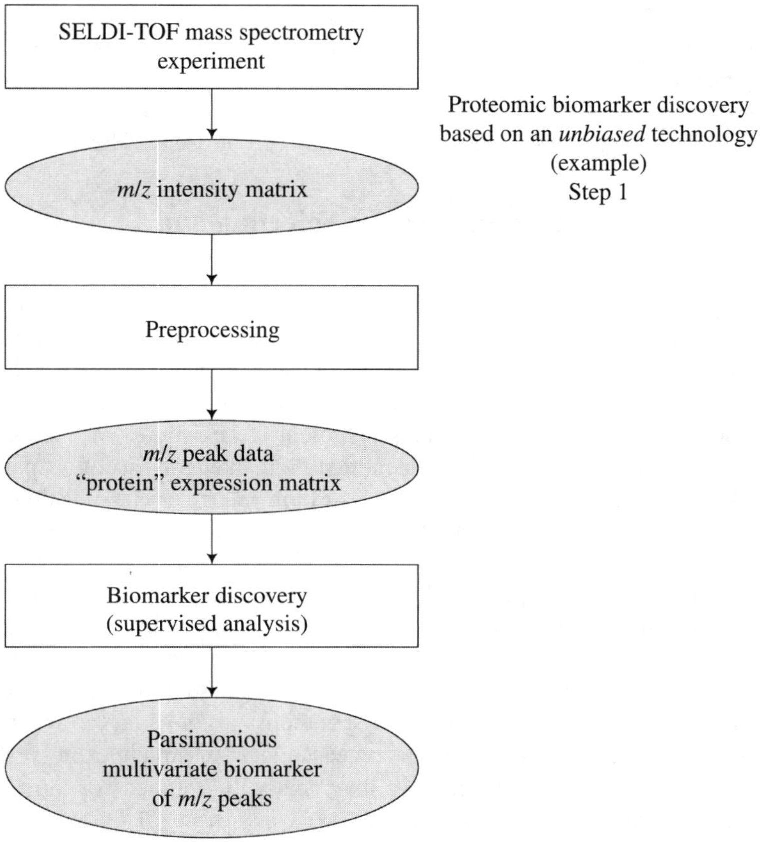

Figure 5.4: Elements of biomarker discovery based on proteomic data generated by an unbiased technology—a SELDI-TOF example; step 1: from a SELDI-TOF experiment to a multivariate biomarker consisting of m/z peaks. After SELDI-TOF mass spectrometry, data is in the raw form of the m/z intensity matrix. Preprocessing of such data may include quality assessment, calibration, baseline correction, smoothing and noise reduction, peak detection, alignment, and normalization. After preprocessing, the data may be represented in the form of a protein expression matrix with the variables (rows) representing the identified m/z peaks and the biological samples (columns) associated with the differentiated classes. The biomarker discovery step, including feature selection and supervised learning, is performed on this protein expression matrix. The result is a parsimonious multivariate biomarker—a small set of m/z peak variables whose joint intensity pattern can significantly separate the discriminated classes.

groups of proteins that have similar functions or are associated with the same biochemical processes. Simultaneous clustering of samples and variables (two-way clustering) may identify subsets of proteins which exhibit distinctive expression patterns only for a subset of samples. Dimensionality reduction techniques (such as PCA) may be used to preserve most of the variation in the data represented by a much smaller number of variables. A low-dimensional representation of the data may be defined either by a subset of original variables or by new variables that are linear combinations of the original variables. However, using original variables allows for more straightforward biological interpretation of the results. Low-dimensional representations of the data—generated by such methods like PCA or self-organizing maps (SOM)—enable visualization of data points and their eventual grouping. For more information on unsupervised learning, cluster analysis, PCA and SOM please read Section 2.8.

5.6.4 Supervised Learning–Feature Selection and Biomarker Discovery

The main goal of feature selection and biomarker discovery is to identify a parsimonious multivariate biomarker consisting of proteins whose joint multi-protein expression pattern can be used for efficient classification of new cases, that is, for accurate prediction of their class membership. When feature selection is performed on the data whose variables only indirectly represent proteins, for example, when the variables are m/z mass spectrometry peaks, the identified biomarker consists of a small set of m/z peaks. Although it is possible to use such biomarkers for classification of new cases, a set of m/z peaks does not allow for direct biological interpretation of classification results. Even if the efficiency of such biomarkers is positively validated on independent test data, their biological validation would be difficult since a set of m/z peaks does not point directly to protein expression patterns underlying class differences. Therefore, it is important that such m/z biomarkers are subject to subsequent experiments and analysis translating them into biomarkers consisting of sets of proteins.

A protein expression matrix may consist of thousands of variables. When biased proteomic technologies allowing for direct measurement of protein expression at the whole-proteome level are developed, we may deal with hundreds of thousands or even a million variables. Similarly, as for gene expression data, the exhaustive search for the best multivariate biomarker is not an option. Recommended feature selection strategies are such heuristic searches that can efficiently identify small subsets of variables associated with local optima in a high-dimensional space of all p variables. Heuristic searches may be implemented as sequential, random, or hybrid ones.

Basic strategies for the sequential heuristic searches are *stepwise forward selection* and *stepwise backward elimination*. The forward selection starts with the empty set and sequentially adds variables, one by one, in a way that maximizes a measure of class separation calculated for the subsets of variables considered for each cardinality. The backward elimination starts with all p variables and sequentially eliminates variables that are multivariately least important for class separation. However, the best results may be achieved when these two strategies are combined into *stepwise hybrid selection*. At each step of the hybrid selection, evaluated are subsets of variables that can be generated from the currently optimal subset by either adding or removing variables. *Heuristic random searches* allow escaping particularly inefficient local optima in the high-dimensional space of p variables. They may randomly select variables at each step of the search. Alternatively, they may randomly select the first variable and then follow either the stepwise forward or the stepwise hybrid strategy.

Whatever search strategy is implemented, the search for an optimal biomarker ends when one of the stopping criteria is achieved. The stopping criteria may be based on the measure of class separation, on the size of the biomarker, on the number of iterations, or upon arriving at a local optimum in the search space.

Feature selection methods may also be classified as implementing either filter or wrapper, or hybrid, or embedded search models. This classification depends on the relationship between the feature selection process and the learning algorithm used to build a classification system. *Filter models* perform feature selection independently of any learning algorithm. In *wrapper models*, a learning algorithm is included in the feature selection process and used to evaluate each of the considered subsets of variables. *Hybrid* models combine the filter and wrapper approaches. First, the filter model is used to identify the optimal biomarker for each of the considered cardinalities. Then, the wrapper model is used to select one of the cardinalities by evaluating each of the optimal biomarkers. In *embedded models*, feature selection is implemented within the learning algorithm. Thus, feature selection is an integral part of the training process.

For a more detailed description of feature selection and biomarker discovery, please refer to Sections 3.2 and 3.1.2 of Chapter 3. Though these sections deal with gene expression analysis, their contents are equally applicable to protein expression analysis.

5.6.5 Supervised Learning–Classification Systems

To build a classification system, we need a training data set and a machine learning algorithm. The training data set has the form of a protein expression matrix and consists of statistical samples representing the populations to be differentiated. The classification system may be designed after the optimal biomarker has been identified at the feature selection step. Alternatively, the learning algorithm may include feature selection (the embedded model of feature selection). In any case, parameters of the classifier are learned from the training data set. Therefore, for the classifier to be generalizable, it is very important that the training set is of high quality and of appropriate size in order to properly represent the underlying populations (the classes to be differentiated).

Many learning algorithms may be used to build a classifier and so far there is no proof that any of them is generally superior over the others. However, depending on the idiosyncrasies of a particular classification problem and the available data, different learning algorithms may yield different results. Unless the differentiated classes are easily separable (in which case any learning algorithm may be able to deliver a satisfactory solution), it is a good idea to try more than one learning algorithm and select the one most appropriate for the project at hand. The very same learning algorithms that are described in Chapter 3 may be considered for designing classifiers based on protein expression data. These algorithms represent classical methods (such as *linear discriminant analysis*), newer and already popular algorithms (for instance, *support vector machines*), and recent and promising approaches (*random forests* and other ensemble-based methods).

Once the classification system is built, it has to be properly validated, which means that its generalization abilities have to be reliably estimated. We always have to remember that for the ultimate estimation of predictive abilities of the classifier, we need to validate it on a large and independent test data set. If such a set is not available, the ensemble-based validation using out-of-bag (OOB) samples is recommended.

More information on classification systems and biomarker discovery as well as details of several learning algorithms are provided in Chapter 3.

5.7 ASSOCIATING BIOMARKER PEAKS WITH PROTEINS

5.7.1 Introduction

A parsimonious multivariate biomarker and a generalizable classifier are all that we need for efficient assignment of new biological samples to one of the differentiated classes. Similarly as for gene expression biomarkers, protein expression biomarkers may be used for diagnostic, prognostic, or therapy selection systems, for drug discovery, and any other applications where biological samples need to be assigned to one of well defined classes.

Since proteins are directly associated with cellular biochemical processes, exploitation of protein expression biomarkers in the area of personalized medicine is more promising and may be more appropriate than the use of gene expression biomarkers. Class membership is defined by the proteomic expression pattern of biomarker variables. If the biomarker variables are proteins, then they (and, eventually, the *Informative Set of Proteins* identified via a sequence of alternative protein expression biomarkers) may be used by domain experts to elucidate biological processes associated with the class differences. However, if the biomarker was identified from mass spectrometry data, its variables are mass-to-charge, or m/z, peaks. Generally, the peaks may correspond to proteins or to peptides. Usually, however, proteomic biomarker discovery studies based on mass spectrometry data utilize SELDI-TOF technology, which generates spectra of protein m/z values.

Protein databases (such as *Swiss-Prot*) can be searched for information about molecular masses of proteins with known amino acid sequences. However, direct comparison of SELDI-TOF m/z peaks with the masses retrieved from the databases is not recommended. Due to intrinsic errors of m/z measurements, the m/z values of the peaks selected into the biomarker (as well as other spectra peaks) represent only the approximate molecular masses of the proteins that generated the peaks. Searching protein databases for one of these m/z values may result in many proteins whose masses are within the error interval of the m/z measurement. As described earlier (in the section on MALDI-TOF mass spectrometry), more robust are database searches utilizing the peptide-mass fingerprint of the protein to be identified. Therefore, in order to identify the biomarker proteins, the sample proteins are often purified and then separated by two-dimensional gel electrophoresis. The gel spots corresponding to the biomarker m/z peaks are then identified, excised from the gel and digested by trypsin. As a result of the digestion process, the spot protein is cleaved into peptides. Subsequently, molecular masses of these peptides are determined by MALDI-TOF mass spectrometry. The resulting peptide-mass fingerprint is characteristic for the biomarker protein. Searching for the identity of this protein is based on

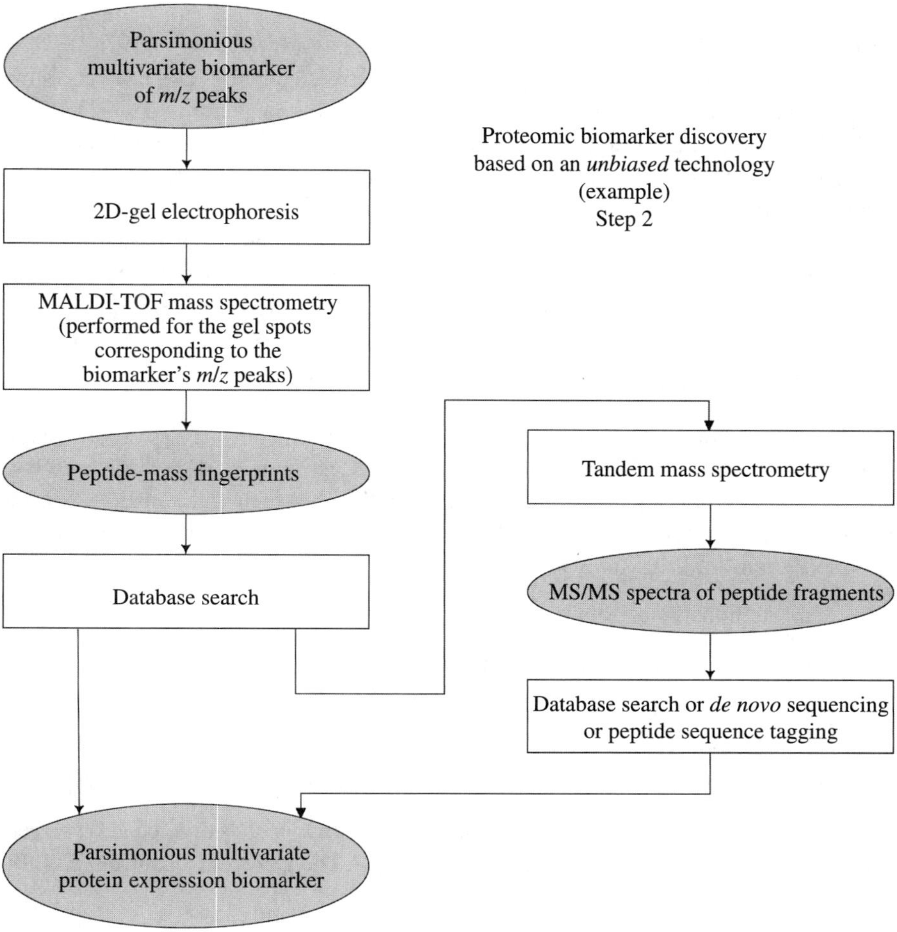

Figure 5.5: Elements of biomarker discovery based on proteomic data generated by an unbiased technology—a SELDI-TOF example; step 2: from a multivariate biomarker of m/z peaks to a multivariate protein expression biomarker. At step 1, an optimal multivariate biomarker has been identified. Such a biomarker, when positively validated on an independent test data set, can be used to build a classification system that is generalizable and allows for classification of new cases with high sensitivity and high specificity. To identify proteins that generated the biomarker's m/z peaks, 2D-gel electrophoresis may be performed to separate sample proteins. The gel spots (proteins) corresponding to each of the biomarker's m/z peaks are then excised and digested by tripsin. MALDI-TOF mass spectrometry is performed on the resulting peptides, and a peptide-mass fingerprint is generated for each of the proteins that produced the biomarker peaks. Subsequently, a protein sequence database may be searched to compare each of the observed fingerprints with theoretical fingerprints calculated for the database proteins (search programs are described in Section 5.7.3). Alternatively (or if a search with peptide-mass fingerprint data is not successful), some or all peptides of a fingerprint may be further fragmented and have m/z ratios determined for the resulting fragments. This may be accomplished with the use of tandem mass spectrometry. A database search, *de novo* sequencing, or peptide sequence tagging (see Section 5.7.4) can be subsequently used to identify the peptides and then the protein.

comparing its fingerprint to peptide masses calculated for theoretical digestion of proteins with known sequences stored in protein databases (see Fig. 5.5).

Because of the many-to-many relationship between peptides and proteins (similarly as for bands and genes in the electrophoresis-based genomic studies of the 1990s), specialized algorithms (and software programs) for the identification of significant associations between the observed peptide-mass fingerprint and theoretical sequences stored in protein databases are necessary. The general idea of such search algorithms is to assess the probability of a protein from a sequence database to generate the observed experimental results. The most widely used protein sequence database is *Swiss-Prot*, which is currently a part of the *Universal Protein Resource*.

5.7.2 The Universal Protein Resource (*UniProt*)

The Universal Protein Resource (The UniProt Consortium 2008) is a freely accessible repository of protein sequences and annotations. The UniProt Consortium (www.uniprot.org) was formed in 2002 as a collaboration between three institutions: the European Bioinformatics Institute (EBI), the Swiss Institute of Bioinformatics (SIB), and the Protein Information Resource (PIR).

UniProt consists of four databases:

- the protein knowledge base (UniProtKB),
- the reference sequence clusters (UniRef),
- the sequence archive (UniParc),
- the repository of metagenomic and environmental sequences (UniMES).

The main part of UniProt is the UniProt knowledge base, which is divided into two sections, Swiss-Prot and TrEMBL (Bairoch et al. 2008; The UniProt Consortium 2008):

Swiss-Prot

Swiss-Prot[21] (UniProtKB/Swiss-Prot) is a cross-referenced database of manually annotated and continuously updated protein sequences. It is a central repository of high-quality information about all publicly available protein sequences. The information is extracted from literature and from results of computational analyses. All information is reviewed, curated and annotated by experts. Each Swiss-Prot sequence entry includes:

- protein name or description,
- amino-acid sequence,
- bibliographical references,
- taxonomic data,
- annotations, which may include information about protein function (or functions), alternate protein products (results of alternative splicing), post-translational modifications, diseases associated with the protein deficiencies, etc.

[21]The first version of Swiss-Prot was released by Amos Bairoch in 1986. Since 1994, Swiss-Prot has been a collaborative project with the European Bioinformatics Institute (Bairoch 2000; Bairoch et al. 2008).

Swiss-Prot entries may also include cross-references and links to nucleotide sequence databases (such as EMBL-Bank, GenBank, and DDBJ), 3D structure databases, enzyme and pathway databases, ontologies, genome annotation databases, gene expression databases (such as ArrayExpress), and many other databases.

TrEMBL

TrEMBL (UniProtKB/TrEMBL) is a supplement to Swiss-Prot that contains computationally analyzed, automatically annotated and unreviewed protein sequences, which await full manual annotation and integration into Swiss-Prot.

5.7.3 Search Programs

While the peptide-mass-fingerprint and the information concerning what enzyme was used for digestion are the main experimental data for protein search, other available information may be helpful in the positive identification of the protein. Examples of such information are the approximate molecular mass of the protein (the m/z value of the SELDI-TOF peak or the mass associated with the gel spot), the isoelectric point of the protein, and the peak intensities associated with the fingerprint's m/z values. To compare the peptide-mass fingerprint to a database sequence, theoretical (*in silico*) digestion of the database sequence—with the same enzyme that was used to obtain the experimental fingerprint—has to be performed. The simplest search algorithms rank sequences in a protein database by the number of theoretical peptide masses matching the fingerprint peptide masses (within a given tolerance). If they do not adjust for the mass of a database sequence, then large proteins will likely score higher. Other algorithms, though also based on counting the number of peptide matches, compensate for the relative abundance of the peptides of a given mass and for the protein size. More sophisticated algorithms add statistical verification of the identified matches. Since different search programs make different assumptions and implement different algorithms, it is a good practice to use more than one search program and look for the overlapping results.

A recent survey of the proteomic literature suggests that Mascot, Ms-Fit and ProFound are the most commonly used programs for the peptide-mass fingerprinting-based protein identification (Damodaran et al. 2007). In addition to these three search programs, we will look at one of the many other algorithms—Aldente, which combines the search with a quite unique recalibration of the empirical masses.

Mascot–Mowse

Mascot[22] (Perkins et al. 1999) implements the Mowse scoring algorithm (Pappin et al. 1993). The Mowse (**Mo**lecular **w**eight **se**arch) algorithm starts with comparing masses of the fingerprint peptides with the theoretical peptide masses calculated for database sequences. The database sequences can be prefiltered by the taxonomy parameter limiting the search to a specified species.[23] A peptide match is identified if the mass

[22] www.matrixscience.com

[23] For example, if we limit a Swiss-Prot search to the human sequences, only 20,408 sequences out of 400,771 Swiss-Prot sequences will be considered (the numbers as of November 14, 2008).

difference falls within a given mass tolerance. The reported sequence-level matches may, however, be limited by the protein mass parameter,[24] which is used as a sliding mass window. Mowse scoring not only counts the peptide matches and adjusts them for the sequence mass, but also uses empirically determined distribution of peptide masses in the database to assign a weight to each peptide match between the fingerprint and database sequences. Then, for each matched sequence, Mascot calculates the probability p that the observed (or better) match between the sequence and the fingerprint is a random event. This p-value is subsequently converted into the reported score S,

$$S = -10 \cdot \log_{10}(p) \tag{5.2}$$

This score, however, is independent of the database size. Therefore, it should be compared to the cut-off score C based on the significance level α adjusted for multiple testing. In Mascot, the adjustment is done by dividing α by N, the number of database sequences considered during the search.

$$C = -10 \cdot \log_{10}\left(\frac{\alpha}{N}\right) \tag{5.3}$$

For example, if the significance level $\alpha = 0.05$ and the number of searched database sequences $N = 50{,}000$, the matches with the Mascot score of $S \geq C = 60$ would be considered statistically significant.

Instead of comparing the score S to the cut-off score C, the E-value (expectation value) can be calculated for each reported sequence match,

$$\begin{aligned} E &= \alpha \cdot 10^{\frac{C-S}{10}} \\ &= p \cdot N \end{aligned} \tag{5.4}$$

If this E-value was restricted to be not greater than 1, it could be interpreted as the p-value of the match adjusted for multiple testing. Without this restriction, it can be interpreted as the expected number of sequences that would be assigned a particular score S (or higher) by chance.[25]

MS-Fit

MS-Fit[26] (Clauser et al. 1999; Baker and Clauser 2008), similarly as Mascot, implements the Mowse scoring algorithm. MS-Fit, however, directly reports the Mowse score and does not provide information on its statistical significance. Reported are all proteins with the number of peptide matches equal to or greater than the user specified minimum number of required peptide matches.

[24]Though called "the protein mass," this parameter is used by Mascot as a sliding mass window. This means that for a sequence to be reported as a match, all of its peptide matches have to be contained within a contiguous stretch of sequence with the mass not greater than the specified protein mass.

[25]Recently, a decoy database search is recommended as a way to estimate the false positive rate or the false discovery rate, especially for MS/MS experiments (see Section 5.7.4) with large numbers of spectra (Elias et al. 2005; Reidegeld et al. 2008). A decoy database is a database consisting of randomized or reversed sequences. The number of matches found in such a decoy database provides an estimate of the number of false positives among the matches found in the real sequence database.

[26]http://prospector.ucsf.edu

ProFound

ProFound[27] (Zhang and Chait 2000) implements a Bayesian approach to searching a protein database for sequences that are likely to generate the observed peptide-mass fingerprint. For each considered database protein k, ProFound calculates the conditional probability $P(k|DI)$ that protein k is the sample protein given the observed fingerprint data D and the available background information I (such as the species, approximate mass of the protein, the enzyme used for protein digestion, peptide mass accuracy, etc.). Proteins are ranked by this Bayesian probability. Let us note that this is the probability that a database protein is the sample protein—this is *not* a p-value, which would be the probability that the match occurred by chance.

Aldente

Aldente,[28] **A**dvanced **L**arge-scale **Ident**ification **E**ngine (Gasteiger et al. 2005), is a search program that compares an experimental peptide-mass fingerprint with the theoretical peptide masses calculated for protein sequences stored in the Swiss-Prot or TrEMBL database. To resolve ambiguities in matching the observed peptide masses with theoretical ones, the Aldente algorithm recalibrates the experimental masses using the Hough line transform (Hough 1962; Duda and Hart 1972). This heuristic technique takes into account the estimated internal precision of the mass spectrometer and identifies a straight line (in the two-dimensional space defined by the experimental and theoretical peptide masses) that maximizes the number of matched masses. During the search, a score is calculated for each protein. The Aldente scoring system can take into account several user supplied parameters, both at the protein level (such as the estimated protein molecular mass and isoelectric point) and at the peptide level (such as the intensities of the fingerprint peaks). If no such parameters are supplied, a protein score is the number of matched peptides. As a result of the search, a list of the best matching proteins is provided. The proteins are listed in decreasing order of their scores. Each score is associated with the p-value, which can be interpreted as the probability of finding—for the same experimental data—a protein sequence with the same or better score in a database of randomly generated sequences.

5.7.4 Tandem Mass Spectrometry

Sometimes, the results of protein identification via peptide-mass fingerprinting may be ambiguous. Different peptides may have similar masses. A mass of a modified peptide may match a theoretical mass of another peptide. A fingerprint peak may be the result of contamination or noise. Supplementing the peptide-mass fingerprint with peptide sequence information would greatly improve chances for correct protein identification. Including sequence information for one or two peptides is often satisfactory for unambiguous protein identification. *Tandem mass spectrometry* (Tandem MS, MS/MS, or MS^2) can be used to obtain information about the sequence of a specific peptide. In

[27] http://prowl.rockefeller.edu

[28] www.expasy.org/tools/aldente

tandem mass spectrometry, a two-step mass analysis is performed.[29] The first step of the analysis is performed for a small range of m/z values focused on the m/z of the selected peptide (*parent ion*). Then the peptide is fragmented[30] and, at the second step of the mass analysis, the ions resulting from its fragmentation (*product ions*) are measured. The final result of tandem mass spectrometry is the MS/MS spectrum with peaks corresponding to m/z values and intensities of the fragment ions (Eidhammer et al. 2007).

Database searching, *de novo* sequencing and p*eptide sequence tagging* are common approaches used to determine the peptide sequence from the MS/MS spectrum of its fragments.

Database Searching

A database search compares the experimental MS/MS spectrum of peptide fragments with the theoretical spectra calculated for peptide sequences stored in a database. For each considered peptide sequence, calculated is the score that indicates the degree of match between this database sequence and the experimental MS/MS spectrum. The best matches are associated with the MS/MS spectrum. To identify the biomarker protein, tandem mass spectrometry and database searches are performed for all peaks from the protein's peptide-mass fingerprint. Then the protein may be identified by analyzing the generated lists of best matching peptide sequences (Eidhammer et al. 2007; Fenyo and Beavis 2008). Mascot (Perkins et al. 1999) and SEQUEST (Eng et al. 1994) are popular database search programs for the peptide identification.

De Novo *Sequencing*

Methods that try to determine a peptide's sequence from a MS/MS spectrum alone are termed *de novo* peptide sequencing. Without going into details of such methods, the idea of *de novo* sequencing is based on ordering the m/z values of fragment ions and trying to match the m/z differences between them to one or more amino acids. A popular approach to *de novo* sequencing is the *spectrum graph* method. A spectrum is represented by a directed acyclic graph, whose nodes correspond to the spectrum peaks arranged in ascending order of their masses. Directed edges (from a lower mass to a higher one) are defined between any two subsequent nodes whose mass difference corresponds to the mass of an amino acid. In an ideal situation, when all possible peptide fragments are represented in the MS/MS spectrum, there will be a path through the graph that will define the peptide sequence (Frank and Pevzner 2005; Xu and Ma 2006). In practice, however, such complete spectra are rare. When some fragments (peaks) are missing, we may allow graph edges to correspond to the sum of masses of two or more amino acids. However, there may be many other departures from the ideal MS/MS spectrum of a peptide (for instance, additional peaks corresponding to contaminations or noise). Various *de novo* sequencing algorithms and programs have been developed to deal with such uncertainties and to generate a ranked list of potential peptide sequences. Among them are Lutefisk (Taylor and Johnson 2001),

[29]Both steps of this analysis can be automatically performed on the same mass spectrometer.

[30]More specifically, fragmented are all ions present in the specified m/z range of the first analysis.

SHERENGA (Dancík et al. 1999), PepNovo (Frank and Pevzner 2005) and PEAKS (Ma et al. 2003). The first three programs implement the spectrum graph approach. PEAKS software works directly on spectrum data, without converting them into a spectrum graph.

Peptide Sequence Tagging

Peptide sequence tagging is another approach to peptide identification. It combines *de novo* sequencing with database searching. Instead of trying to determine the complete sequence of a peptide, the *de novo* step of this approach infers from the MS/MS spectrum only short partial sequences—*sequence tags*. A sequence tag divides the peptide into three sections, the region preceding the tag, the sequence tag and the region trailing the tag. The combination of the sequence tag with molecular masses of these two regions defines *a peptide sequence tag*, which is used to search for matching peptide sequences in the database (Mann and Wilm 1994). Algorithms that implement the peptide sequence tagging approach may use the spectrum graph method to generate a number of short tags (typically three amino acids), and then may assign to each tag the probability that its sequence is correct. The goal is to use a small set of tags that includes at least one correct tag. Therefore, only tags with high probabilities are used for the database search. This allows for efficient database filtering as well as for high probability of success (Frank et al. 2005). InsPecT (Tanner et al. 2005) and GutenTag (Tabb et al. 2003) are examples of programs implementing peptide sequence tagging.

5.8 SUMMARY

Protein expression analysis is likely to become one of the main sources of new biomarkers for personalized medicine, which may include early medical diagnosis, tailoring therapy selection to the prediction of individual response to available treatment modalities, and assessing treatment progression and drug efficacy. Multivariate approaches to feature selection coupled with large and good quality training data sets will lead to the identification of parsimonious proteomic biomarkers representing multi-protein expression patterns characteristic for the differentiated classes. Proteomic biomarkers will be routinely used during drug discovery and development to assess efficacy, safety, and toxicity of drug candidates.

When biased proteomic technologies achieve a high-throughput whole-proteome level of direct measurement of relative protein expression, they are likely to replace the currently dominating unbiased proteomic approaches. Two-dimensional gel electrophoresis and mass spectrometry are examples of popular unbiased proteomic technologies. Their advantage is that they can be used to identify any protein or sequence, whether known or unknown. This advantage will, however, be disappearing as our knowledge of the human proteome (or human proteomes, depending on the definition used) will increase. High throughput biased technologies, such as antibody or protein microarrays, will be more and more popular, especially when they become cost- and time-efficient. *Whole-proteome* microarrays would enable direct and simultaneous measurement of expression levels of all proteins of a proteome as large and complex as the human proteome. Functional microarrays would enable elucidation of protein functions on a whole-proteome scale.

Whatever technology is used to generate protein expression data, the methods that should be used to analyze such data depend on goals of the study. For example, unsupervised learning methods should be used for studies aimed at new taxonomic knowledge, but supervised learning has to be the main approach for biomarker discovery.

Preprocessing of raw proteomic data depends on the technology that generated the data. Nevertheless, after low-level preprocessing we can represent any protein expression data in the form of a *protein expression matrix*. The variables of this matrix can represent proteins either directly (as in the case of antibody microarrays) or indirectly (for instance, SELDI-TOF m/z variables). In either case, the higher-level analysis of such protein expression data can be performed with the use of the same data mining methods that we are using to analyze gene expression data. If the goal of our analysis is biomarker discovery, we try to identify a small set of variables whose joint expression pattern can significantly separate the differentiated classes. When these variables are m/z peaks rather that proteins, additional analysis is necessary to associate the biomarker variables with proteins. For example, SELDI-TOF analysis may be followed by 2D gel electrophoresis and MALDI-TOF mass spectrometry in order to identify peptide-mass fingerprints for proteins associated with the SELDI-TOF m/z peaks selected into the biomarker. Specialized search programs may be then used to associate the fingerprints with protein sequences stored in protein databases. If this does not lead to the positive identification of all biomarker proteins, tandem mass spectrometry may be subsequently applied. The fingerprint peptides are further fragmented and the masses of these secondary fragments are used to identify the peptides via database search, *de novo* sequencing, or sequence tagging.

Heuristic approaches to multivariate feature selection, efficient supervised learning algorithms, improving generalization by introducing randomness to training data sets and to algorithms, and validation of the generated classifiers with the use of independent test data sets are the elements of biomarker discovery approaches that can efficiently analyze data sets with a very large number of variables, such as the whole-proteome level protein expression data.

CHAPTER 6

SKETCHES FOR SELECTED EXERCISES

6.1 INTRODUCTION

In this chapter, we draw sketches of selected approaches to various stages of analysis of gene expression data. It is important to indicate that no single method works optimally in all situations. With the possible exception of very simple cases when the differentiated classes are particularly easy to separate and almost any method can yield plausible results, it is a good idea to try more than one method and compare the results. One may notice our preferences, such as a heuristic search approach for feature selection. Our primary method used for feature selection and biomarker discovery is the stepwise hybrid search driven by Lawley–Hotelling T^2 measure of class separation as described in Chapter 3. For validation (when we do not have an independent test set or when we want to estimate generalizability of a classifier before testing it on independent data), we use the *modified bagging* schema (see Chapter 3). This schema generates a large number of classifiers built on random subsets of the original training samples. However, we do not treat these classifiers as an ensemble of classifiers used for prediction. We use each of them to classify its out-of-bag (OOB) samples—the samples that were not selected for the training set used for feature selection performed for this particular classifier. Averaging the results of OOB classification over hundreds or thousands of classifiers gives us an estimate of the generalization abilities of an optimal biomarker built from the entire training set. To facilitate biological interpretation of class differences, we generate a sequence of alternative multivariate markers and use them for the identification of the *Informative Set of Genes*. This method has been introduced in Chapter 4. Furthermore, by analyzing the distribution of the informative set variables among classifiers built by the *modified bagging* schema, we identify the *primary expression patterns* and their *frequent primary* genes, which can be used for a more focused biological interpretation of class differences and for the identification of more robust multivariate biomarkers.

Let us stress here that we are not endorsing or recommending any software package or solution. Though we perform feature selection, the identification of the *Informative Set of Genes*, and biomarker optimization and validation with the *modified bagging* schema and on independent data using the *MbMD* biomarker discovery

Data Mining for Genomics and Proteomics. By Darius M. Dziuda

software,[1] various algorithms and software solutions may be successfully used for these tasks as long as they use multivariate methods implementing a supervised learning approach.

We feel that it is also important to indicate that a promising data set does not always lead to the sought after results. In real-life biomedical studies, it is sometimes necessary to conclude (after various approaches and methods are tried) that the data at hand is not sufficient for the identification of reliable and efficient biomarkers, and that either a larger or more representative training set is required, or maybe data of a different type should be gathered and analyzed (for instance, protein expression data instead of gene expression data). We would discourage our readers from reporting, in such situations, perfect or near-perfect reclassification or internal cross-validation results. With a large number of variables and much fewer samples, such results are often easy to achieve,[2] but they should not be reported as the only results of a study if they are not supported by more reliable indications of good generalization of biomarker classification abilities.

6.2 MULTICLASS DISCRIMINATION (EXERCISE 3.2)

6.2.1 Data Set Selection, Downloading, and Consolidation

ALL3 data (http://www.stjuderesearch.org/data/ALL3/dataFiles.html) from St. Jude Children's Research Hospital (Memphis, TN) is an example of a good quality multiclass data set (Ross et al. 2003). It includes 132 samples representing seven groups of pediatric acute lymphoblastic leukemia (ALL)—six subtypes of ALL (T-ALL, E2A-PBX1, TEL-AML1, MLL rearrangement, BCR-ABL, and hyperdiploid karyotypes with more than 50 chromosomes) and the seventh group of "other" ALL samples. The samples were hybridized on the Affymetrix HG-U133 set of microarrays (HG-U133A and HG-U133B).

Download ALL3 Data Files from the St. Jude Website

- The data set consists of 14 files; there are two files for each of the seven groups of samples: one file with the HG-U133A expression data (*Chip A*) and one with the HG-U133B data (*Chip B*).
- Each *Chip A* file has 22,285 rows, and each *Chip B* file has 22,647 rows (included in the numbers are two header rows in each file).

Check Whether Arrays are Scaled to the Same Target. Rescale if Necessary

- For each of the 14 downloaded files, calculate a four percent trimmed mean for each array (column).

  Question: Why do we use the four percent trimmed mean (see Chapter 2)?

[1] www.MultivariateBiomarkers.com

[2] Such results can be achieved even for data that represent randomly generated noise.

Excel Tip

To calculate the trimmed means in a *Chip A* file:

- in the spreadsheet cell C22286, define function "TRIMMEAN(C3:C22285, 0.04)"
- highlight the three cells: C22286 to E22286,
- copy them and then paste to the remaining columns of row C22286,
- calculate the MIN and MAX values of the trimmed means in row C22286. Both of them should be very close to the target intensity value (500).

- If we prefer to work with data scaled to a different target value, we will multiply each expression data value by a rescaling factor. For example, to change the target intensity to 100, we use the rescaling factor of 0.2 (see Chapter 2).

Consolidate the Data into a Single File (Gene Expression Matrix)

- In each of the 14 data files, each biological sample is represented by three columns: expression level, detection call, and *p*-value of the detection call. If we use the default *p*-value ranges for detection calls, the columns with the detection *p*-values may be removed. We may also remove the second and the last column (probe set descriptions).

  Question: What are the default p-value ranges for the Present, Marginal, and Absent detection calls (See Chapter 2)?

- For each of the seven groups, append the *Chip B* data to the bottom of the *Chip A* data. For each of the seven combined files, check that the sample names in the columns of the appended *Chip B* file correspond to those of the *Chip A* columns.

Excel Tip

- Cut the two header rows of the appended *Chip B* data and insert them below row 2.
- Insert a new row below row 4 and using the Excel IF function check whether the sample names in rows 1 and 3 are the same for all columns (in C5, enter " =IF(C1=C3,0,1)", copy this function to all cells of row 5 and then sum the results—the sum should be zero).
- Remove rows 2 to 5. We now should have 44,930 rows in each of the seven resulting Excel files: two header rows and 44,928 probe set rows. Each sample is represented by two columns: the intensity and the detection call.

- Remove the control probe sets: 68 probe sets with the names starting with AFFX and the set of 100 normalization controls (200000_s_at to 200099_s_at). Since they are repeated on the U133A and 133B arrays, we are removing 336 probe sets (if we have not changed the order of probe sets, these control probe sets are the first 168 rows in the *Chip A* and *Chip B* data). Each of the seven files should now include 44,592 probe sets.

- Since we will filter probe sets by the fraction of Present calls in each class, it is convenient to calculate this fraction now, when we still have seven separate class files. In each of them, we add two columns.
    - A column that calculates the number of Present calls in a row. For example, for the first probe set of the BCR-ABL file, which has the data for its 15 samples in columns B3 to AE3, this would be calculated as "=COUNTIF(B3:AE3,"P")".
    - A column that calculates the percent of Present calls in a row.
- After replacing the functions (in the newly added last two columns) with values, we may remove all columns with the detection call information. Hence, each of the seven files will now have one column for each sample.
- Now we can combine all seven files. Starting with one of the files, we will be appending—one by one—columns from the remaining six files. It is a good idea to compare—after each append—the order of probe sets.

  Excel Tip

    - When the columns of a file are appended to the right of the current data, the first appended column includes the probe set names. Cut this column and insert it after the first column.
    - Insert an empty column after the two, and define there a function checking if the probe set names are the same in the first two columns. For example, for row 3, the function would be: "=IF(A3=B3,0,1)".
    - Copy this function to the remaining cells of column 3.
    - Check if the sum of values in column 3 is zero. If so, columns 2 and 3 can be removed, and the next file appended to the right.
- The last two columns of each of the combined seven files may be moved to the very right of the spreadsheet. The spreadsheet would now consist of the gene expression matrix with 14 additional columns holding the information about the number and percent of Present calls in each class.

6.2.2 Filtering Probe Sets

The main goal of filtering is to eliminate the probe sets (i.e., rows of the gene expression matrix) whose expression measurements are not reliable or represent experimental noise. Since this data set includes detection call information, we may filter variables by the *fraction of Present calls* in a class. This should remove a significant portion of unreliable measurements. In addition, we may apply filtering based on the intensity level, for example filtering by the *range of expression values*.

Filtering by the Fraction of Present Calls in a Class

Since numbers of samples per class are relatively small (four out of seven classes have fewer than 20 samples), we may filter out probe sets that have less than 50 percent of Present calls in each of the seven classes. There are 28,388 such probe sets.

Filtering by the Range of Expression Values

Assuming that we rescaled the data to the trimmed mean of 100 (for each array), we may decide to remove all probe sets for which the amplitude of their intensity values is less than 100 (Max–Min <100). There are 9980 such probe sets. However, 9473 of them are already included in the set of probe sets to be removed by the fraction of Present calls filter. Therefore, this filtering by the *range of expression values* removes only 507 additional probe sets.

> *Question: Explain the fact that the majority of probe sets targeted by this filter have already been included in the set of variables marked for removal by the fraction of Present calls filter (See Chapter 2)?*

After filtering and removing the columns with filter information, we have the expression data ready for analysis. The gene expression matrix includes 15,697 probe sets and 132 samples assigned to seven classes.

The unfiltered data consisted of 5,886,144 intensity values (44,592 probe sets times 132 samples). Only 28.09 percent of these values were associated with Present calls. After filtering, we have 2,072,004 intensity values, and 71.63 percent of them are associated with Present calls.

6.2.3 Designing a Multistage Classification Schema

There are several approaches to multiclass differentiation. One is to try and build a single classifier to differentiate all of the classes simultaneously. For example, Clemmensen and colleagues applied this approach to a subset of six classes selected from an older version of the St. Jude ALL data (Yeoh et al. 2002). Using sparse discriminant analysis, they identified a single multivariate marker consisting of twenty five variables (Clemmensen et al. 2008). In the original study of ALL3 data (Ross et al. 2003), two decision tree schemas were designed. One of them implemented a parallel approach, in which each leukemia subtype was discriminated against all other cases of the training set. The second, called "*the differential diagnosis decision tree approach*," implemented a multistage schema with predetermined sequence of differentiated classes. All three are valid approaches to multiclass classification. We suggest a multistage schema with a *data-driven* (rather than predetermined) sequence of differentiated classes. Furthermore, individual classifiers of the schema do not have to be limited to binary discrimination. When training data supports it, the multistage schema may include multiclass models.

When the class membership is the predominant factor differentiating biological samples of the training data, unsupervised approaches (such as hierarchical clustering or principal component analysis) may be used to decide on the sequence of models of a multistage schema. However, this does not need to be the case. Supervised methods represent a more general approach to the identification of sets of classes that should be differentiated at the subsequent stages of a multistage classification schema.

Stage 1: Seven Classes

To decide which of the seven classes (represented in the ALL3 data set) should be discriminated and which should be combined into a relatively homogeneous group at the first stage of the classification schema, we use a supervised approach. Though various supervised learning algorithms may be used for this purpose, we prefer those that provide a measure of class separation and overlaps. Ideally, a numerical measure describing class relationships would be accompanied by a low-dimensional projection of the discriminatory space. To accomplish this, we may use linear discriminant analysis (LDA) coupled with the stepwise hybrid feature selection driven by the T^2 measure of discriminatory power (as described in Chapter 3).

We begin by building a model that tries to separate all seven classes. We expect that one or two classes will dominate the separation, causing the remaining classes to overlap—this is a common situation in multiclass discrimination. When we run a T^2-driven feature selection, the heuristic search stops after a set of 11 variables is identified—the stopping criterion for the discriminatory power, $T^2 \geq 100$, has been achieved. A projection of the six-dimensional discriminatory space indicates that T-ALL and E2A-PBX1 are totally separated, and that the remaining five classes overlap (see Fig. 6.1).

To investigate whether the observed separation of T-ALL and E2A-PBX1 is not due to chance, we may build an ensemble of some 10 to 100 classifiers. Following the *modified bagging* schema (see Chapter 3), each of the classifiers is based on a randomly selected subset of the training samples. For example, each of the randomly generated training sets may include 80 percent of all training samples. This would leave 20 percent OOB samples that can be used as the test data for each classifier. Note that we are not interested in finding a classifier yet. The goal of this *modified*

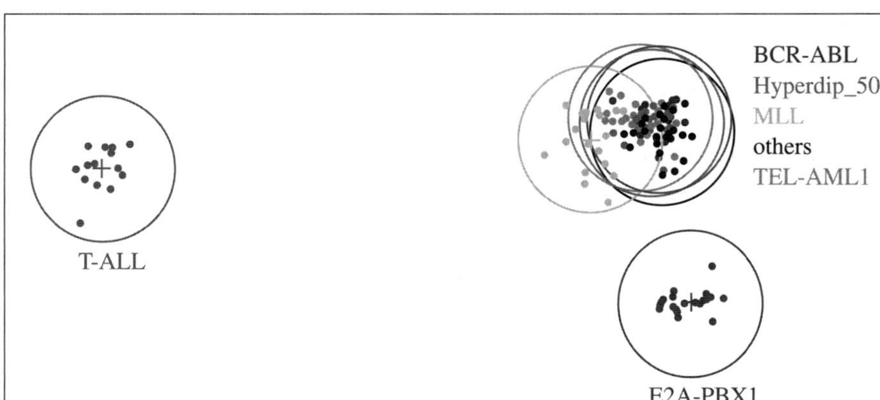

Figure 6.1: The classification model built on a set of 11 variables. When seven classes are differentiated, the discriminatory space has six dimensions (see Chapter 3). The projection onto a two-dimensional space of the first two linear discriminant functions (with the two largest eigenvalues) presented here represents 97.96 percent of the discriminatory information, which allows drawing conclusions about class separation in the six-dimensional discriminatory space. The circles represent the two-dimensional projections of the six-dimensional hyperspheres enclosing 95 percent of the probability in each class. The points are projections of the six-dimensional vectors representing training samples.

bagging experiment is to verify the assumption that the data support separation of T-ALL and E2A-PBX1 from the other five classes.

By investigating different cardinality sets identified during the T^2-driven heuristic search, we can see that subsets of three to five variables may have satisfactory discriminatory power (to at least separate one class from the others). We may decide to use the *modified bagging* schema to generate classifiers based on sets of five variables. For each classifier, an independent feature selection is performed and a multivariate marker of five variables is identified.

This experiment indicates that:

- T-ALL is separated from all other classes,
- in the majority of models, both T-ALL and E2A-PBX1 are totally separated from the other classes (see Fig. 6.2).

Based on the results of these experiments, we may decide on one of two plausible designs for the first stage of classification:

- to build a model that differentiates three classes: T-ALL, E2A-PBX1, and the remaining five groups treated as a single class, or
- to build two binary models—the first one separating T-ALL from the six remaining groups, and the second one to separate E2A-PBX1 from the remaining five groups.

Note:

For either scenario, biomarkers with very few variables can be found.

Stage 1: Three Classes

Assume that at stage 1 we differentiate three classes: T-ALL, E2A-PBX1, and all other samples combine into the third class. To verify this assumption, we first modify the meta-data in our gene expression matrix to include only three classes (all samples not belonging to T-ALL or E2A-PBX1 will be associated with the third class that we can call *5-Remaining*) and then use it as the base training data set to build an ensemble of 1000 classifiers. As before, the *modified bagging* schema may be implemented, which means that:

- each classifier is based on its own training set that includes about 80 percent of the samples that have been randomly selected from the base training data set,
- the remaining about 20 percent of the samples are OOB samples that can be used as test samples for this particular classifier.

To have about 20 percent of the OOB samples selected from each class, we assume the proportion of the OOB samples $\gamma_{OOB} = 0.2$ and calculate the number of OOB samples from class j as

$$nOOB_j = \text{int}(\gamma_{OOB} \cdot n_j + 0.99), \quad j = 1, \ldots, J \tag{6.1}$$

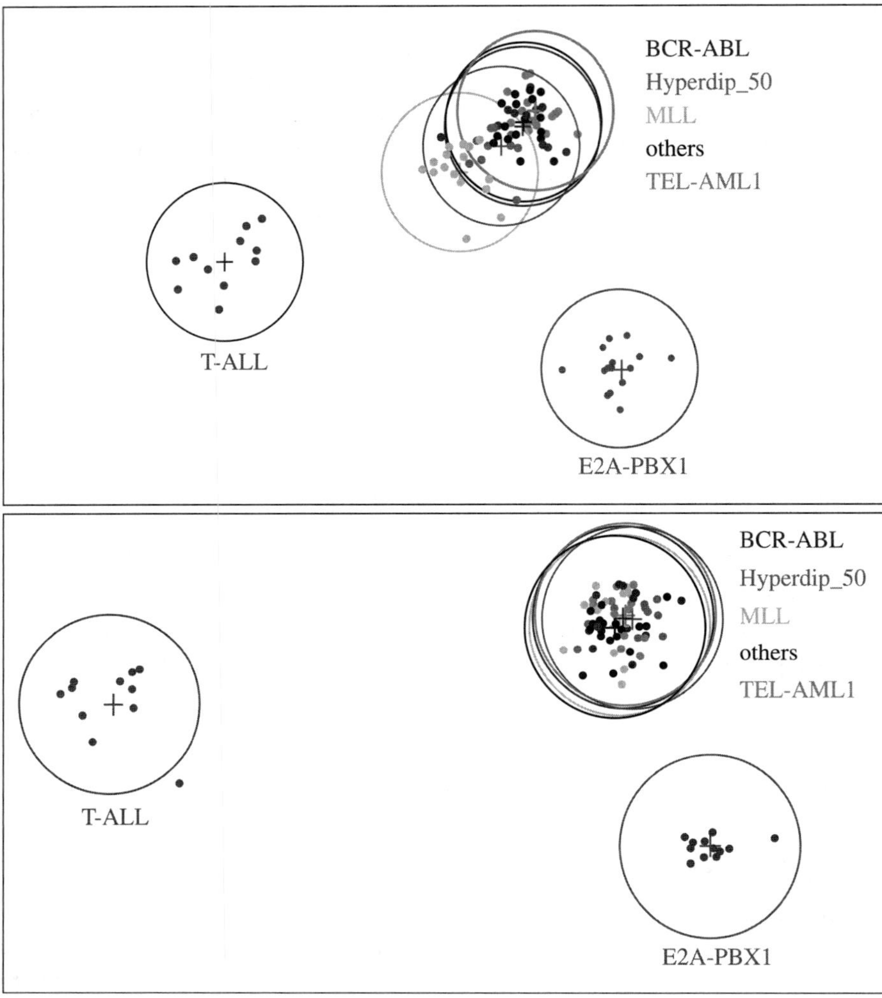

Figure 6.2: Two examples of classifiers based on multivariate sets of five variables. T-ALL and E2A-PBX1 are totally separated from the other classes. The remaining five classes heavily overlap. These two projections represent 92.9 percent and 98.5 percent of the discriminatory information in their respective six-dimensional discriminatory spaces.

This way, the total number of randomly selected OOB samples for each classifier can be calculated as

$$nOOB = \sum_{j=1}^{J} nOOB_j \tag{6.2}$$

For our training data with 14 T-ALL samples, 18 E2A-PBX1 samples and 100 samples in the *5-Remaining* class, each of the 1000 randomly generated training sets will include 11 T-ALL samples, 14 E2A-PBX1 samples, and 80 samples from the *5-Remaining* class. The size of each OOB set will be 27, with three T-ALL and four E2A-PBX1 samples. Please note that the number of different training sets that can be generated by the *modified bagging* schema is much larger than 1000, and can be calculated as

$$\prod_{j=1}^{J} \binom{n_j}{nOOB_j} = \binom{14}{3} \cdot \binom{18}{4} \cdot \binom{100}{20} \approx 5.97 \cdot 10^{26} \tag{6.3}$$

For each of the 1000 classifiers, a separate feature selection is performed. We may decide to stop each selection process when a set of five variables is selected (from all 15,697 variables represented in the training data). If the T^2-driven heuristic search is used for feature selection, an additional level of randomness may be added by starting each search from a randomly selected variable (see Chapter 3).

Note that we are still not in a biomarker building phase. All we aim to do now is to verify our initial assumption about differentiating the three classes at the first stage of the classification schema. From this experiment, we seek the following:

a) to determine if the three classes can be well separated by a multivariate marker of five variables,

b) an initial estimate of the accuracy for a classifier based on a marker of five variables.

To verify our assumption, we look at the discriminatory spaces of classifiers based on the randomly selected training sets and at the estimate of the misclassification error rate provided by classification of their OOB samples.

Results:

- 98.9 percent of the OOB samples are classified correctly (99.7 percent for T-ALL, 99.3 percent for E2A-PBX1, and 98.7 percent for the samples belonging to the *5-Remaining* class). This result is based on the classification of 27,000 samples (27 OOB samples for each of the 1000 classifiers).
- The discriminatory spaces of the models show that the three classes are totally separated (see an example in Fig. 6.3). Note that this only means that the *training samples* are well separated. Whether new samples can be correctly assigned to their true classes depends on how well the training data represents the underlying populations. Nevertheless, these results suggest that, when we enter the phase of building an optimal biomarker, we can consider biomarkers with fewer than five variables.

Stage 2: Five Classes

The next stage of multiclass differentiation includes only the five groups that previously belonged to the *5-Remaining* class. Since samples belonging to the subtypes addressed at stage 1 (T-ALL and E2A-PBX1) were excluded from consideration, our training set changes and we need to revisit probe set filtering. Using the same approach as before, that is, filtering by the *fraction of Present calls* in a class and filtering by the *range of expression values*, we now end up with 14,537 variables.

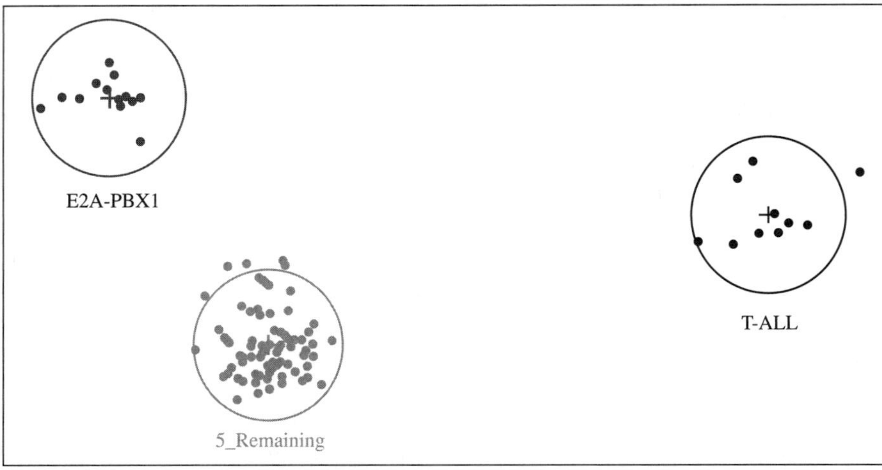

Figure 6.3: An example of a discriminatory space typical for classifiers built on five-gene sets selected from training sets generated by the *modified bagging* schema. Since three classes are differentiated here, the discriminatory space has only two dimensions. The circles represent constant density boundaries enclosing 95 percent of the probability in each class. The points represent the training samples.

As before, to check which group or groups dominate separation, we start with five classes, each class representing one group. When feature selection utilizing the T^2-driven heuristic search is performed, the process stops when a set of 14 variables with $T^2 \geq 100$ (strong indication of overfitting) is identified (Fig. 6.4). The examination of models with 1 to 14 variables identified during this process indicates that sets of four to eight variables may have satisfactory discriminatory power.

To decide, which combination of classes should be separated at this stage, we build an ensemble of 100 classifiers, utilizing the *modified bagging* schema with random selection of about 80 percent of samples into each bootstrap training set.

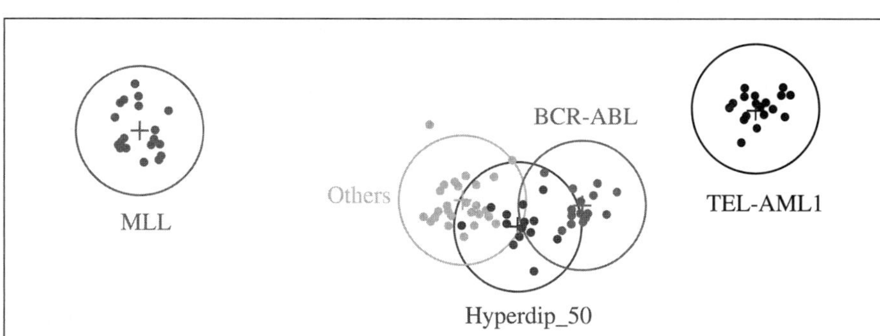

Figure 6.4: Discriminatory space of the classifier built on a multivariate set of 14 variables. This two-dimensional projection of the four-dimensional discriminatory space represents 98.6 percent of the discriminatory information.

This time we stop feature selection for each model when a set of eight variables is identified.

The discriminatory spaces of all identified models indicate a good separation of MLL from the other four classes. Many models separate two classes: MLL and TEL-AML1. Hence, we may also consider models that separate either three or two classes at this stage.

Stage 2: Three Classes

First consider separation of three classes: MLL, TEL-AML1, and the remaining three groups of samples treated as one class. We build a single model based on a marker of five variables (see Fig. 6.5). All three classes are well separated in its discriminatory space.

To check whether a model of five variables has a chance for good generalization, we use the *modified bagging* approach to build 1000 classifiers based on training sets consisting of randomly selected 80 percent of all training samples. When OOB samples are classified, only 92.8 percent of them are assigned to their true classes.

Stage 2: Two Classes

Now let's look at the discrimination of two classes only—MLL versus the remaining four groups treated as a single class. First, we build a single model with five variables and confirm that the training samples of the two classes are totally separated.

Next, we build 1000 classifiers based on the randomized training sets generated by the *modified bagging* schema. This time, 97.8 percent of the OOB samples are correctly classified, making this design more likely to deliver a generalizable classifier.

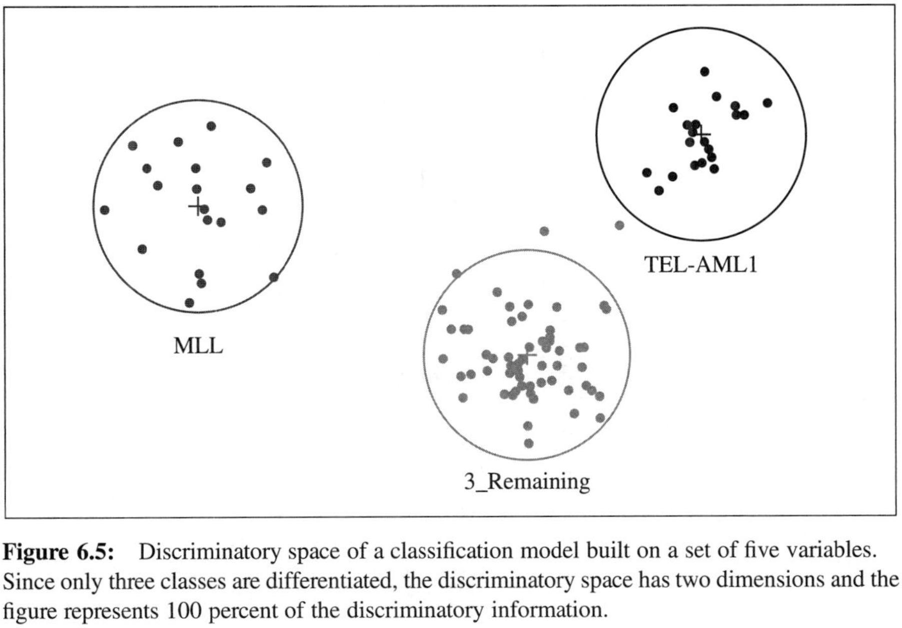

Figure 6.5: Discriminatory space of a classification model built on a set of five variables. Since only three classes are differentiated, the discriminatory space has two dimensions and the figure represents 100 percent of the discriminatory information.

TABLE 6.1: Multistage Classification of Seven ALL Classes

Stage	Model	Number of classes	Differentiated classes						
			T-ALL	E2A-PBX1	MLL	TEL-AML1	Hyperdip > 50	BCR-ABL	"Other"
1	Model 1	3	**T-ALL**	**E2A-PBX1**	**5-Remaining**				
2	Model 2	2			**MLL**	**4-Remaining**			
3	Model 3	2				**TEL-AML1**	**3-Remaining**		
4	Model 4	2					**Hyperdip > 50**	**2-Remaining**	
5	Model 5	2						**BCR-ABL**	**"Other"**

Based on these results, at stage 2 we may decide to differentiate two classes—MLL against all remaining samples combined into the *4-Remaining* class.

Remaining Stages of the Classification Schema

Following the approach described for the first two stages, we are designing the five-stage classification schema presented in Table 6.1.

6.3 IDENTIFYING THE *INFORMATIVE SET OF GENES* (EXERCISES 4.2–4.6)

We will illustrate the identification of the *Informative Set of Genes* on the example of Model 2 of the multistage classification schema designed for ALL3 data. This model differentiates MLL from the *4-Remaining* class that includes four groups of samples[3] (TEL-AML1, Hyperdiploid $>$ 50, BCR-ABL, and "Others"). The previously performed *modified bagging* experiments indicate that a classification model based on a marker of five variables may be expected to properly classify about 97.8 percent of the samples not seen by the model. Actually, we may anticipate a better performance out of a classifier built on the entire training set since the 97.8 percent average accuracy was based on models built from training sets including only 80 percent of the samples from the base training data. Furthermore, using the *Informative Set of Genes* for identification of robust biomarkers may also lead to a better than average performance of such biomarkers (see Chapter 4).

At this point, one may consider additional filtering of noise by criteria specific for two-class differentiation. Examples of such filters are as follows.

- Filtering by the *average expression level* in a class. For this data, we could consider retaining only those probe sets whose average intensity level exceeds a specified threshold (e.g., 100) in at least one of the two classes (McClintick and Edenberg 2006). This would constitute much more aggressive filtration than that already applied using the *range of expression values* (Max − Min $<$ 100). Here, it would remove an additional 2699 probe sets.
- Filtering by the *fold change*. In Chapter 2, we did not recommend using this as the only filtration method. Here, we add an additional warning against filtering by the fold change with the commonly encountered thresholds of 1.5 or greater. Filtering with such thresholds may add a strong univariate bias to the analysis. Genes whose average expression is not very different in the two classes may have expression patterns complementary to the expression patterns of some of the other genes. Their combined expression changes may define a strong multivariate pattern significantly separating the classes, even when genes contributing to the pattern are not individually discriminatory. However, with much lower thresholds, we may use the fold change filter for the additional elimination of noise (especially at lower expression levels).

[3]Hence, this *Informative Set of Genes* will contain discriminatory information significant for MLL subtype prediction.

Assume that we applied the *fold change* filter and eliminated probe sets whose ratio of the average expressions in the two classes was less than 1.1. Our gene expression matrix now includes 8950 variables.

To illustrate the identification and then verification of the *Informative Set of Genes*, we will use T^2-based stepwise hybrid feature selection and the *modified bagging* schema with LDA (details of these methods are described in Chapter 3). However, any supervised learning approach that can identify parsimonious multivariate markers and provide a measure of their discriminatory power may be used for this purpose.

6.3.1 The *Informative Set of Genes*

To identify the *Informative Set of Genes*, we generate a sequence of alternative multivariate biomarkers and use them to build alternative classification models (see Chapter 4 for a description of the method). Each alternative marker is a result of feature selection performed on a training set that does not include genes already selected into the previously identified markers. The stepwise hybrid search driven by the T^2 measure of class separation is used to identify alternative markers of five variables. The discriminatory power of subsequent alternative markers (and models) has a strong decreasing tendency that can be approximated by a logarithmic trend line (Fig. 6.6).

To decide at which point significant discriminatory information of the training data is exhausted, we investigate alternative models with different levels of

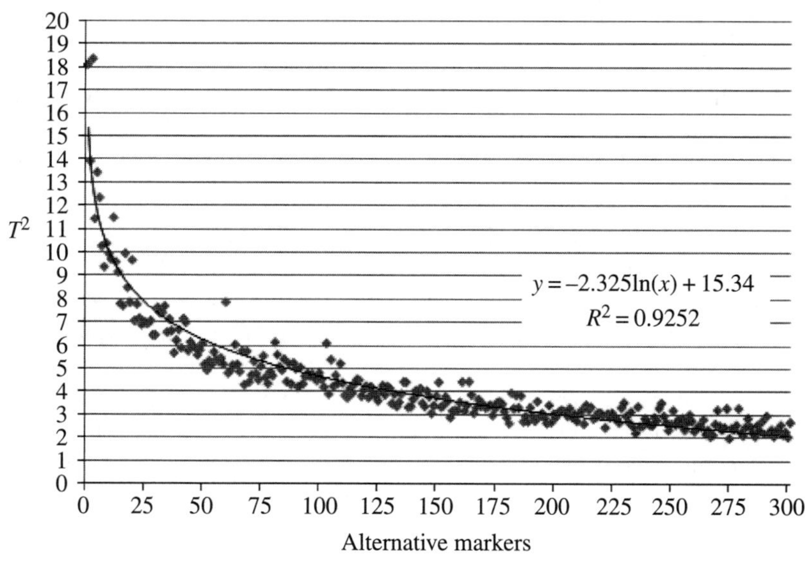

Figure 6.6: Decreasing discriminatory power of the subsequently identified alternative markers.

discriminatory power T^2. For this data, we decide[4] that a plausible cut-off is at $T^2 = 5$. The logarithmic trend line crosses this level in the neighborhood of the alternative marker 84 (M_{84}). We define the *Informative Set of Genes* as the set of genes included in these of M_0 to M_{84} markers[5] which have $T^2 \geq 5$. Since there are seventy one such markers, our *Informative Set of Genes* includes 355 genes. Let us call this set INF-355.

> ***Note:***
>
> If we decided on a cut-off at $T^2 = 4$, the informative set would include 600 genes selected into the first 120 markers with $T^2 \geq 4$.

To verify our selection of the *Informative Set of Genes*, we compare the estimated MLL sensitivity of the classifiers of three ensembles built on:

- the entire training set of 8950 variables,
- the informative set INF-355,
- the training set without variables included in INF-355.

In each case, the *modified bagging* schema is used to build 1000 classifiers based on randomly selecting 80 percent of the samples from each class. Please recall that separate feature selection is performed for each classifier. MLL sensitivity is estimated by the classification of the OOB samples.

The results are summarized in Table 6.2. When the training data include only the INF-355 variables, MLL sensitivity is 98.2 percent. This sensitivity drops about 21 percent (to 77.1 percent) when the classifiers are built using the data including all variables but those belonging to INF-355. Note that MLL sensitivity estimated for classifiers selecting variables from all 8950 variables is about 3 percent lower than the one for classifiers considering only the INF-355 variables. A plausible explanation of this fact is that the classifiers built on the *Informative Set of Genes* are less likely to fit noise.

6.3.2 Primary Expression Patterns of the Informative Set

We assume that the *Informative Set of Genes* includes gene expression patterns important for separation of MLL samples from samples belonging to the *4-Remaining* class. To identify these patterns, we start with clustering the INF-355 variables. Various clustering approaches may be used to identify clusters of genes with similar expression patterns. Here, we use a self-organizing map (SOM) with 16 neurons of the output layer organized into a 4×4 rectangular grid. Pearson correlation is used as a distance measure between the input gene expression vectors and the weight vectors

[4]We consider not only the T^2 discriminatory power of the models, but also look at their discriminatory spaces to evaluate the distribution of training data points in relation to the boundaries enclosing 95 percent of the probability in each class. Models with a T^2 of 5 and above look like they provide good discrimination. Models with a T^2 of 3 or 4 look like they provide a borderline class separation. This indicates that significant discriminatory information may be exhausted somewhere between $T^2 = 5$ and $T^2 = 4$.

[5]Marker M_0 is the optimal multivariate biomarker identified from the entire training set (including all 8950 variables).

TABLE 6.2: Summary of the Modified Bagging Experiments

Variables in training set	Number of variables to select	Number of classifiers	% OOB samples	MLL sensitivity %	Specificity %	Accuracy %	Average T^2	Number of perfect OOB classifiers
All 8950	5	1000	20	95.3	98.5	97.8	21.3	641
INF-355 only	5	1000	20	98.2	99.1	98.9	20.3	778
8950 minus INF-355	5	1000	20	77.1	93.2	90.0	6.7	107
8950 minus INF-600	5	1000	20	72.7	90.1	86.6	5.4	44
8950 minus top 100 univariate	5	1000	20	87.7	96.7	94.9	9.5	362

In each experiment, 1000 classifiers are built. Each of them is based on its own training set that includes 80 percent of the samples in each class by random selection. The remaining 20 percent of the samples are not used for training a classifier, and serve as OOB test samples to estimate the classification efficiency of the classifier. When classifiers are built from a training set including only the INF-355 variables, the maximum MLL sensitivity of 98.2 percent is achieved. When the 355 variables of the *Informative Set of Genes* are removed from the training data of 8950 variables, the MLL sensitivity is only about 77 percent. Removing the 600 variables that would constitute the informative set if it was defined for the cutoff of $T^2 = 4$ results in only about a 4 percent further decrease of the MLL sensitivity. This is a strong indication that there is very little MLL-specific discriminatory information left in the data after the INF-355 variables are removed. When 100 variables with top univariate scores are removed from the training data, MLL sensitivity of about 88 percent indicates that there is still some significant discriminatory information left in the data. The term *Perfect OOB classifiers* (in the last column) refers to the classifiers that correctly classify all of their OOB samples (see Chapter 4). Although averaging over 1000 classifiers yields quite stable results, differences of 0.5 percent or so should be considered meaningless since they are likely to occur due to the two levels of randomness built into the implemented *modified bagging* schema (the random selection of training samples and the random start of heuristic feature selection).

representing cluster prototypes. As a result, we have 16 clusters of genes with cluster sizes varying from 4 to 55 genes. Although very small clusters may suggest that their genes were selected into alternative markers by chance, we do not evaluate the importance of gene expression patterns solely on the size of clusters representing the patterns.

To identify primary expression patterns, we look at the distribution of each cluster's genes among the classifiers built with the *modified bagging* schema. We use two ensembles of 1000 classifiers. One ensemble is built on the entire training set of 8950 variables, the other on 355 variables of the INF-355 set only (see Table 6.2). The *average use* of cluster genes by the perfect OOB classifiers of these ensembles (i.e., the classifiers that correctly classify all of their respective OOB samples) is our measure of the importance of gene expression patterns represented by the clusters. The results (see Table 6.3) indicate that the genes of the same group of four clusters are most often used by the perfect classifiers of both ensembles. Gene expression

TABLE 6.3: Use of Clusters and Their Genes Among the Perfect OOB Classifiers of Two Ensembles[6]

		Ensemble built on 355 variables		Ensemble built on 8950 variables	
Cluster	Size	Cluster Use	Average Use	Cluster Use	Average Use
12	10	694	69.40	569	56.90
13	25	497	19.88	408	16.32
1	55	813	14.78	573	10.42
15	16	235	14.69	181	11.31
3	4	54	13.50	8	2.00
11	12	140	11.67	67	5.58
14	24	238	9.92	116	4.83
5	38	346	9.11	162	4.26
6	24	202	8.42	80	3.33
16	40	304	7.60	142	3.55
2	35	245	7.00	140	4.00
7	11	29	2.64	11	1.00
8	23	56	2.43	33	1.43
9	4	8	2.00	4	1.00
4	28	23	0.82	10	0.36
10	6	4	0.67	4	0.67

One ensemble of 1000 classifiers was built on the training set including all 8950 variables. The other ensemble of 1000 classifiers was built on the training set limited to the 355 genes of the *Informative Set of Genes*. The perfect OOB classifiers of both ensembles most often tap into the same four clusters. The *Cluster Use* column shows the number of times genes of a cluster are used in the classifiers. The *Average Use* is the average number of times a gene from the cluster is selected into the classifiers. Note that Cluster 1 is used 813 times in the 778 perfect OOB classifiers built on the INF-355 variables. This means that some of these classifiers were built on multivariate markers that included more than one gene from this cluster.

[6]Instead of using only the perfect OOB classifiers, we could use all classifiers of both ensembles (see Chapter 4).

patterns represented by these four *primary clusters* are called the *primary expression patterns*. We assume that these patterns represent the most important biological processes associated with the class differences. Hence, the interpretation of differences between MLL and the other four subtypes represented in this experiment may be based on the four identified primary expression patterns.

6.3.3 The Most Frequently Used Genes of the Primary Expression Patterns

The four primary clusters representing the *primary expression patterns* include 106 genes. These genes are not necessarily equally important for classification and for biological interpretation of class differences. We may assume that genes that are most often selected into the perfect OOB classifiers generated by the *modified bagging* schema are more important for class differentiation than other genes. To identify

TABLE 6.4: List of 21 *Frequent Primary* Genes of the Informative Set INF-355

Probe set	Cluster	Gene use in perfect OOB classifiers of the ensemble built on 355 variables	Gene use in perfect OOB classifiers of the ensemble built on 8950 variables	GenBank accession	Chip
203434_s_at	12	652	542	AI433463	A
210487_at	13	463	389	M11722	A
226415_at	1	371	295	AA156723	B
219686_at	15	191	142	NM_018401	A
219463_at	1	140	123	NM_012261	A
231899_at	1	76	36	AB051513	B
226546_at	1	50	33	BG477064	B
203076_s_at	15	32	36	U65019	A
209905_at	1	21	20	AI246769	A
203435_s_at	12	17	10	NM_007287	A
202603_at	1	16	11	N51370	A
204069_at	1	16	9	NM_002398	A
223046_at	12	12	11	AL117352	B
206875_s_at	13	11	13	NM_014720	A
218566_s_at	1	11	8	NM_012124	A
220668_s_at	1	22	5	NM_006892	A
219988_s_at	1	18	3	NM_018150	A
223467_at	13	15	4	AF069506	B
203375_s_at	1	11	2	NM_003291	A
202365_at	1	10	3	BC004815	A
219036_at	1	9	3	NM_024491	A

The 15 genes at the top of the list are the *most frequent primary* genes (selected into at least one percent of the perfect OOB classifiers of the two ensembles, one built on the training set including all 8950 genes, the other built on the INF-355 genes only).

such genes, we will revisit the two ensembles of 1000 classifiers (one built on all 8950 variables and the other on the 355 genes of the informative set). This time, however, we will look at the distributions of the primary cluster genes among the five-gene markers used in the perfect OOB classifiers. By defining the frequently used genes as those that are selected into at least one percent of the perfect OOB classifiers of the ensemble built on the training set including only the INF-355 variables, we identify 21 genes that are considered the frequently used genes of the primary clusters.[7] We call these genes the *frequent primary* genes. Fifteen of these 21 genes are used in at least one percent of the perfect OOB classifiers of both ensembles; they are the *most frequent primary* genes of the informative set. Information about all frequent primary genes is presented in Table 6.4.

6.4 USING THE *INFORMATIVE SET OF GENES* TO IDENTIFY ROBUST MULTIVARIATE MARKERS (EXERCISE 4.8)

The main goal of searching for the *Informative Set of Genes*, its primary patterns and their most frequently used genes is to facilitate biological interpretation of class differences. Before we identified the informative set, we could already have a parsimonious multivariate biomarker built on the training set including all 8950 variables. However, such a marker may have a higher chance of overfitting the training data than a marker built from the genes representing the primary expression patterns associated with class differentiation. When we use the *Informative Set of Genes* or its subset of the *frequent primary* genes, we may expect not only a more robust biomarker, but also one that may be associated with more plausible biological

[7]Please check that some of these genes are quite far on the univariately ordered list of 8950 genes.

TABLE 6.5: The Multivariate Marker of Five Genes Identified for Model 2 (of the Multistage Classification Schema) Differentiating MLL from the Four Other ALL Subtypes

Probe set	Cluster	GenBank accession	Chip
203434_s_at	12	AI433463	A
210487_at	13	M11722	A
226415_at	1	AA156723	B
226546_at	1	BG477064	B
204069_at	1	NM_002398	A

Three of the four primary clusters are represented in this biomarker. Three genes of this biomarker are selected from the largest 55-gene Cluster 1. This suggests that there are at least three subpatterns in this cluster that are joined at the level of similarity used by applied SOM clustering. For biological interpretation, we may either try to interpret the super pattern of Cluster 1 or split it into a few subpatterns (by further clustering the gene expressions of Cluster 1).

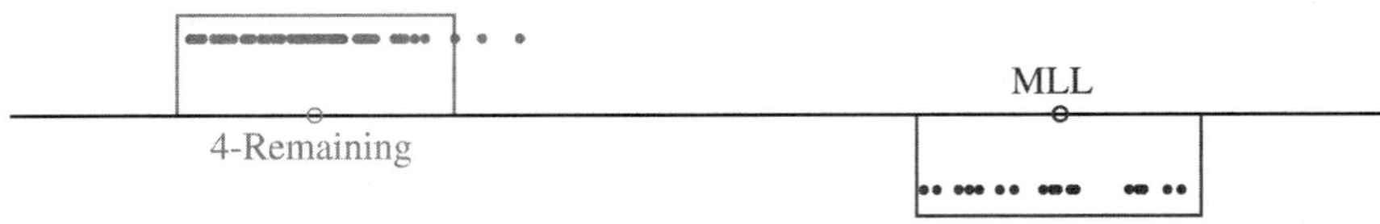

Figure 6.7: Discriminatory space of the five-gene biomarker for Model 2 differentiating MLL from the four other ALL subtypes. Since two classes are differentiated, the discriminatory space is one-dimensional and is represented by the horizontal line. The vertical offset is added only to improve the visibility. The classes are represented by their centroids and the segments (of the discriminatory direction) that include 95 percent of the probability in each class. The points (or more precisely, their horizontal coordinates) represent the training samples. The fact that the training samples are well separated in this discriminatory space is only a necessary but not satisfactory condition for the marker to be generalizable. However, the fact that this marker is based on the *most frequent primary* genes of the *Informative Set of Genes* suggests that it may be more robust than similar markers based on all the variables of the original training set.

interpretation. When we perform T^2-based stepwise hybrid feature selection using only the 15 *most frequent primary* genes, we identify the multivariate biomarker of five genes presented in Table 6.5. Figure 6.7 shows the discriminatory space of an LDA classifier built on this marker.

> *Question: Use the INF-355 training set with other multivariate methods that can identify a biomarker consisting of a small number of genes and can estimate its generalization. One such method is Recursive Feature Elimination often implemented with support vector machines or random forests learning algorithms. Do the resulting biomarkers prefer genes that are on the short list of the 21 frequent primary genes?*

6.5 VALIDATING BIOMARKERS ON AN INDEPENDENT TEST DATA SET (EXERCISE 4.8)

To validate the five-gene biomarker identified in the previous section, we search public repositories for a data set with samples representing subtypes of pediatric ALL and processed on the same or a compatible microarray platform. In Gene Expression Omnibus we can find a recently published data set GSE13351 (Den Boer et al. 2009). This data set is also available in ArrayExpress under the accession number E-GEOD-13351. It includes 107 cases representing the seven ALL subtypes. To use it as a test set for a Model 2 classifier, we will use a subset of 90 samples (four MLL samples and 86 samples representing the four subtypes of the *4-Remaining* class). The samples were processed on Affymetrix GeneChip HG-U133 Plus 2.0 microarrays. Though this is a different array than that used for the ALL3 data, the two designs are compatible in such a way that all probe sets represented on the HG-U133A and HG-U133B arrays are also represented on the Plus 2.0 array.

TABLE 6.6: Testing Six Classifiers on an Independent Test Set

			Classification of independent test data					
Feature selection method	Marker size	Learning algorithm	MLL	Sensitivity %	4-R	Specificity %	Accuracy %	Notes
T^2 search	5	LDA	4 of 4	100	86 of 86	100.0	100.0	337072 close to the decision boundary
Top 5 of 21 frequent primary	5	LDA	4 of 4	100	86 of 86	100.0	100.0	337072 close to the decision boundary
Recursive feature elimination	5	SVM	4 of 4	100	85 of 86	98.8	98.9	337072 misclassified
		LDA	4 of 4	100	86 of 86	100.0	100.0	337072 close to the boundary
		D-LDA	4 of 4	100	84 of 86	97.7	97.8	337066, 337072 misclassified
		KNN ($k = 3$)	4 of 4	100	86 of 86	100.0	100.0	

Three multivariate markers of five genes are used to build the classifiers. T^2-driven stepwise hybrid feature selection (T^2 search) has been performed on the training set that included only the 15 *most frequent primary* genes to identify the marker shown in Table 6.5. Then this marker is used to build an LDA classifier. Another LDA classifier uses the top five genes of the set of 21 *frequent primary* genes of the *Informative Set of Genes* (these five genes represent all four primary clusters). Support vector machine (SVM) is used to perform recursive feature elimination (RFE) on the training set including all 355 genes of the *Informative Set of Genes*. SVM-RFE identifies the third marker of five genes. Not surprisingly, all genes selected into this marker belong to the set of 21 frequent primary genes. This third marker of five genes is then tested using Linear SVM and three other learning algorithms. D-LDA is the Diagonal Linear Discriminant Analysis that ignores correlations among the variables and works on the diagonal covariance matrices. Here, the *k*-Nearest Neighbor (KNN) method is represented by the 3-Nearest Neighbor algorithm (which performs very well on this data). A closer look at the results of the testing on the independent data indicates that all these learning algorithms performed comparably well. The only differences are related to classification of two samples (GSE337072 and GSE337066), which are always close or very close to a decision boundary (see an example in Fig. 6.8). This may suggest that these samples are either outliers in their class or members of a subpopulation underrepresented in the training data.

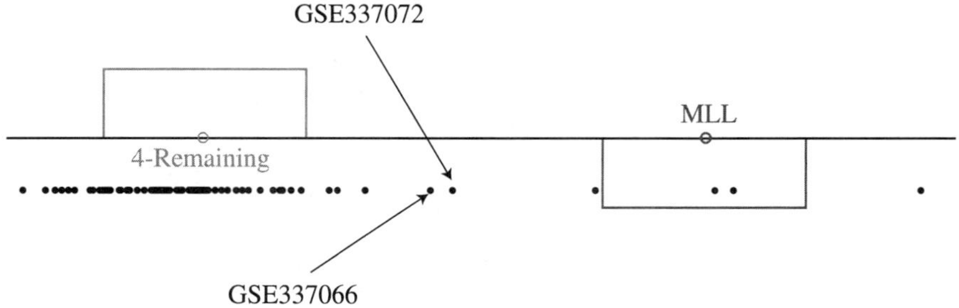

Figure 6.8: The results of the classification of the independent test set (GSE13351) by the LDA classifier based on the five-gene biomarker identified by the T^2-driven stepwise hybrid feature selection. As in Figure 6.7, the discriminatory space is one dimensional and is represented by the horizontal line; the vertical offset is added only to improve the visibility of the results. The points represent the classification results of the biological samples of the independent test set. Since the classifying software does not know the true classes of the test samples, it paints all points with the same vertical offset. All four MLL samples are correctly classified to the MLL class. All 86 samples representing the four subtypes of the *4-Remaining* class are also correctly classified into their true class. However, two of these 86 samples (GSE337072 and GSE337066) are close to the decision boundary.

Take the following steps.

- Either download the probe set intensity data and rescale each array to the same target as used for the ALL3 data, or download the CEL files and preprocess them using the Affymetrix *Expression Console* software.
- Log2 the preprocessed (or rescaled) test data.

Table 6.6 summarizes the results of testing several classifiers built from the training set including only the 355 genes of the *Informative Set of Genes*. Each classifier is used to classify samples of the independent test set represented by the GSE13351 data.

6.6 USING A TRAINING SET THAT COMBINES MORE THAN ONE DATA SET (EXERCISES 3.5 AND 4.1–4.8)

In this exercise, we will combine two pediatric ALL data sets. One is the ALL3 set used in the previous exercises. The other is the GSE13425 data set (Den Boer et al. 2009) recently uploaded to GEO (also available in ArrayExpress as E-GEOD-13425). The samples of the GSE13425 data set have been processed on Affymetrix HG-U133A 2.0 microarrays, which are compatible with the HG-U133 chips used for ALL3. The GSE13425 data set includes 190 samples associated with the same seven ALL subtypes that are represented in ALL3. Since the two data sets originate not only in different medical centers but also in different countries, we may hope that the combined training set will better represent the underlying populations of the ALL subtypes.

This may lead to more robust biomarkers and classifiers. However, there is also a disadvantage in combining these particular sets. The GSE13425 data is limited to *Chip A* probe sets, which means that we will need to discard the discriminatory information contained in the *Chip B* probe sets of the ALL3 data set.

To be able to compare the results of this exercise with the previous results, we may prepare and then use the combined training set to identify the *Informative Set of Genes* and biomarkers for the same stage of the multistage classification schema (i.e., Model 2 that differentiates MLL from other four ALL subtypes).

6.6.1 Combining the Two Data Sets into a Single Training Set

Since neither the GSE13425 nor the E-GEOD-13425 probe set level expression data include the detection call information that we want to use for filtering unreliable variables, we will download the CEL files and perform low-level preprocessing using the Affymetrix *Expression Console* software.

- From *ArrayExpress*, download the 190 CEL files of the GSE13425 (E-GEOD-13425) data set.
- From *ArrayExpress*, download the *Detailed sample annotation* (or SDRF file) for this data set, and save it as an Excel file.
- From Affymetrix, download and install the *Expression Console* software.
- Open the *Expression Console* application and:
  - download the library and annotation files for the HG-U133A microarray (this will define report controls and thresholds),
  - create a new study,
  - add the 190 CEL files to the study,
  - define a new advanced configuration of the MAS5 algorithm and specify a target intensity of 100,
  - run the analysis by executing this advanced MAS5 configuration,
  - set the report options to include *Signal* and *Detection Calls*,
  - export the probe set data to a text file,
  - save the study.
- Open the exported data in Excel.
  - Using the *Detailed sample annotation*, associate each column with a biological sample and its ALL subtype.
  - Remove the control probe sets: the 68 probe sets with names starting with AFFX and the set of 100 normalization controls (200000_s_at to 200099_s_at). The file should now include 22,115 probe sets.
  - Append, to the right of the spreadsheet, the *Chip A* data of the ALL3 file (including the *Signal* and *Detection Call* columns). Be sure that the header rows of both files represent the same type of information and that the order of the probe sets is identical.

- Since in this exercise we are interested in differentiating MLL from the other four ALL subtypes (as identified for Model 2 of the multistage classification schema), remove the T-ALL and the E2A-PBX1 samples.
- Rearrange the column order so that the samples of the same subtype are together.

6.6.2 Filtering Probe Sets of the Combined Data

- Filter by the *fraction of Present calls* and remove the 13,132 probe sets that have no class with at least 50 percent of Present calls.
- Filter by the *range of expression values* and remove the additional 131 probe sets with amplitude of expression level below 100 (Max − Min < 100). There are 5951 such probe sets but almost all of them are already removed by the fraction of Present calls filter.
- Filter by the *fold change* and remove the additional 1962 probe sets whose ratio of the average expressions in the two classes is less than 1.1.

After removing the Detection Call columns, we should have the data in the form of a gene expression matrix with 6890 variables and 241 samples (24 MLL samples and 217 samples of the *4-Remaining* class).

6.6.3 Assessing the Discriminatory Power of the Biomarkers and Their Generalization

If we had not already performed exercises for Model 2, we would assess here the approximate size of a multivariate biomarker by performing feature selection experiments and checking the discriminatory power and the class separation of various cardinality markers. However, the previous results allow us to expect that a marker of five or so variables still should be sufficient to separate the classes well. Due to the increased heterogeneity of the training data, we may also expect new markers to be more robust, that is, less fit to the training data but still well generalizable. When we design an ensemble of 1000 classifiers based on the randomized training sets generated with the use of the *modified bagging* schema, we notice that the average T^2 measure of discriminatory power of these classifiers is significantly lower than when only the ALL3 data were used. However, they provide similar MLL sensitivity, specificity, and accuracy in the classification of their OOB samples.

6.6.4 Identifying the *Informative Set of Genes*

As described in Chapter 4, identification of the *Informative Set of Genes* is based on a sequence of alternative markers (models) resulting from feature selection performed on the training data from which all the variables selected into the previously identified alternative markers have been excluded. Here, we build a sequence of alternative markers of eight variables.[8] The discriminatory information in the data is quickly

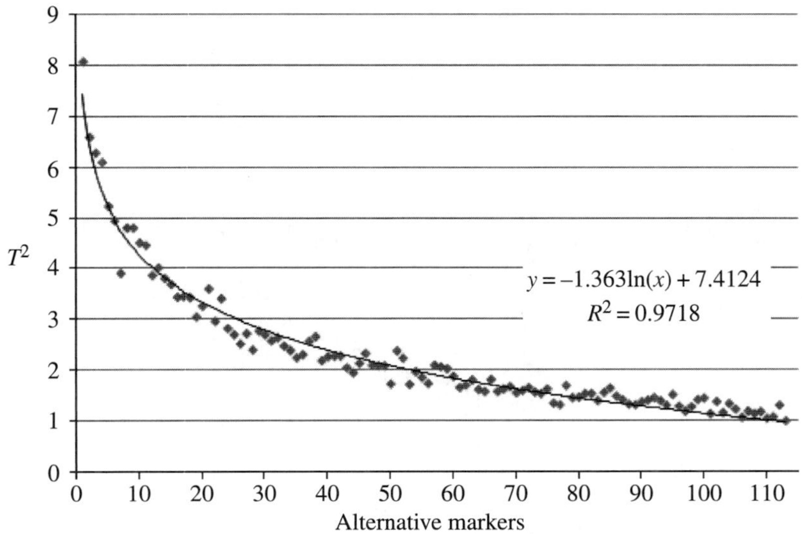

Figure 6.9: Alternative models built on the training set combining the ALL3 and the GSE13425 data sets. The quickly decreasing discriminatory power of subsequent alternative markers is approximated by a logarithmic trend line. The variables selected into the alternative markers represented by the points above the horizontal line $T^2 = 3$ and to the left of the point in which the trend line crosses this horizontal line are included in the *Informative Set of Genes*. There are 22 such alternative markers. Hence, the *Informative Set of Genes* includes 176 genes.

exhausted and we define the *Informative Set of Genes* as the set of 176 genes included in the 22 markers with $T^2 \geq 3$ (see Fig. 6.9). We will call this set INF-176.

To verify our selection of the *Informative Set of Genes*, we estimate the MLL sensitivity using three ensembles of 1000 classifiers built on:

- the entire training set of 6890 variables,
- the informative set INF-176,
- the training set without the variables included in INF-176.

In each case, the classifiers are based on randomized training sets generated by the *modified bagging* schema (which includes independent feature selection performed for each classifier). They are then used to classify their respective OOB samples.

The results are summarized in Table 6.7. Classifiers based on the multivariate markers selected from the INF-176 set correctly classify 95.1 percent of the MLL OOB samples. When the INF-176 variables are excluded from the training set, the MLL sensitivity is only 76.2 percent. This is a clear indication that no significant discriminatory information is left in this set. Figure 6.10 presents a heat map of the *Informative Set of Genes*.

[8]We are increasing the size of the alternative markers to account for the decreased discriminatory power of a single marker and for the decreased amount of discriminatory information in the data (since only *Chip A* data are included).

TABLE 6.7: Summary of the *Modified Bagging* Experiments on the Training Set Combining ALL3 and GSE13425 Data Sets

Variables in training set	Number of variables to select	Number of classifiers	% OOB samples	MLL sensitivity %	Specificity %	Accuracy %	Average T^2	Number of perfect OOB classifiers
All 6890	5	1000	20	94.2	99.3	98.8	6.2	548
INF-176 only	5	1000	20	95.1	99.4	99.0	6.1	607
6890 minus INF-176	5	1000	20	76.2	95.5	93.5	2.1	27
6890 minus top 100 univariate	5	1000	20	84.8	97.6	96.3	2.9	134

The MLL sensitivity results indicate that the INF-176 *Informative Set of Genes* includes the most significant MLL-specific discriminatory information. When 1000 classifiers are built from the INF-176 set, the MLL sensitivity is 95.1 percent. Furthermore, over 60 percent of these classifiers provide perfect classification of their OOB samples. When the training set includes all but the INF-176 variables, the MLL sensitivity drops to about 76 percent and only 2.7 percent of the classifiers can correctly classify all of their OOB samples. This indicates that very little MLL-specific discriminatory information is left in the data after the INF-176 variables are removed.

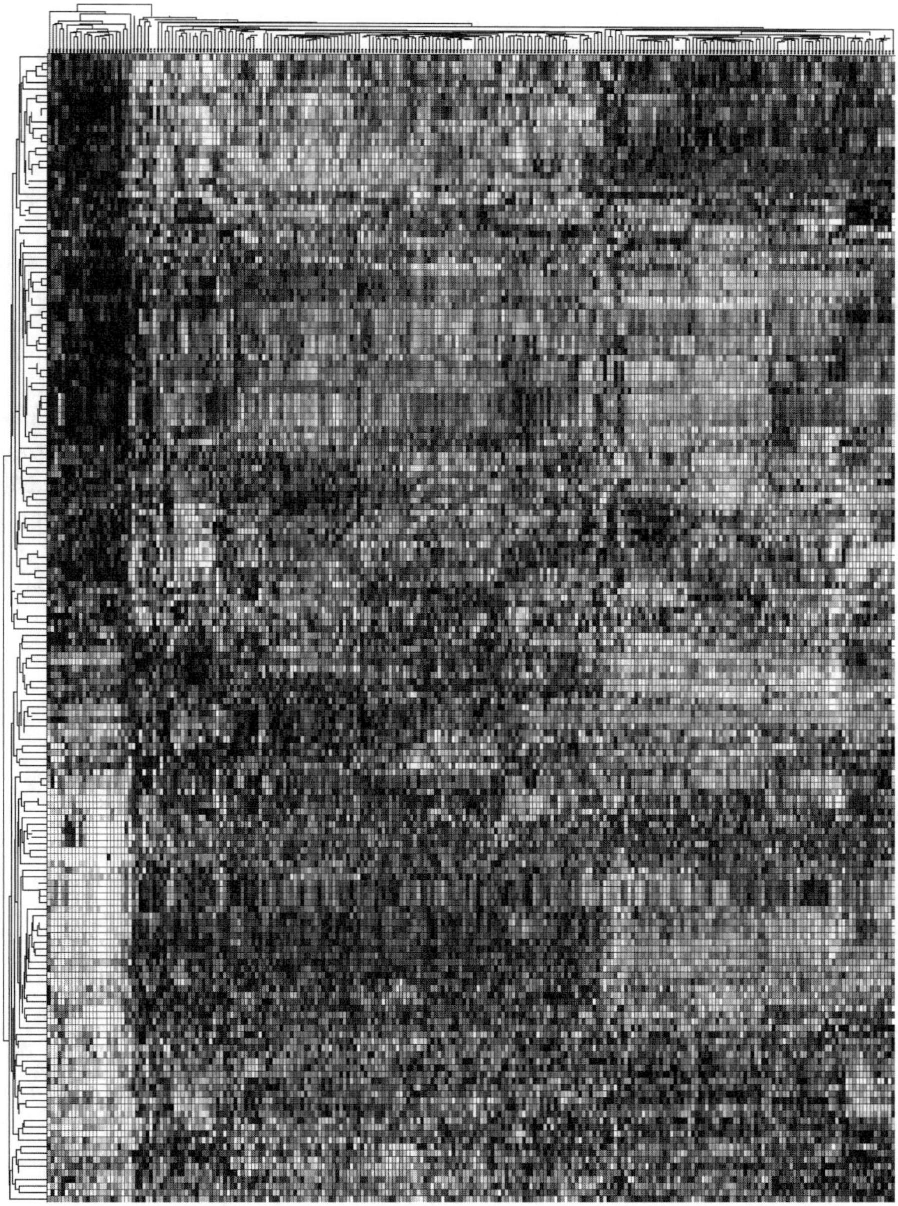

Figure 6.10: A heat map representing the two-way clustering of the INF-176 *Informative Set of Genes*. The dendrogram at the top of the image represents the results of hierarchical clustering of the training samples. The leftmost cluster of this dendrogram includes all 24 MLL samples. The dendrogram at the left side of the image represents the results of clustering genes. Since the multivariate approach was used to identify the *Informative Set of Genes*, some of them are far from the top of a univariately ordered list of genes. Individually, they do not discriminate the two classes. However, their expression patterns are complementary to the patterns of some other genes in a way that a joint expression pattern of a small subset of such complementary genes provides good separation of the MLL samples from the samples of the other class. (See color insert.)

6.6.5 Primary Expression Patterns of the *Informative Set of Genes*

To identify groups of similar expression patterns included in the *Informative Set of Genes*, we cluster the 176 variables of the INF-176 set. We use a SOM with a 3×3 rectangular topology of the output layer and with Pearson correlation as a distance measure between the vectors representing gene expressions and the weight vectors of cluster prototypes. The resulting nine clusters include 1 to 44 genes.

To identify the primary expression patterns, we check how often the genes of these nine clusters are used by the two ensembles of 1000 classifiers. One of these ensembles is built on the training set with all 6890 variables, the other—on the 176 variables of the *Informative Set of Genes* only (see Table 6.7). We examine only the perfect OOB classifiers, that is, the classifiers that correctly classify all their OOB samples. The genes of the same group of four clusters are most often selected into the perfect OOB classifiers of the two ensembles. These clusters also have the highest average use of their genes (the results are summarized in Table 6.8). These four primary clusters represent the *primary expression patterns* (see Fig. 6.11)—the patterns associated with the most important biological processes underlying the differences between MLL and the other four ALL subtypes considered by Model 2.

TABLE 6.8: Use of Clusters and Their Genes Among the Perfect OOB Classifiers of Two Ensembles

		Ensemble built on 176 variables		Ensemble built on 6890 variables	
Cluster	Size	Cluster Use	Average Use	Cluster Use	Average Use
9	51	1728	33.88	1453	28.49
4	16	463	28.94	413	25.81
2	16	218	13.63	214	13.38
1	44	437	9.93	203	4.61
6	11	78	7.09	35	3.18
8	10	37	3.70	19	1.90
7	22	42	1.91	31	1.41
3	5	4	0.80	3	0.60

One ensemble of 1000 classifiers was built on the training set that included all 6890 variables, the other—on the training set limited to the 176 genes of the *Informative Set of Genes*. In both ensembles, the perfect OOB classifiers most often used genes of the same four clusters. The *Cluster Use* column shows the number of times the genes of a cluster were selected into these classifiers. The *Average Use* represents the average number of times a gene from the cluster was selected into the classifiers. Note that Cluster 9 is used more than 548 times by the perfect OOB classifiers of the ensemble built on all 6890 variables, and more than 607 times by the perfect OOB classifiers of the ensemble built on the *Informative Set of Genes*. This means that many of these classifiers use more than one gene from this cluster. This may prompt further clustering of the 51 genes of this cluster in order to identify subpatterns in its expression pattern. Note also that Cluster 5 is not shown in the table; its single gene has not been selected into any of the perfect OOB classifiers.

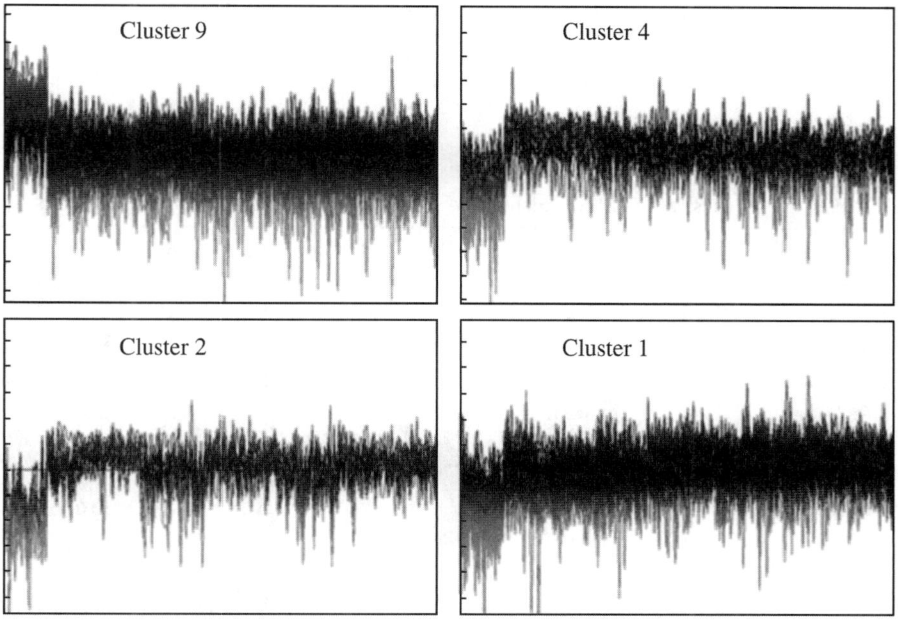

Figure 6.11: The four *primary expression patterns* of the *Informative Set of Genes* identified from the combined training data set (ALL3 + GSE13425). (See color insert.)

6.6.6 The Most Frequently Used Genes of the Primary Expression Patterns

The four primary clusters representing the *primary expression patterns* include 127 genes. To find out which of them are the most representative for class discrimination and for biological interpretation of the class differences, we examine how often these genes are selected into the perfect OOB classifiers of the same two ensembles that were used to identify these primary patterns (one ensemble of 1000 classifiers selecting from all 6890 variables, and the other ensemble of 1000 classifiers selecting variables from the INF-176 set only).

We define the *frequent primary* genes as those genes of the primary clusters that are selected into at least one percent of the perfect OOB classifiers of the ensemble built on the *Informative Set of Genes*. The *most frequent primary* genes are those that are selected into at least one percent of the perfect OOB classifiers in both ensembles. The four primary clusters include 27 frequent primary genes.[9] Eighteen of them are the most frequent primary genes. Information about these genes is presented in Table 6.9. Figure 6.12 shows expressions of the 18 most frequent primary genes of the informative set.

Please note that it is a common practice for gene expression studies to use the terms *probe set* and *gene* as synonyms. However, more than one probe set may be associated with the same gene. For example, 27 probe sets listed in Table 6.9

[9]Please check that some of these genes are quite far on a univariately ordered list of 6890 genes.

TABLE 6.9: List of 27 *Frequent Primary* Genes of the *Informative Set of Genes*

Probe set	Cluster	Gene use in perfect OOB classifiers of the ensemble built on 176 variables	Gene use in perfect OOB classifiers of the ensemble built on 6890 variables	GenBank accession	Gene symbol
219463_at	9	596	528	NM_012261	C20orf103
209905_at	9	434	374	AI246769	HOXA9
203434_s_at	4	408	386	NM_007287	MME
201153_s_at	9	273	248	N31913	MBNL1
209354_at	1	242	134	BC002794	TNFRSF14
219686_at	2	207	204	NM_018401	STK32B
204069_at	9	144	123	NM_002398	MEIS1
201151_s_at	9	76	44	BF512200	MBNL1
209038_s_at	1	71	22	AL579035	EHD1
211066_x_at	9	53	60	BC006439	PCDHGC3
211126_s_at	9	52	33	U46006	CSRP2
208729_x_at	1	48	8	D83043	HLA-B
212237_at	9	32	11	N64780	ASXL1
208112_x_at	1	24	11	NM_006795	EHD1
206492_at	4	22	7	NM_002012	FHIT
200989_at	1	17	12	NM_001530	HIF1A
220448_at	9	9	12	NM_022055	KCNK12
200953_s_at	2	8	7	NM_001759	CCND2
212762_s_at	1	13	5	AI375916	TCF7L2
208779_x_at	4	13	5	L20817	DDR1
208690_s_at	4	6	5	BC000915	PDLIM1
219821_s_at	1	10	4	NM_018988	GFOD1
204304_s_at	9	12	3	NM_006017	PROM1
203795_s_at	9	12	3	NM_020993	BCL7A
218764_at	4	7	2	NM_024064	PRKCH
203837_at	9	11	1	NM_005923	MAP3K5
213150_at	9	6	1	BF792917	HOXA10

The 18 genes at the top of the list are the *most frequent primary* genes (selected into at least one percent of the perfect OOB classifiers of the two ensembles, one built on the training set including all 6890 genes, the other on the INF-176 genes only).

correspond to 25 unique gene symbols. Although each of these 27 probe sets has a different GenBank accession,[10] the probe sets 201151_s_at and 201153_s_at are associated with gene symbol MBNL1, whereas probe sets 208112_x_at and 209038_s_at with EHD1. Such "redundant" probe sets are associated with different

[10]The GenBank database accepts all submitted sequences, which means multiple GenBank accessions may represent the same gene. Only after curation by experts a sequence may be elevated to the status of a reference sequence (Gibson and Muse 2009).

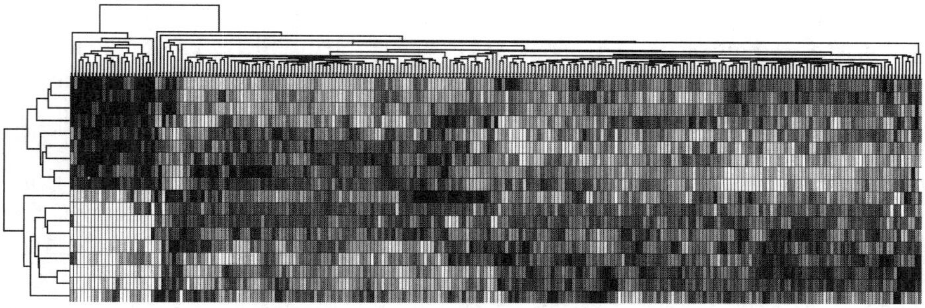

Figure 6.12: A heat map of the expression data of the 18 *most frequent primary* genes of the *Informative Set of Genes*. The leftmost cluster of the top dendrogram includes all 24 MLL samples of the combined training set. (See color insert.)

regions of the targeted gene. If those regions belong to different exons, the probe sets may represent different splice variants of the gene.

Biological Interpretation of the Most Frequent Primary Genes (Contributed by Dr. Preethi Gunaratne from Baylor College of Medicine)

The presented multivariate approach identified a set of 176 informative genes that are important for discrimination of MLL from four other B-cell ALL subtypes (TEL-AML1, Hyperdiploid > 50, BCR-ABL, and "Others"). Eighteen of these genes are most frequently used by a large number of bootstrap-based classifiers and they constitute the best starting point for biological interpretation of the results. This set of 18 most frequent primary genes consists of genes with diverse functions and cellular locations. Among them are genes that regulate a number of important cellular processes that have been linked with cancer, including cellular differentiation, cell adhesion, and regulation of cell proliferation and survival. This is very illuminating from the point of view that cancer cells display properties that overlap with cells that have either undergone developmental arrest or retained stem cell characteristics. In that context, the presence of at least four key developmental genes (HOXA9, MEIS1, ASXL1, and CSRP2) in the set suggests that one of the reasons for poor prognosis of MLL, as opposed to other B-cell ALL subtypes, is likely due to premature termination of the developmental cascades in these cells. The other genes that may have then contributed to the malignancy include genes that promote cell proliferation (CCND2, HIF1A) and cell survival (TNFRSF14). Finally, cell–cell communication and adhesion properties are also significantly different in differentiated cells as compared to stem cells. The former are characterized by increased cell-to-cell communication and adhesion, and the latter by decreased adhesion required for stem cells self-renewal. But for a few other genes (MBNL1—an RNA-binding protein, STK32B—a serine/threonine kinase, and KCNK12—a potassium channel gene), the majority of these 18 genes have a direct link to stem cell related processes.

Genes Linked to Cancer Related Pathways

Developmental Genes The first cellular process that is striking in the signature is a set of genes involved in the specification and development of the hematopoietic lineage. These include the homeodomain proteins HOXA9 and MEIS1, a polycomb group (PcG) repressor ASXL1, and a LIM domain protein CSRP2. HOX proteins are a family of DNA binding transcription factors that play a central role in the specification of segmental body patterns during development. MEIS1 is a lineage-specific transcription factor that is expressed in hematopoietic stem and progenitor cells. HOXA9 and Hox-cofactor MEIS1 have been highly linked in MLL-rearranged leukemias (Dou and Hess 2008; Faber et al. 2009; Li et al. 2009). Increased HOXA9 is linked with coordinate up-regulation of HOXA10, MEIS1, PBX3, and MEF2C, which together lead to increased cell survival and poor prognosis (Faber et al. 2009). ASXL1 is a member of the PcG group of proteins that play an important role in the maintenance of the pluripotent embryonic stem cells in self-renewal, through the stable repression of homeotic loci. Mutations in this locus have been associated with myelodysplastic syndrome and chronic myelomonocytic leukemia (Gelsi-Boyer et al. 2009). This gene has also been identified as a fusion partner with PAX5 resulting in the inactivation of this gene in human cancers (An et al. 2008). Another gene, CSRP2, is a member of a family of genes containing LIM domains. This group of proteins has been highly implicated in cellular differentiation in the developing embryo. Collectively, this group of genes is likely responsible for the developmental arrest of hematopoietic development that is in part responsible for the MLL phenotype.

Cell Proliferation, Survival and Adhesion CCND2 activity is required for cell cycle G1/S transition through its regulation of CDK4 or CDK6. This gene has been implicated in a large number of cancers including hepatocellular carcinoma, gastric, and ovarian cancer (Goode et al. 2009; Hirata et al. 2009; Moribe et al. 2009; Yasuda et al. 2009). It is likely that this gene plays a major role in the increased proliferative potential of this very aggressive leukemia. Two transcription factors, FHIT and HIF-1, are also included in the set. FHIT is a fragile histidine triad gene that is also a tumor suppressor gene found to be rearranged in infant acute lymphoblastic leukemia (Stam et al. 2006). Promoter hypermethylation of FHIT has been postulated to be linked to disease progression in ALL (Yang et al. 2006) and myelodysplastic syndrome MDS (Iwai et al. 2005). In addition, decreased expression of FHIT has been shown to be correlated with poor prognosis in patients with diffuse large B-cell lymphoma (Jais et al. 2008). Hypoxia-inducible factor-1 (HIF-1) is a transcription factor that controls homeostatic responses to hypoxia. HIF-1 has been found to induce survival or death in different contexts. Hypoxia in general has been found to lead to survival of leukemic cells when transplanted into mice. HIF-1A associates with RUNX1. RUNX1 is highly linked to myeloid leukemia (Munker et al. 2009) and AML-linked with MDS (Harada and Harada 2009). TNFRSF14 is a member of the tumor necrosis factor receptor that is primarily known for its role in mediating the entry of herpesvirus into T cells to serve in the costimulatory pathway to promote T-cell survival. Disruptions of this gene have been found in association with synovial cell carcinoma (Yu et al. 1999; Costello et al. 2003).

Genes involved in cell–cell communication are also found among these 18 genes. Membrane metallo-endopeptidase (MME) is a transmembrane glycoprotein that cleaves and inactivates several peptide hormones. It is also a common cell surface marker for acute lymphocytic leukemia (ALL), and has been shown to predict progression-free survival of diffuse large B-cell lymphoma (DLBCL) in a multivariate model (Jais et al. 2008). EHD1 plays an important role in protein-protein interactions with regard to intracellular sorting. The protein encoded by this gene is thought to play a role in the endocytosis of IGF1 receptors. EHD1 is one among seven biomarkers that are able to distinguish between neuroblastoma, non-Hodgkin lymphoma, rhabdomyosarcoma, and Ewing sarcoma that share similar histology and therefore often present a confusing clinical picture (Pal et al. 2007). PCDHGC3 belongs to the cadherin-like cell adhesion proteins found to play an important role in the establishment of cell–cell connections.

Other Genes STK32B is a serine/threonine kinase of which not much is known; however, other members of this kinase family, such as GSK3, have been linked to poor prognosis human leukemia. KCNK12 belongs to the superfamily of potassium channel proteins containing two pore-forming P domains. In addition, MBNL1—a double stranded RNA binding protein that specifically binds to expanded CUG triplet repeat sequences—is also included in the set of frequent primary genes. This protein has been linked with muscular dystrophy but not with leukemia, so it is yet unclear what its role is in this disease.

6.6.7 Using the *Informative Set of Genes* to Identify Robust Multivariate Markers

Although we identify the *Informative Set of Genes* to facilitate the biological interpretation of a parsimonious multivariate biomarker that we may have already found and initially validated, the genes included in the *primary expression patterns* of this set—and especially the *frequent primary* genes—may serve as a basis for the identification of potentially more robust biomarkers. Furthermore, the fact that such biomarkers will only tap into the primary expression patterns of the *Informative Set of Genes* should allow their easier and more meaningful biological interpretation.

TABLE 6.10: Multivariate Marker of Five Genes Identified for Model 2 (of the Multistage Classification Schema) When Feature Selection is Performed on the Training Data Only Including the 18 *Most Frequent Primary* Genes

Probe set	Cluster	GenBank accession	Gene symbol
219463_at	9	NM_012261	C20orf103
201153_s_at	9	N31913	MBNL1
203434_s_at	4	NM_007287	MME
209905_at	9	AI246769	HOXA9
209354_at	1	BC002794	TNFRSF14

Three of the five genes selected into this marker belong to Cluster 9 that includes 51 genes. This suggests that the expression super-pattern represented by this cluster includes some complementary subpatterns that together are important for separating the MLL subtype from the other four subtypes.

TABLE 6.11: Testing Five Classifiers on an Independent Test Set

			Classification of test data					
Feature selection method	Marker size	Learning algorithm	MLL	Sensitivity %	4-R	Specificity %	Accuracy %	Notes
T^2 search	5	LDA	4 of 4	100.0	85 of 86	98.8	98.9	337066 misclassified
Recursive Feature Elimination	5	SVM	4 of 4	100.0	86 of 86	100.0	100.0	
		LDA	4 of 4	100.0	86 of 86	100.0	100.0	
		D-LDA	4 of 4	100.0	86 of 86	100.0	100.0	
		KNN ($k = 3$)	4 of 4	100.0	86 of 86	100.0	100.0	

The classifiers are built from the *Informative Set of Genes* identified from the training set combining two data sets—ALL3 and GSE13425. The stepwise hybrid feature selection (T^2 search) has been performed on the 18 *most frequent primary* genes of the *Informative Set of Genes*. These are the genes that are most frequently used by the perfect OOB classifiers of the two ensembles (one built on the variables of the INF-176 set, the other—on all 6890 variables). This time, the identified biomarker includes the top five frequent primary genes, which represent three of the four primary clusters. An LDA classifier is built on this marker. Another marker of five genes is selected from the 176 genes of the *Informative Set of Genes* by Recursive Feature Elimination with linear SVM. All genes selected into this marker belong to the *frequent primary* genes. Four learning algorithms (SVM, LDA, Diagonal LDA, and k-Nearest Neighbors) are used to build classifiers based on this second multivariate biomarker. All five classifiers perform similarly well; the only difference is the misclassification of one sample (GSM337066) by the classifier built on the marker identified with the parametric LDA-based method.

When T^2-driven stepwise hybrid feature selection is performed to select an optimal subset of the eighteen *most frequent primary* genes, we identify the multivariate biomarker of the five genes presented in Table 6.10.

6.6.8 Validating Biomarkers on an Independent Test Data Set

We will validate our biomarker on the independent test set available in GEO as GSE13351 (and in ArrayExpress as E-GEOD-13351). This is the same test set we used in the exercise where only the ALL3 data were used for training. The fact that this test set includes samples from a different country[11] than the training set and that it was processed on a different (though compatible) microarray[12] makes the test results more reliable. The GSE13351 data set consists of 107 samples representing seven ALL subtypes. Ninety of these samples are associated with the subtypes

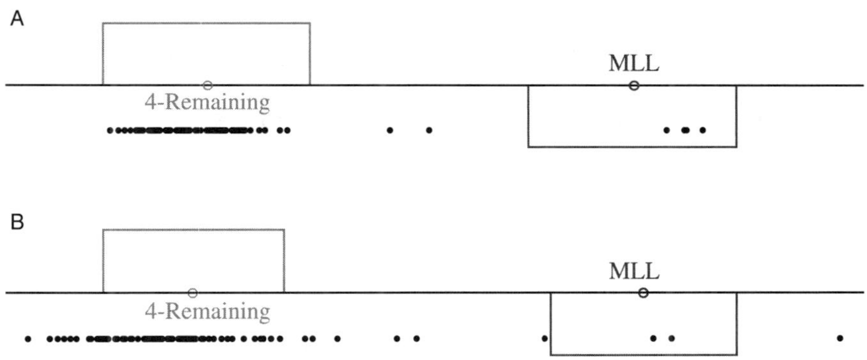

Figure 6.13: Comparing two LDA classifiers built on different training sets. Both classifiers are based on the multivariate markers of five genes identified by the stepwise hybrid feature selection driven by the T^2 measure of discriminatory power. For both, the one-dimensional discriminatory spaces (the vertical offset is added only to improve the visibility) and results of classification of the independent test samples (GSE13351) are shown. Classifier A (top figure) is based on the training set combining two data sets (ALL3 and GSE13425). Classifier B (bottom figure) is based on the ALL3 training data only. Both of them were identified by performing feature selection on the *most frequent primary* genes of their respective *Informative Sets of Genes*. As could be expected, the larger and more heterogeneous training set allows for more precise estimation of the distributions of the differentiated classes (top figure). All but two samples of the independent test set are classified within the segments (of the one-dimensional discriminatory space) representing 95 percent of the probability of their true classes. Only the same two samples that were previously suspected of being outliers are outside of such segment of their class (and one of them is misclassified).

[11]The combined training set (ALL3 plus GSE13425) includes samples of patients from USA and Germany, the test set GSE13351 consists of samples from Dutch patients.

[12]The test set samples were processed on Affymetrix GeneChip HG-U133 Plus 2.0 microarrays, the ALL3 part of the training set used HG-U133A chips, and the GSE13425 part of the training set HG-U133A 2.0 arrays.

differentiated by Model 2 and they will constitute our test data. They include four MLL samples and 86 samples of the *4-Remaining* class including four ALL subtypes.

To prepare the test data we may download the CEL files and preprocess them using the *Expression Console* software. Alternatively, we may download the preprocessed probe set intensity data and rescale it to the target value used for the training data.

Table 6.11 summarizes classification of the test samples by classifiers based on multivariate biomarkers identified from the training set including only 176 genes of the *Informative Set of Genes*.

Questions:

What could be a plausible explanation for the fact that the sample GSM337066 (which was previously suspected of being one of the two outliers) is correctly classified when a multivariate marker is identified by a nonparametric learning algorithm (SVM-based), but misclassified when a marker is selected by a parametric algorithm (LDA-based)?

What are advantages and disadvantages of the parametric and nonparametric approaches?

(See Chapter 3 and the discussion under Fig. 6.13.)

REFERENCES

Abe, S. (2005). *Support vector machines for pattern classification*. London, Springer.

Affymetrix (2001). *Statistical Algorithms Reference Guide*, Affymetrix, Inc., Santa Clara, CA. www.affymetrix.com

Affymetrix (2002). *Statistical Algorithms Description Document*, Affymetrix, Inc., Santa Clara, CA.

Affymetrix (2003). *Design and Performance of the GeneChip® Human Genome U133 Plus 2.0 and Human Genome U133A 2.0 Arrays – Technical Note*, Affymetrix, Inc., Santa Clara, CA.

Affymetrix (2005a). *GeneChip® Exon Array Design – Technical Note*, Affymetrix, Inc., Santa Clara, CA.

Affymetrix (2005b). *Guide to Probe Logarithmic Intensity Error (PLIER) Estimation*, Affymetrix, Inc., Santa Clara, CA.

Affymetrix (2006). *Affymetrix Expression Console™ Software Version 1.0—User Guide*, Affymetrix, Inc., Santa Clara, CA.

Affymetrix (2007). *GeneChip® Gene 1.0 ST Array Design – Technical Note*, Affymetrix, Inc., Santa Clara, CA.

Affymetrix (2008a). *GeneChip® HT HG-U133+PM Array Plate*, Affymetrix, Inc., Santa Clara, CA.

Affymetrix (2008b). *GeneChip® Human Exon 1.0 ST Array – Design Statistics Summary*, Affymetrix, Inc., Santa Clara, CA.

Ahrens, H. and J. Läuter (1974). *Mehrdimensionale Varianzanalyse: Hypothesenprüfung, Dimensionserniedrigung, Diskrimination bei multivariaten Beobachtungen*. Berlin: Akademie-Verlag.

Alizadeh, A. A., M. B. Eisen, R. E. Davis, C. Ma, I. S. Lossos, A. Rosenwald (2000). Distinct types of diffuse large B-cell lymphoma identified by gene expression profiling. *Nature* **403**(6769): 503–511.

Amaldi, E. and V. Kann (1998). On the approximability of minimizing nonzero variables or unsatisfied relations in linear systems. *Theoretical Computer Science* **209**: 237–260.

Amaratunga, D. and J. Cabrera (2004). *Exploration and analysis of DNA microarray and protein array data*. Hoboken, NJ: Wiley.

Ambroise, C. and G. J. McLachlan (2002). Selection bias in gene extraction on the basis of microarray gene-expression data. *Proceedings of the National Academy of Science* **99**(10): 6562–6566.

An, Q., S. L. Wright, Z. J. Konn, E. Matheson, L. Minto, A. V. Moorman (2008). Variable breakpoints target PAX5 in patients with dicentric chromosomes: a model for the basis of unbalanced translocations in cancer. *Proceedings of the National Academy of Science USA* **105**(44): 17050–17054.

Data Mining for Genomics and Proteomics. By Darius M. Dziuda

Anderson, T. W. (2003). *An introduction to multivariate statistical analysis*. Hoboken, NJ: Wiley-Interscience.

Archer, K. J. and R. V. Kimes (2008). Empirical characterization of random forest variable importance measures. *Computational Statistics & Data Analysis* **52**: 2249–2260.

Astle, J. M. and T. Kodadek (2008). Microarray approaches to autoantibody profiling. In: Van Eyk, J. E. and M. J. Dunn, editors. *Clinical proteomics: From diagnosis to therapy*. Weinheim: Wiley-VCH, pp. 511–532.

Avery, O. T., C. M. MacLeod and M. McCarty (1944). Studies on the chemical nature of the substance inducing transformation of pneumococcal types. Inductions of transformation by a desoxyribonucleic acid fraction isolated from pneumococcus type III. *Journal of Experimental Medicine* **149**(2): 297–326.

Bader, J. S., Y. Liu and D. M. Dziuda (2006). *Methods for classifying nucleic acids and polypeptides. US Patent 7016788*, 21 March 2006.

Bairoch, A. (2000). Serendipity in bioinformatics, the tribulations of a Swiss bioinformatician through exciting times! *Bioinformatics* **16**(1): 48–64.

Bairoch, A., R. Apweiler and C. H. Wu (2008). *UniProt Knowledgebase: Swiss-Prot Protein Knowledgebase, TrEMBL Protein Database. User Manual; Release 14.4 of 04-Nov-2008*, Swiss Institute of Bioinformatics and European Bioinformatics Institute. www.expasy.ch

Baker, M. (2005). In biomarkers we trust. *Nature Biotechnology* **23**(3): 297–304.

Baker, P. R. and K. R. Clauser (2008). *ProteinProspector/MS-Fit*, UCSF Mass Spectrometry Facility. http://prospector.ucsf.edu

Balboni, I., S. M. Chan, M. Kattah, J. D. Tenenbaum, A. J. Butte and P. J. Utz (2006). Multiplexed protein array platforms for analysis of autoimmune diseases. *Annual Review of Immunology* **24**: 391–418.

Ball, C. A., A. Brazma, H. Causton, S. Chervitz, R. Edgar, P. Hingamp (2004). Submission of microarray data to public repositories. *PLoS Biology* **2**(9): 1276–1277.

Ball, C. A., G. Sherlock, H. Parkinson, P. Rocca-Sera, C. Brooksbank, H. Causton (2002). An open letter to the scientific journals. *Bioinformatics* **18**(11): 1409.

Barnes, M., J. Freudenberg, S. Thompson, B. Aronow and P. Pavlidis (2005). Experimental comparison and cross-validation of the Affymetrix and Illumina gene expression analysis platforms. *Nucleic Acids Research* **33**(18): 5914–5923.

Barrett, T., T. O. Suzek, D. B. Troup, S. E. Wilhite, W.-C. Ngau, P. Ledoux (2005). NCBI GEO: mining millions of expression profiles—database and tools. *Nucleic Acids Research* **33**: D562–D566.

Barrett, T., D. B. Troup, S. E. Wilhite, P. Ledoux, D. Rudnev, C. Evangelista (2007). NCBI GEO: mining tens of millions of expression profiles—database and tools update. *Nucleic Acids Research* **35**: D760–D765.

Bates, D. W. (1998). Drugs and adverse drug reactions. How worried should we be? *Journal of American Medical Association* **279**(15): 1216–1217.

Bellman, R. E. (1961). *Adaptive control processes: a guided tour*. Princeton, NJ: Princeton University Press.

Benjamini, Y. and Y. Hochberg (1995). Controlling the false discovery rate: A practical and powerful approach to multiple testing. *Journal of the Royal Statistical Society. Series B (Methodological)* **57**(1): 289–300.

Bennett, K. P. and C. Campbell (2000). Support vector machines: Hype or hallelujah. *SIGKDD Explorations* **2**(2): 1–13.

Beranova-Giorgianni, S. (2003). Proteome analysis by two-dimensional gel electrophoresis and mass spectrometry: strengths and limitations. *Trends in Analytical Chemistry* **22**(5): 273–281.

Berrar, D. P., M. Granzow and W. Dubitzky (2007). Introduction to genomic and proteomic data analysis. In: Dubitzky, W., M. Granzow and D. P. Berrar, editors. *Fundamentals of data mining in genomics and proteomics*. New York: Springer, pp. 1–37.

Bertail, P. (1997). Second-order properties of an extrapolated bootstrap without replacement under weak assumptions. *Bernoulli* **3**(2): 149–179.

Bertucci, F. and A. Goncalves (2008). Clinical proteomics and breast cancer: strategies for diagnostic and therapeutic biomarker discovery. *Future Oncology* **4**(2): 271–287.

Beyer, S., Y. Walter, J. Hellmann, P. J. Kramer, A. Kopp-Schneider, M. Kroeger (2006). Comparison of software tools to improve the detection of carcinogen induced changes in the rat liver proteome by analyzing SELDI-TOF-MS spectra. *Journal of Proteome Research* **5**(2): 254–261.

Bickel, P. J. and D. A. Freedman (1984). Asymptotic normality and the bootstrap in stratified sampling. *The Annals of Statistics* **12**(2): 470–482.

Bickel, P. J., P. Götze and W. R. van Zwet (1994). *Resampling fewer than n observations: gains, losses, and remedies for losses*, Technical Report 419, Department of Statistics, University of California, Berkeley.

Bio-Rad Laboratories (2008). *ProteinChip Arrays*, Bio-Rad Laboratories, Inc., Hercules, CA, http://www.bio-rad.com

Bishop, C. M. (1995). *Neural networks for pattern recognition.* New York: Clarendon Press.

Bolstad, B. M. (2004). *Low-level analysis of high-density oligonucleotide array data: Background, normalization and summarization*, Ph.D. thesis, University of California, Berkeley.

Bolstad, B. M. (2007). Pre-processing DNA microarray data. In: Dubitzky, W., M. Granzow and D. P. Berrar, editors. *Fundamentals of data mining in genomics and proteomics*. New York: Springer: pp. 51–78.

Bolstad, B. M. (2008). Preprocessing and normalization for Affymetrix GeneChip expression microarrays. In: Stafford, P, editors. *Methods in microarray normalization.* Boca Raton, FL: CRC Press, pp. 41–59.

Bolstad, B. M., R. A. Irizarry, M. Astrand and T. P. Speed (2003). A comparison of normalization methods for high density oligonucleotide array data based on variance and bias. *Bioinformatics* **19**(2): 185–193.

Boser, B. E., I. Guyon and V. N. Vapnik (1992). *A training algorithm for optimal margin classifiers*. Fifth Annual Workshop on Computational Learning Theory, Pittsburgh, ACM, 144–152.

Box, G. E. P. and D. R. Cox (1964). An analysis of transformations. *Journal of the Royal Statistical Society. Series B (Methodological)* **26**(2): 211–252.

Brazma, A., P. Hingamp, J. Quackenbush, G. Sherlock, P. Spellman, C. Stoeckert (2001). Minimum information about a microarray experiment (MIAME) – toward standards for microarray data. *Nature Genetics* **29**(4): 365–371.

Breiman, L. (1996a). Bagging predictors. *Machine Learning* **24**: 123–140.

Breiman, L. (1996b). Heuristics of instability and stabilization in model selection. *The Annals of Statistics* **24**(6): 2350–2383.

Breiman, L. (1996c). *Out-of-bag estimation*. ftp://ftp.stat.berkeley.edu/pub/users/breiman/OOBestimation.ps.Z

Breiman, L. (1998). ARCing classifiers. *The Annals of Statistics* **26**(3): 801–849.

Breiman, L. (2001). Random forests. *Machine Learning* **45**(5): 5–32.

Breiman, L. (2002a). *WALD lecture I: Machine Learning*. 277th meeting of the Institute of Mathematical Statistics, Banff, Alberta, Canada, 1–39.

Breiman, L. (2002b). *WALD lecture II: Looking Inside the Black Box*. 277th meeting of the Institute of Mathematical Statistics, Banff, Alberta, Canada, 1–35.

Breiman, L. (2003). *Manual—Setting up, using, and understanding random forests V4.0*, berkeley.edu: 1–33.

Breiman, L. and A. Cutler (2004). Random forests: Classification/clustering. www.math.usu.edu/~adele

Breiman, L., J. Friedman, R. Olshen and C. Stone (1984). *Classification and regression trees*. Belmont, California, Wadsworth International Group.

Brettschneider, J., F. Collin, B. M. Bolstad and T. P. Speed (2008). Quality assessment for short oligonucleotide arrays. *Technometrics* **50**(3): 241–264, 279–283.

Bühlmann, P. and B. Yu (2002). Analyzing bagging. *The Annals of Statistics* **30**(4): 927–961.

Campbell, A. M. and L. J. Heyer (2007). *Discovering genomics, proteomics, and bioinformatics*. San Francisco, CSHL Press: Pearson/Benjamin Cummings.

Campbell, M. K. and S. O. Farrell (2006). *Biochemistry*. Belmont, CA: Thomson.

Carroll, W. L., D. Bhojwani, D. J. Min, E. Raetz, M. Relling, S. Davies (2003). Pediatric acute lymphoblastic leukemia. *Hematology (Am Soc Hematol Educ Program)* **2003**: 102–131.

Chan, D., S. M. Bridges and S. C. Burgess (2007). An ensemble method for identifying robust features for biomarker discovery. In: Liu, H. and H. Motoda, editors. *Computational methods of feature selection*. Boca Raton, FL: Taylor & Francis, pp. 377–392.

Charboneau, L., H. Scott, T. Chen, M. Winters, E. F. Petricoin III, L. A. Liotta (2002). Utility of reverse phase protein arrays: Applications to signalling pathways and human body arrays. *Briefings in Functional Genomics and Proteomics* **1**(3): 305–315.

Chernick, M. R. (2008). *Bootstrap methods: a guide for practitioners and researchers*. Hoboken, NJ: Wiley.

Chipman, H., T. Hastie and R. Tibshirani (2003). Clustering microarray data. In: Speed, T. P., editors. *Statistical analysis of gene expression microarray data*. Boca Raton, FL: Chapman & Hall/CRC, pp. 159–200.

Chipman, H. and R. Tibshirani (2006). Hybrid hierarchical clustering with applications to microarray data. *Biostatistics* **7**(7): 286–301.

Chipman, H. and R. Tibshirani (2008). *HybridHclust: Hybrid hierarchical clustering software for R*. http://math.acadiau.ca/chipmanh/hybridHclust/index.html

Chu, G., B. Narasimhan, R. Tibshirani and V. Tusher (2007). *SAM. "Significance Analysis of Microarrays". Users guide and technical document*. http://www-stat.stanford.edu/~tibs/SAM/sam.pdf

Clamp, M., B. Fry, M. Kamal, X. Xie, J. Cuff, M. F. Lin (2007). Distinguishing protein-coding and noncoding genes in the human genome. *Proceedings of the National Academy of Sciences* **104**(49): 19428–19433.

Clauser, K. R., P. Baker and A. L. Burlingame (1999). Role of accurate mass measurement (+/− 10 ppm) in protein identification strategies employing MS or MS/MS and database searching. *Analytical Chemistry* **71**(14): 2871–2882.

Clemmensen, L., T. Hastie and B. Ersbøll (2008). *Sparse Discriminant Analysis*, DTU Informatics, Technical University of Denmark, Lyngby, pp. 1–25.

Coombes, K. R., K. A. Baggerly and J. S. Morris (2007). Pre-processing mass spectrometry data. In: Dubitzky, W., M. Granzow and D. P. Berrar, editors. *Fundamentals of data mining in genomics and proteomics*. New York: Springer, pp. 79–102.

Coombes, K. R., H. A. Fritsche, C. Clarke, J.-N. Chen, K. A. Baggerly, J. S. Morris (2003). Quality control and peak finding for proteomics data collected from nipple aspirate fluid by surface-enhanced laser desorption and ionization. *Clinical Chemistry* **49**(10): 1615–1623.

Coombes, K. R., S. Tsavachidis, J. S. Morris, K. A. Baggerly, M.-C. Hung and H. M. Kuerer (2005). Improved peak detection and quantification of mass spectrometry data acquired from surface-enhanced laser desorption and ionization by denoising spectra with the undecimated discrete wavelet transform. *Proteomics* **5**(16): 4107–4117.

Cope, L. M., R. A. Irizarry, H. A. Jaffee, Z. Wu and T. P. Speed (2004). A benchmark for Affymetrix GeneChip expression measures. *Bioinformatics* **20**(3): 323–331.

Costello, R., F. Mallet, B. Barbarat, J. Schiano De Colella, D. Sainty and R. Sweet (2003). Stimulation of non-Hodgkin's lymphoma via HVEM: an alternate and safe way to increase Fas-induced apoptosis and improve tumor immunogenicity. *Leukemia* **17**(12): 2500–2507.

Cox, D. R. and N. J. H. Small (1978). Testing multivariate normality. *Biometrika* **65**(2): 263–272.

Crick, F. (1958). *On protein synthesis*. Symposium of the Society for Experimental Biology, London, 138–163.

Crick, F. (1970). Central dogma of molecular biology. *Nature* **227**: 561–563.

Crick, F. (1988). *What mad pursuit: a personal view of scientific discovery*. New York: Basic Books.

Cristianini, N. (2001). *Support vector and kernel machines*. http://www.support-vector.net/icml-tutorial.pdf

Cristianini, N. and J. Shawe-Taylor (2000). *An introduction to support vector machines: and other kernel-based learning methods*. New York: Cambridge University Press.

Cruz-Marcelo, A., R. Guerra, M. Vannucci, Y. Li, C. Lau and T.-K. Man (2008). Comparison of algorithms for pre-processing of SELDI-TOF mass spectrometry data. *Bioinformatics* **24**(19): 2129–2136.

Damodaran, S., T. D. Wood, P. Nagarajan and R. A. Rabin (2007). Evaluating peptide mass fingerprinting-based protein identification. *Genomics, Proteomics & Bioinformatics* **5**(3–4): 152–157.

Dancík, V., T. A. Addona, K. R. Clauser, J. E. Vath and P. A. Pevzner (1999). De novo peptide sequencing via tandem mass spectrometry. *Journal of Computational Biology* **6**(3–4): 327–342.

Das, S. (2001). *Filters, wrappers and a boosting based hybrid for feature selection*. Proceedings of the Eighteenth International Conference on Machine Learning, Williams College, 74–81.

Den Boer, M. L., M. van Slegtenhorst, R. X. De Menezes, M. H. Cheok, J. G. Buijs-Gladdines, S. T. Peters (2009). A subtype of childhood acute lymphoblastic leukaemia with poor treatment outcome: a genome-wide classification study. *Lancet Oncology* **10**(2): 125–134.

DePalma, A. (2003). Capturing protein using antibody arrays. *Genomics and Proteomics* **3**(3): 40–43.

Diaz-Uriarte, R. and S. Alvarez de Andres (2006). Gene selection and classification of microarray data using random forest. *BMC Bioinformatics* **7**: 3.

Donoghue, P. M., M. Stastna and M. J. Dunn (2008). Protein separation by two-dimensional electrophoresis. In: Van Eyk, J. E. and M. J. Dunn, editors. *Clinical proteomics: From diagnosis to therapy*. Weinheim: Wiley-VCH, pp. 13–29.

Dou, Y. and J. L. Hess (2008). Mechanisms of transcriptional regulation by MLL and its disruption in acute leukemia. *International Journal of Hematology* **87**(1): 10–18.

Drăghici, S. (2003). *Data analysis tools for DNA microarrays*. Boca Raton, FL: Chapman & Hall/CRC.

Dreyfus, G. (2005). *Neural networks: methodology and applications*. New York: Springer.

Duan, K., J. Rajapakse, H. Wang and F. Azuaje (2005). Multiple SVM-RFE for gene selection in cancer classification with expression data. *IEEE Transactions on Nanobioscience* **4**(3): 228–234.

Dubitzky, W., M. Granzow and D. P. Berrar (2007). *Fundamentals of data mining in genomics and proteomics*. New York: Springer.

Duda, R. O. and P. E. Hart (1972). Use of the Hough Transformation to detect lines and curves in pictures. *Communications of the ACM* **15**(1): 11–15.

Duda, R. O., P. E. Hart and D. G. Stork (2001). *Pattern classification*. New York: Wiley.

Dudoit, S. and M. J. van der Laan (2008). *Multiple testing procedures with applications to genomics*. New York: Springer.

Dudoit, S., Y. H. Yang, M. J. Callow and T. P. Speed (2002). Statistical methods for identifying differentially expressed genes in replicated cDNA microarray experiments. *Statistica Sinica* **12**: 111–139.

Durbin, B. P., J. S. Hardin, D. M. Hawkins and D. M. Rocke (2002). A variance-stabilizing transformation for gene-expression microarray data. *Bioinformatics* **18**(Supplement 1): S105–S110.

Dziuda, D. (1990). Specialized PC software package for creation of computer systems supporting partial diagnostics based on numerical results of medical examinations. *Medical Informatics (London)* **15**(4): 319–326.

Dziuda, D. (2007). *Multivariate biomarkers and the Informative Set of Genes*. 5th Conference of International Drug Discovery Science and Technology, Shanghai, China.

Dziuda, D. (2008). *Informative Set of Genes: The missing link between diagnostic biomarkers and biological processes underlying differentiation*, Biomarker World Congress 2008, Philadelphia, PA.

Dziuda, D. and M. J. Czar (2006). *Advanced informatics in biomedicine – individualized biomedicine*, Technical Report, Virginia Bioinformatics Institute.

Dziuda, D. and Y. Zhou (2007). *Multivariate method for identification of the informative set of genes*. Proceedings of the 6th Annual Hawaii International Conference on Statistics, Mathematics and Related Fields, Honolulu, Hawaii, 283–290.

Efron, B. (1979). Bootstrap methods: another look at the jackknife. *The Annals of Statistics* **7**(1): 1–26.

Efron, B. (1983). Estimating the error rate of a prediction rule: improvement on cross-validation. *Journal of the American Statistical Association* **78**(382): 316–331.

Efron, B. and R. Tibshirani (1993). *An introduction to the bootstrap*. New York: Chapman & Hall.

Efron, N. and N. Intrator (2004). *The effect of noisy bootstrapping on the robustness of supervised classification of gene expression data*. IEEE International Workshop on Machine Learning for Signal Processing, Brazil, 411–420.

Eidhammer, I., K. Flikka, L. Martens and S.-O. Mikalsen (2007). *Computational methods for mass spectrometry proteomics*. Hoboken, NJ: Wiley.

Eisen, M. B., P. T. Spellman, P. O. Brown and D. Botstein (1998). Cluster analysis and display of genome-wide expression patterns. *Proceedings of the National Academy of Science USA* **95**: 14863–14868.

Ekins, R. P. (1989). Multi-analyte immunoassay. *Journal of Pharmaceutical and Biomedical Analysis* **7**(2): 155–168.

Elias, J. E., W. Haas, B. K. Faherty and S. P. Gygi (2005). Comparative evaluation of mass spectrometry platforms used in large-scale proteomics investigations. *Nature Methods* **2**(9): 667–675.

Elvidge, G. (2006). Microarray expression technology: from start to finish. *Pharmacogenomics* **7**(1): 123–134.

Eng, J. K., A. L. McCormack and J. R. Yates (1994). An approach to correlate tandem mass spectral data of peptides with amino acid sequences in a protein database. *Journal of the American Society for Mass Spectrometry* **5**(11): 976–989.

Faber, J., A. V. Krivtsov, M. C. Stubbs, R. Wright, T. N. Davis, M. van den Heuvel-Eibrink (2009). HOXA9 is required for survival in human MLL-rearranged acute leukemias. *Blood* **113**(11): 2375–2385.

Falk, R. (1986). What is a gene? *Studies in History and Philosophy of Science Part A* **17**(2): 133–173.

Fenyo, D. and R. C. Beavis (2008). Informatics development: challenges and solutions for MALDI mass spectrometry. *Mass Spectrometry Reviews* **27**: 1–19.

Fielding, A. (2007). *Cluster and classification techniques for the biosciences*. New York: Cambridge University Press.

Flicek, P., B. L. Aken, K. Beal, B. Ballester, M. Caccamo, Y. Chen (2008). Ensembl 2008. *Nucleic Acids Research* **36**(Database Issue): D707–D714.

Flury, B. and H. Riedwyl (1988). *Multivariate statistics: a practical approach*. New York: Chapman & Hall.

Fodor, S. P. A., J. Leighton Read, M. C. Pirrung, L. Stryer, A. T. Lu, and D. Solas (1991). Light-directed, spatially addressable parallel chemical synthesis. *Science* **251**(4995): 767–773.

Frank, A. and P. Pevzner (2005). PepNovo: De Novo Peptide Sequencing via Probabilistic Network Modeling. *Analytical Chemistry* **77**: 964–973.

Frank, A., S. Tanner, V. Bafna and P. Pevzner (2005). Peptide Sequence Tags for Fast Database Search in Mass-Spectrometry. *Journal of Proteome Research* **4**: 1287–1295.

Frank, R. (1992). Spot-synthesis: an easy technique for the positionally addressable, parallel chemical synthesis on a membrane support. *Tetrahedron* **48**(42): 9217–9232.

Frank, R. (2002). The SPOT-synthesis technique. Synthetic peptide arrays on membrane supports—principles and applications. *Journal of Immunological Methods* **267**: 13–26.

Franklin, R. E. and R. G. Gosling (1953). Molecular Configuration in Sodium Thymonucleate. *Nature* **171**(4356): 740–741.

Freund, Y. and R. E. Shapire (1997). A decision-theoretic generalization of on-line learning and an application to boosting. *Journal of Computer and System Sciences* **55**(1): 119–139.

Friedman, J. H. (1989). Regularized discriminant analysis. *Journal of the American Statistical Association* **84**(405): 165–175.

Ganten, D. and K. Ruckpaul (2006). *Encyclopedic reference of genomics and proteomics in molecular medicine*. New York: Springer.

Gao, X., J. P. Pellois, Y. Na, Y. Kim, E. Gulari and X. Zhou (2004). High density peptide microarrays. In situ synthesis and applications. *Molecular Diversity* **8**(3): 177–187.

Garrett, R. and C. M. Grisham (2007). *Biochemistry*. Belmont, CA: Thomson Brooks/Cole.

Gasteiger, E., C. Hoogland, A. Gattiker, S. Duvaud, M. R. Wilkins, R. D. Appel (2005). Protein Identification and Analysis Tools on the ExPASy Server. In: Walker, J. M., editor. *The Proteomics Protocols Handbook*. New York: Humana Press, pp. 571–607.

Ge, H. (2000). Protein arrays immobilizing entire proteins are used for functional assays and detection of protein enzymatic activities. *Nucleic Acids Research* **28**(2): e3.

Ge, Y., S. Dudoit and T. P. Speed (2003). *Resampling-based multiple testing for microarray data analysis*, Technical report 633, Department of Statistics, University of California, Berkeley.

Gelsi-Boyer, V., V. Trouplin, J. Adélaïde, J. Bonansea, N. Cervera, N. Carbuccia (2009). Mutations of polycomb-associated gene ASXL1 in myelodysplastic syndromes and chronic myelomonocytic leukaemia. *British Journal of Haematology* **145**(6): 788–800.

Gerstein, M. B., C. Bruce, J. S. Rozowsky, D. Zheng, J. Du, J. O. Korbel (2007). What is a gene, post-ENCODE? History and updated definition. *Genome Research* **17**(6): 669–681.

Gibson, G. and S. V. Muse (2009). *A primer of genome science*. Sunderland, MA: Sinauer Associates.

Gingeras, T. R. (2007). Origin of phenotypes: Genes and transcripts. *Genome Research* **17**(6): 682–690.

Glaser, V. (2007). Protein microarray market scope broadens: Focused protein profiling strategies and reverse-phase arrays drive growth. *Genetic Engineering & Biotechnology News* **27**(6).

Gobom, J. (2006). Mass spectrometry: MALDI. In: Ganten, D. and K. Ruckpaul, editors. *Encyclopedic reference of genomics and proteomics in molecular medicine, Volume 2*, New York: Springer, pp. 1023–1027.

Goode, E. L., B. L. Fridley, R. A. Vierkant, J. M. Cunningham, C. M. Phelan, S. Anderson (2009). Candidate gene analysis using imputed genotypes: cell cycle single-nucleotide polymorphisms and ovarian cancer risk. *Cancer Epidemiology Biomarkers and Prevention* **18**(3): 935–944.

Gross, S. T. (1980). *Median estimation in sample surveys*. Proceedings of the Survey Research Methods Section, American Statistical Association, 181–184.

Gulmann, C., K. M. Sheehan, E. W. Kay, L. A. Liotta and E. F. Petricoin, III (2006). Array-based proteomics: mapping of protein circuitries for diagnostics, prognostics, and therapy guidance in cancer. *Journal of Pathology* **208**(5): 595–606.

Gunderson, K. L., S. Kruglyak, M. S. Graige, F. Garcia, B. G. Kermani, C. Zhao (2004). Decoding randomly ordered DNA arrays. *Genome Research* **14**(5): 870–877.

Guyon, I. and A. Elisseeff (2003). An introduction to variable and feature selection. *Journal of Machine Learning Research* **3**(7/8): 1157–1182.

Guyon, I., S. Gunn, M. Nikravesh and L. A. Zadeh (2006). *Feature extraction: foundations and applications*. New York: Springer-Verlag.

Guyon, I., J. Weston, S. Barnhill and V. N. Vapnik (2002). Gene selection for cancer classification using support vector machines. *Machine Learning* **46**(1–3): 389–422.

Hall, M. A. (1999). Correlation-based feature selection for machine learning, Ph.D. thesis, Department of Computer Science, The University of Waikato.

Hall, M. A. and L. A. Smith (1999). Feature selection for machine learning: Comparing a correlation-based filter approach to the wrapper. *Florida Artificial Intelligence Symposium. AAAI Press*: 235–239.

Hall, P. (1992). *The bootstrap and Edgeworth expansion*. New York: Springer-Verlag.

Hamdan, M. and P. G. Righetti (2005). *Proteomics today: protein assessment and biomarkers using mass spectrometry, 2D electrophoresis, and microarray technology*. Hoboken, NJ: Wiley.

Han, J. and M. Kamber (2006). *Data mining: concepts and techniques*. Amsterdam: Elsevier.

Hand, D. J., H. Mannila and P. Smyth (2001). *Principles of data mining*. Cambridge, MA: MIT Press.

Harada, Y. and H. Harada (2009). Molecular pathways mediating MDS/AML with focus on AML1/RUNX1 point mutations. *Journal of Cellular Physiology* **220**(1): 16–20.

Hart, P. E. (1968). The condensed nearest neighbor rule. *IEEE Transactions on Information Theory* **14**: 515–516.

Hartigan, J. A. (1972). Direct clustering of data matrix. *Journal of the American Statistical Association* **67**(337): 123–129.

Hastie, T., R. Tibshirani, M. Eisen, A. A. Alizadeh, R. Levy, L. M. Staudt (2000). "Gene shaving" as a method for identifying distinct sets of genes with similar expression patterns. *Genome Biology* **1**(2): 1–21.

Hastie, T., R. Tibshirani and J. H. Friedman (2001). *The elements of statistical learning: data mining, inference, and prediction*. New York: Springer.

Hastie, T., R. Tibshirani and J. H. Friedman (2009). *The elements of statistical learning, second edition: data mining, inference, and prediction*. New York: Springer.

Haykin, S. S. (2008). *Neural networks and learning machines*. Upper Saddle River, NJ: Prentice Hall.

Hebb, D. O. (1949). *The organization of behavior; a neuropsychological theory*. New York: Wiley.

Hellman, M. E. (1970). The nearest neighbor classification rule with a reject option. *IEEE Transactions on Systems Science and Cybernetics* **6**(3): 179–185.

Henzel, W. J., T. M. Billeci, J. T. Stults, S. C. Wong, C. Grimley and C. Watanabe (1993). Identifying proteins from two-dimensional gels by molecular mass searching of peptide fragments in protein sequence databases. *Proceedings of the National Academy of Sciences of USA* **90**(11): 5011–5015.

Hilario, M., A. Kalousis, C. Pellegrini and M. Müller (2006). Processing and classification of protein mass spectra. *Mass Spectrometry Reviews* **25**(3): 409–449.

Hirata, Y., N. Ogasawara, M. Sasaki, T. Mizushima, T. Shimura, T. Mizoshita (2009). BCL6 degradation caused by the interaction with the C-terminus of pro-HB-EGF induces cyclin D2 expression in gastric cancers. *British Journal of Cancer* **100**(8): 1320–1329.

Holm, S. (1979). A simple sequentially rejective multiple test procedure. *Scandinavian Journal of Statistics* **6**: 65–70.

Hotelling, H. (1933). Analysis of a complex of statistical variables into principal components. *Journal of Educational Psychology* **24**(67): 417–441, 498–520.

Hotelling, H. (1951). *A generalized T test and measure of multivariate dispersion*. Proceedings of the Second Berkeley Symposium on Mathematical Statistics and Probability (July 31–August 12, 1950), Berkeley, CA: University of California Press, pp. 23–41.

Hough (1962). *Method and means for recognizing complex patterns. US Patent 3069654*, 18 December 1962.

Huang, S., A. A. Yeo, L. Gelbert, X. Lin, L. Nisenbaum and K. G. Bemis (2004). At what scale should microarray data be analyzed? *American Journal of Pharmacogenomics* **4**(2): 129–139.

Huber, W., A. von Heydebreck, H. Sültmann, A. Poustka and M. Vingron (2002). Variance stabilization applied to microarray data calibration and to the quantification of differential expression. *Bioinformatics* **18**(Supplement 1): S96–S104.

Huberty, C. J. and S. Olejnik (2006). *Applied MANOVA and discriminant analysis*. Hoboken, NJ: Wiley.

HUGO Gene Nomenclature Committee (2008). *HGNC Guidelines*, Human Genome Organisation (HUGO). www.genenames.org

Hutchens, T. W. and T.-T. Yip (1993). New desorption strategies for the mass spectrometric analysis of macromolecules. *Rapid Communications in Mass Spectrometry* **7**(7): 576–580.

Illumina (2008a). *Product Guide 2008 – Simplifying Genetic Analysis*, Illumina, Inc., San Diego, CA. www.illumina.com

Illumina (2008b). *An update to the 2008 Illumina Product Guide*, Illumina, Inc., San Diego, CA. www.illumina.com

Ingelman-Sundberg, M. (2008). Pharmacogenomics biomarkers for prediction of severe adverse drug reactions. *New England Journal of Medicine* **358**(6): 637–639.

International Human Genome Sequencing Consortium (2001). Initial sequencing and analysis of the human genome. *Nature* **409**(6822): 860–921.

International Human Genome Sequencing Consortium (2004). Finishing the euchromatic sequence of the human genome. *Nature* **431**(7011): 931–945.

Irizarry, R. A., B. M. Bolstad, F. Collin, L. M. Cope, B. Hobbs and T. P. Speed (2003). Summaries of Affymetrix GeneChip probe level data. *Nucleic Acids Research* **31**(4): e15.

Irizarry, R. A., B. Hobbs, F. Collin, Y. D. Beazer-Barclay, K. J. Antonellis, U. Scherf (2003). Exploration, normalization, and summaries of high density oligonucleotide array probe level data. *Biostatistics* **4**(2): 249–264.

Irizarry, R. A., Z. Wu and H. A. Jaffee (2006). Comparison of Affymetrix GeneChip expression measures. *Bioinformatics* **22**(7): 789–794.

Iwai, M., H. Kiyoi, K. Ozeki, T. Kinoshita, I. N. Em, R. Ohno and T. Naoe (2005). Expression and methylation status of the FHIT gene in acute myeloid leukemia and myelodysplastic syndrome. *Leukemia* **19**(8): 1367–1375.

Izenman, A. J. (2008). *Modern multivariate statistical techniques: regression, classification, and manifold learning*. New York: Springer.

Jain, A. K., J. Mao and K. M. Mohiuddin (1996). Artificial neural networks: A tutorial. *Computer* **29**(3): 31–44.

Jais, J., C. Haioun, T. J. Molina, D. S. Rickman, A. de Reynies, F. Berger (2008). The expression of 16 genes related to the cell of origin and immune response predicts survival in elderly patients with diffuse large B-cell lymphoma treated with CHOP and rituximab. *Leukemia* **22**(10): 1917–1924.

John, G. H., R. Kohavi and K. Pfleger (1994). *Irrelevant features and the subset selection problem*. Machine Learning: Proceedings of the Eleventh International Conference, Morgan Kaufmann, pp. 121–129.

Johnson, R. A. and D. W. Wichern (2007). *Applied multivariate statistical analysis*. Upper Saddle River, NJ: Prentice Hall.

Jolliffe, I. T. (2002). *Principal component analysis*. New York: Springer.

Jordan, M. I. (2004). *The kernel trick*, Department of Statistics, University of California, Berkeley. www.cs.berkeley.edu/~jordan

Jungblut, P. R. (2006). Two-dimensional gel electrophoresis. In: Ganten, D. and K. Ruckpaul, editors. *Encyclopedic reference of genomics and proteomics in molecular medicine*, Volume 2, New York: Springer, pp. 1942–1944.

Kaku, M. (1994). *Hyperspace: a scientific odyssey through parallel universes, time warps, and the tenth dimension*. New York: Oxford University Press.

Kaufman, L. and P. J. Rousseeuw (1990). *Finding groups in data: an introduction to cluster analysis*. New York: Wiley.

Kohonen, T. (1982a). Self-organized formation of topologically correct feature maps. *Biological Cybernetics* **43**(1): 59–69.

Kohonen, T. (1982b). Analysis of a simple self-organizing process. *Biological Cybernetics* **44**(2): 135–140.

Kohonen, T. (2001). *Self-organizing maps*. New York: Springer.

Koller, D. and M. Sahami (1996). *Toward optimal feature selection*. Proceedings of the 13th International Conference on Machine Learning (ICML), Morgan Kaufmann, pp. 284–292.

Korf, U. (2006). Protein microarrays as a tool for protein profiling and functional proteomics. In: Ganten, D. and K. Ruckpaul, editors. *Encyclopedic reference of genomics and proteomics in molecular medicine*, Volume 2, New York: Springer, pp. 1521–1525.

Kreil, D. P., N. A. Karp and K. S. Lilley (2004). DNA microarray normalization methods can remove bias from differential protein expression analysis of 2D difference gel electrophoresis results. *Bioinformatics* **20**(13): 2026–2034.

Kultima, K., B. Scholz, H. Alm, K. Sköld, M. Svensson, A. R. Crossman (2006). Normalization and expression changes in predefined sets of proteins using 2D gel electrophoresis: A proteomic study of L-DOPA induced dyskinesia in an animal model of Parkinson's disease using DIGE. *BMC Bioinformatics* **7**: 475.

Kung, L. A. and M. Snyder (2006). Proteome chips for whole-organism assays. *Nature Reviews: Molecular Cell Biology* **7**: 617–622.

Larose, D. T. (2005). *Discovering knowledge in data: an introduction to data mining*. Hoboken, NJ: Wiley.

Larose, D. T. (2006). *Data mining methods and models*. Hoboken, NJ: Wiley.

Lawley, D. N. (1938). A generalization of Fisher's z test. *Biometrika* **30**(1/2): 180–187, Correction 467–469.

Lazzeroni, L. and A. O. Owen (2002). Plaid model for gene expression data. *Statistica Sinica* **12**: 61–86.

Lee, Y., Y. Lin and G. Wahba (2004). Multicategory support vector machines: Theory and application to the classification of microarray data and satellite radiance data. *Journal of the American Statistical Association* **99**(465): 67–81.

Li, Z., R. Luo, S. Mi, M. Sun, P. Chen, J. Bao (2009). Consistent deregulation of gene expression between human and murine MLL rearrangement leukemias. *Cancer Research* **69**(3): 1109–1116.

Liaw, A., M. R. P. Wiener, L. Breiman and A. F. O. Cutler (2008). *The randomforest package: Breiman and Cutler's random forests for classification and regression.* http://cran.r-project.org

Lin, S. M., P. Du, W. Huber and W. A. Kibbe (2008). Model-based variance-stabilizing transformation for Illumina microarray data. *Nucleic Acids Research* **36**(2): e11.

Liu, H. and H. Motoda (2007a). *Computational methods of feature selection.* Boca Raton, FL: Taylor & Francis.

Liu, H. and H. Motoda (2007b). Less is more. In: Liu, H. and H. Motoda, editors. *Computational methods of feature selection.* Boca Raton, FL: Taylor & Francis, pp. 3–17.

Liu, H. and L. Yu (2005). Toward integrating feature selection algorithms for classification and clustering. *IEEE Transactions on Knowledge and Data Engineering* **17**(4): 491–502.

Lloyd, S. P. (1957). *Least squares quantization in PCM*, Technical report, Bell Laboratories. (Portions presented at the Institute of Mathematical Statistics Meeting Atlantic City, New Jersey, September 1957. Published in 1982 in *IEEE Transactions on Information Theory* **28**, 129–137.)

Lunetta, K., L. B. Hayward, J. Segal and P. Van Eerdewegh (2004). Screening large-scale association study data: exploiting interactions using random forests. *BMC Genetics* **5**(32).

Ma, B., K. Zhang, C. Hendrie, C. Liang, M. Li, A. Doherty-Kirby (2003). PEAKS: Powerful software for peptide de novo sequencing by MS/MS. *Rapid Communications in Mass Spectrometry* **17**(20): 2337–2342.

MacNaughton-Smith, P., W. T. Williams, M. B. Dale and L. G. Mockett (1964). Dissimilarity analysis: A new technique of hierarchical sub-division. *Nature* **202**: 1034–1035.

MacQueen, J. (1967). *Some methods for classification and analysis of multivariate observations.* Proceedings of the Fifth Berkeley Symposium on Mathematical Statistics and Probability (June 21–July 18, 1965 and December 27, 1965–January 7, 1966), Berkeley, CA: University of California Press, 281–297.

Madeira, S. C. and A. L. Oliveira (2004). Biclustering algorithms for biological data analysis: a survey. *IEEE Transactions on Computational Biology and Bioinformatics* **1**(1): 23–45.

Maercker, C. (2005). Protein arrays in functional genome research. *Bioscience Reports* **25**(1/2): 57–70.

Mann, M. and M. Wilm (1994). Error-tolerant identification of peptides in sequence databases by peptide sequence tags. *Analytical Chemistry* **66**: 4390–4399.

Mardia, K. V. (1970). Measures of multivariate skewness and kurtosis with applications. *Biometrika* **57**(3): 519–530.

Mazzatti, D. J., G. Pawelec, R. Longdin, J. R. Powell and R. J. Forsey (2007). SELDI-TOF-MS ProteinChip array profiling of T-cell clones propagated in long-term culture identifies human profilin-I as a potential bio-marker of immunosenescence. *Proteome Science* **5**: 7.

McClintick, J. N. and H. J. Edenberg (2006). Effects of filtering by present call on analysis of microarray experiments. *BMC Bioinformatics* **7**(49).

McCulloch, W. and W. Pitts (1943). A logical calculus of the ideas immanent in nervous activity. *Bulletin of Mathematical Biophysics* **5**: 115–133.

Meyers, L. S., G. Gamst and A. J. Guarino (2006). *Applied multivariate research: design and interpretation.* Los Angeles: Sage Publications.

MGED_Society (2008a). *MicroArray and gene expression – MAGE*, MGED Society. www.mged.org

MGED_Society (2008b). *Minimum information about a microarray experiment – MIAME 2.0*, MGED Society. www.mged.org

Miller, D. (2008). Central Connecticut State University, Department of Mathematical Sciences, New Britain, CT. Personal Communication, May 6, 2008.

Min, D. and M. Mrksich (2004). Peptide arrays: towards routine implementation. *Current Opinion in Chemical Biology* **8**: 554–558.

Mitchell, T. M. (1997). *Machine learning*. New York: McGraw-Hill.

Modrek, B. and C. J. Lee (2003). Alternative splicing in the human, mouse and rat genomes is associated with an increased frequency of exon creation and/or loss. *Nature Genetics* **34**(2): 177–180.

Moribe, T., N. Iizuka, T. Miura, N. Kimura, S. Tamatsukuri, H. Ishitsuka (2009). Methylation of multiple genes as molecular markers for diagnosis of a small, well-differentiated hepatocellular carcinoma. *International Journal of Cancer* **125**(2): 388–397.

Morring, B. D. and T. R. Martinez (2004). Weighted instance typicality search (WITS): A nearest neighbor data reduction algorithm. *Intelligent Data Analysis* **8**(1): 61–78.

Morrison, D. F. (2005). *Multivariate statistical methods*. Belmont, CA: Thomson.

Moulines, E. (2008). *Kernels*, Ecole Nationale Supérieure des Télécommunications (ENST). www.tsi.enst.fr/~moulines

Munker, R., M. Nordberg, D. Veillon, B. J. Williams, A. Roggero, W. Kern (2009). Characterization of a new myeloid leukemia cell line with normal cytogenetics (CG-SH). *Leuk Res*. Available online May 2009.

O'Farrell, P. H. (1975). High resolution two-dimensional electrophoresis of proteins. *The Journal of Biological Chemistry* **250**(10): 4007–4021.

Oja, M., J. Nikkila, P. Toronen, G. Wong, E. Castren and S. Kaski (2006). Exploratory clustering of gene expression profiles of mutated yeast strains. In: Zhang, W. and I. Shmulevich, editors. *Computational and statistical approaches to genomics*. New York: Springer, pp. 61–74.

Pal, N., K. Aguan, A. Sharma and S. Amari (2007). Discovering biomarkers from gene expression data for predicting cancer subgroups using neural networks and relational fuzzy clustering. *BMC Bioinformatics* **8**: 5.

Pang, S., J. Smith, D. Onley, J. Reeve, M. Walker and C. Foy (2005). A comparability study of the emerging protein array platforms with established ELISA procedures. *Journal of Immunological Methods* **302**(1–2): 1–12.

Pappin, D. J., P. Hojrup and A. J. Bleasby (1993). Rapid identification of proteins by peptide-mass fingerprinting. *Current Biology* **3**(6): 327–332.

Parkinson, H., M. Kapushesky, N. Kolesnikov, G. Rustici, M. Shojatalab, N. Abeygunawardena (2009). ArrayExpress update—from an archive of functional genomics experiments to the atlas of gene expression. *Nucleic Acids Research* **37** (Database issue): D868–D872.

Parkinson, H., M. Kapushesky, M. Shojatalab, N. Abeygunawardena, R. Coulson, A. Farne (2007). ArrayExpress—a public database of microarray experiments and gene expression profiles. *Nucleic Acids Research* **35**: D747–D750.

Paweletz, C. P., L. Charboneau, V. E. Bichsel, N. L. Simone, T. Chen, J. W. Gillespie (2001). Reverse phase protein microarrays which capture disease progression show activation of pro-survival pathways at the cancer invasion front. *Oncogene* **20**: 1981–1989.

Pearson, H. (2008). Biologists initiate plan to map human proteome. *Nature* **452**(24): 920–921.

Pearson, K. (1901). On line and planes of closest fit to systems of points in space. *The London, Edinburgh and Dublin Philosophical Magazine and Journal of Science* **2**(6): 559–572.

Pellois, J. P., X. Zhou, O. Srivannavit, T. Zhou, E. Gulari and X. Gao (2002). Individually addressable parallel peptide synthesis on microchips. *Nature Biotechnology* **20**: 922–926.

Pennisi, E. (2007). DNA study forces rethink of what it means to be a gene. *Science* **316**: 1556–1557.

Perkins, D. N., D. J. C. Pappin, D. M. Creasy and J. S. Cottrell (1999). Probability-based protein identification by searching sequence databases using mass spectrometry data. *Electrophoresis* **20**(18): 3551–3567.

Perkins, S., K. Lacker and J. Theiler (2003). Grafting: Fast, incremental feature selection by gradient descent in function space. *Journal of Machine Learning Research* **3**: 1333–1356.

Pesole, G. (2008). What is a gene? An updated operational definition. *Gene* **417**(1–2): 1–4.

Phimister, B. (1999). Going global. *Nature Genetics* **21**(Jan 1999 Supplement): 1.

Politis, D. N. and J. P. Romano (1994). Large sample confidence regions based on subsamples under minimal assumptions. *The Annals of Statistics* **22**(4): 2031–2050.

Pollard, H. B., O. Eidelman, M. Srivastava, C. Joswik, S. Rothwell, G. P. Mueller (2008). Antibody and reverse capture protein microarrays for clinical proteomics. In: Van Eyk, J. E. and M. J. Dunn, editors. *Clinical proteomics: From diagnosis to therapy.* Weinheim: Wiley-VCH, pp. 549–569.

Poon, T. C. W. (2007). Opportunities and limitations of SELDI-TOF-MS in biomedical research: practical advices. *Expert Review of Proteomics* **4**(4): 51–65.

Prados, J., A. Kalousis, J.-C. Sanchez, L. Allard, O. Carrette and M. Hilario (2004). Mining mass spectra for diagnosis and biomarker discovery of cerebral accidents. *Proteomics* **4**(8): 2320–2332.

Press, W. H., S. A. Teukolsky, W. T. Vetterling and B. P. Flannery (2007). *Numerical recipes: the art of scientific computing*. New York: Cambridge University Press.

Pruitt, K. D., T. Tatusova and D. R. Maglott (2007). NCBI reference sequences (RefSeq): a curated non-redundant sequence database of genomes, transcripts and proteins. *Nucleic Acids Research* **35**(Database Issue): D61–D65.

Rayner, T. F., P. Rocca-Serra, P. T. Spellman, H. C. Causton, A. Farne, E. Holloway (2006). A simple spreadsheet-based, MIAME-supportive format for microarray data: MAGE-TAB. *BMC Bioinformatics* **7**: 489.

Reddy, G. and E. A. Dalmasso (2003). SELDI ProteinChip array technology: Protein-based predictive medicine and drug discovery applications. *Journal of Biomedicine and Biotechnology* **4**: 237–241.

Reddy, G., R. Rubin and S. Weinberger (2006). Mass Spectrometry: SELDI. In: Ganten, D. and K. Ruckpaul, editors. *Encyclopedic reference of genomics and proteomics in molecular medicine*, Volume 2, New York: Springer, pp. 1034–1036.

Reidegeld, K. A., M. Eisenacher, I. M. Koh, D. Chamrad, G. Körting, M. Blüggel (2008). An easy-to-use Decoy Database Builder software tool, implementing different decoy strategies for false discovery rate calculation in automated MS/MS protein identifications. *Proteomics* **8**(6): 1129–1137.

Rencher, A. C. (2002). *Methods of multivariate analysis*. New York: Wiley & Sons.

Rheinberger, H.-J. and S. Müller-Wille (2008). *Gene*, The Stanford Encyclopedia of Philosophy (Fall 2008 Edition), Edward N. Zalta (ed.). http://plato.stanford.edu

Ripley, B. D. (1996). *Pattern recognition and neural networks*. New York: Cambridge University Press.

Rosenblatt, F. (1958). The perceptron: a probabilistic model for information storage and organization in the brain. *Psychological Review* **65**(6): 386–408.

Ross, M. E., X. Zhou, G. Song, S. A. Shurtleff, K. Girtman, W. K. Williams (2003). Classification of pediatric acute lymphoblastic leukemia by gene expression profiling. *Blood* **102**(8): 2951–2959.

Saeed, A. I., V. Sharov, J. White, J. Li, W. Liang, N. Bhagabati (2003). TM4: a free, open-source system for microarray data management and analysis. *BioTechniques* **34**(2): 374–378.

Saeys, Y., I. Inza and P. Larranaga (2007). A review of feature selection techniques in bioinformatics. *Bioinformatics* **23**(19): 2507–2517.

Schena, M., D. Shalon, R. W. Davis and P. O. Brown (1995). Quantitative monitoring of gene expression patterns with a complementary DNA microarray. *Science* **270**(5235): 467–470.

Schèolkopf, B. and A. J. Smola (2002). *Learning with kernels: support vector machines, regularization, optimization, and beyond*. Cambridge, MA, MIT Press.

Schweitzer, B., P. Predki and M. Snyder (2003). Microarrays to characterize protein interactions on a whole-proteome scale. *Proteomics* **3**(11): 2190–2199.

Seidel, C. (2008). Introduction to DNA microarrays. In: Emmert-Streib, F. and M. Dehmer, editors. *Analysis of microarray data: a network-based approach*. Weinheim: Wiley-VCH, pp. 1–26.

Seo, J. and E. P. Hoffman (2006). Probe set algorithms: is there a rational best bet? *BMC Bioinformatics* **7**(395).

Service, R. F. (2008). Proteomics ponders prime time. *Science* **321**: 1758–1761.

Shawe-Taylor, J. and N. Cristianini (2004). *Kernel methods for pattern analysis*. New York: Cambridge University Press.

Sidak, Z. (1967). Rectangular confidence regions for the means of multivariate normal distributions. *Journal of the American Statistical Association* **62**(318): 626–633.

Singleton, P. (2008). *Dictionary of DNA and genome technology*. Malden, MA: Blackwell.

Smith, S. P. and A. K. Jain (1988). Test to determine the multivariate normality of a data set. *IEEE Transactions on Pattern Analysis and Machine Intelligence* **10**(5): 757–761.

Smyth, G. K. (2004). Linear models and empirical bayes methods for assessing differential expression in microarray experiments. *Statistical Applications in Genetics and Molecular Biology* **3**(1).

Spellman, P. T., M. Miller, J. Stewart, C. Troup, U. Sarkans, S. Chervitz (2002). Design and implementation of microarray gene expression markup language (MAGE-ML). *Genome Biology* **3**(9): research0046.1–0046.9.

Spurrier, B., P. Honkanen, A. Holway, K. Kumamoto, M. Terashima, S. Takenoshita (2008). Protein and lysate array technologies in cancer research. *Biotechnology Advances* **26**: 361–369.

Srivastava, M. S. (2002). *Methods of multivariate statistics*. New York: Wiley.

Stam, R., M. den Boer, M. Passier, G. E. Janka-Schaub, S. E. Sallan, S. A. Armstrong (2006). Silencing of the tumor suppressor gene FHIT is highly characteristic for MLL gene rearranged infant acute lymphoblastic leukemia. *Leukemia* **20**(2): 264–271.

Stamm, S. (2006). Alternative Splicing. In: Ganten, D. and K. Ruckpaul, editors. *Encyclopedic reference of genomics and proteomics in molecular medicine*. New York: Springer, pp. 45–48.

Stein, L. D. (2004). End of the beginning. *Nature* **431**(7011): 915–916.

Szekely, G. J. and M. L. Rizzo (2005). A new test for multivariate normality. *Journal of Multivariate Analysis* **93**: 58–80.

Tabachnick, B. G. and L. S. Fidell (2007). *Using multivariate statistics*. Boston: Pearson.

Tabb, D. L., A. Saraf and J. R. Yates (2003). GutenTag: High-throughput sequence tagging via an empirically derived fragmentation model. *Analytical Chemistry* **75**: 6415–6421.

Tamayo, P., D. Slonim, J. Mesirov, Q. Zhu, S. Kitareewan, E. Dmitrovsky (1999). Interpreting patterns of gene expression with self-organizing maps: Methods and application to hematopoietic differentiation. *Proceedings of the National Academy of Science USA* **96**: 2907–2912.

Tanaka, G., H. Funabashi, M. Mie and E. Kobatake (2006). Fabrication of an antibody microwell array with self-adhering antibody binding protein. *Analytical Biochemistry* **350**(2): 298–303.

Tanner, S., H. Shu, A. Frank, L.-C. Wang, E. Zandi, M. Mumby (2005). InsPecT: Identification of posttranslationally modified peptides from tandem mass spectra. *Analytical Chemistry* **77**: 4626–4639.

Taylor, J. A. and R. S. Johnson (2001). Implementation and uses of automated de novo peptide sequencing by tandem mass spectrometry. *Analytical Chemistry* **73**(11): 2594–2604.

The ENCODE Project Consortium (2007). Identification and analysis of functional elements in 1% of the human genome by the ENCODE pilot project. *Nature* **447**(7146): 799–816.

The UniProt Consortium (2008). The Universal Protein Resource (UniProt). *Nucleid Acid Research* **36**: D190–D195.

Theodoridis, S. and K. Koutroumbas (2006). *Pattern recognition*. Boston: Academic Press.

Therneau, T. M. and K. V. Ballman (2008). What does PLIER really do? *Cancer Informatics* **2008**(6): 423–431.

Tibshirani, R., T. Hastie, M. Eisen, D. Ross, D. Botstein and P. Brown (1999). *Clustering methods for the analysis of DNA microarray data*, Technical report, Department of Health Research and Policy, and Statistics, Stanford University, Stanford, CA. http://www-stat.stanford.edu/~tibs/

Tibshirani, R., T. Hastie, B. Narasimhan and G. Chu (2002). Diagnosis of multiple cancer types by shrunken centroids of gene expression. *Proceedings of the National Academy of Science* **99**(10): 6567–6572.

Turney, P. (1995). Technical Note: Bias and the quantification of stability. *Machine Learning* **20**(1–2): 23–33.

Tusher, V., R. Tibshirani and G. Chu (2001). Significance analysis of microarrays applied to transcriptional responses to ionizing radiation. *Proceedings of the National Academy of Science USA* **98**: 5116–5121.

Tuv, E. (2006). Ensemble Learning. In: Guyon, I., S. Gunn, M. Nikravesh and L. A. Zadeh, editors. *Feature extraction: foundations and applications*. New York: Springer-Verlag, pp. 187–204.

Tuv, E., A. Borisov and K. Torkkola (2007). Ensemble-based variable selection using independent probes. In: Liu, H. and H. Motoda, editors. *Computational methods of feature selection*. Boca Raton, FL: Taylor & Francis, pp. 131–145.

Uhlen, M. (2007). Mapping the human proteome using antibodies. *Molecular and Cellular Proteomics* **6**(8): 1455–1456.

Unlü, M., M. E. Morgan and J. S. Minden (1997). Difference gel electrophoresis: a single gel method for detecting changes in protein extracts. *Electrophoresis* **18**(11): 2071–2077.

VanMeter, A., M. Signore, M. Pierobon, V. Espina, L. A. Liotta and E. F. Petricoin (2007). Reverse-phase protein microarrays: application to biomarker discovery and translational medicine. *Expert Review of Molecular Diagnostics* **7**(5): 625–633.

Vapnik, V. N. (1998). *Statistical learning theory*. New York: Wiley.

Vapnik, V. N. (2000). *The nature of statistical learning theory*. New York: Springer.

Venter, J. C., M. D. Adams, E. W. Myers, P. W. Li, R. J. Mural, G. G. Sutton (2001). The sequence of the human genome. *Science* **291**(5507): 1304–1351.

Vlahou, A. and M. Fountoulakis (2005). Proteomic approaches in the search for disease biomarkers. *Journal of Chromatography B* **814**: 11–19.

Volkmer-Engert, R. (2006). Peptide chips. In: Ganten, D. and K. Ruckpaul, editors. *Encyclopedic reference of genomics and proteomics in molecular medicine*, Volume 2, New York: Springer, pp. 1372–1377.

Wain, H. M., E. A. Bruford, R. C. Lovering, M. J. Lush, M. W. Wright and S. Povey (2002). Guidelines for human gene nomenclature. *Genomics* **79**(4): 464–470.

Wang, Y., I. V. Tetko, M. A. Hall, E. Frank, A. Facius, K. F. X. Mayer and H. W. Mewes (2005). Gene selection from microarray data for cancer classification—a machine learning approach. *Computational Biology and Chemistry* **29**: 37–46.

Warner, R. M. (2008). *Applied statistics: from bivariate through multivariate techniques*. Los Angeles: Sage Publications.

Watson, J. D. and F. H. C. Crick (1953a). Molecular structure of nucleic acids; a structure for deoxyribose nucleic acid. *Nature* **171**(4356): 737–738.

Watson, J. D. and F. H. C. Crick (1953b). Genetical implications of the structure of deoxyribonucleic acid. *Nature* **171**(4361): 964–967.

Westfall, P. H. and S. S. Young (1993). *Resampling-based multiple testing: examples and methods for P-value adjustment*. New York: Wiley.

Weston, J., A. Elisseeff and B. Schèolkopf (2003). Use of zero-norm with linear models and kernel methods. *Journal of Machine Learning Research* **3**: 1439–1461.

Wheeler, D. A., M. Srinivasan, M. Egholm, Y. Shen, L. Chen, A. Mcguire (2008). The complete genome of an individual by massively parallel DNA sequencing. *Nature* **452**(7189): 872–876.

Wilkins, M. H. F., A. R. Stokes and H. R. Wilson (1953). Molecular structure of nucleic acids: Molecular structure of deoxypentose nucleic acids. *Nature* **171**(4356): 738–740.

Wilming, L. G., J. G. Gilbert, K. Howe, S. Trevanion, T. Hubbard and J. L. Harrow (2008). The vertebrate genome annotation (Vega) database. *Nucleic Acids Research* **36**(Database Issue): D753–D760.

Witten, I. H. and E. Frank (2005). *Data mining: practical machine learning tools and techniques*. Boston: Morgan Kaufman.

Wright, G. L. J. (2002). SELDI proteinchip MS: a platform for biomarker discovery and cancer diagnosis. *Expert Review of Molecular Diagnostics* **2**(6): 549–563.

Wu, Z. and R. A. Irizarry (2005). Stochastic models inspired by hybridization theory for short oligonucleotide arrays. *Journal of Computational Biology* **12**(6): 882–893.

Wu, Z., R. A. Irizarry, R. Gentleman, F. Martinez-Murillo and F. Spencer (2004). A model-based background adjustment for oligonucleotide expression arrays. *Journal of the American Statistical Association* **99**(468): 909–917.

Xing, E. P. (2003). Feature selection in microarray analysis. In: Berrar, D. P., W. Dubitzky and M. Granzow, editors. *A practical approach to microarray data analysis*. Boston: Kluwer Academic Publishers, pp. 110–131.

Xing, E. P., M. I. Jordan and R. M. Karp (2001). *Feature selection for high-dimensional genomic microarray data*. Proceedings of the Eighteenth International Conference on Machine Learning (ICML2001).

Xu, C. and B. Ma (2006). Software for computational peptide identification from MS–MS data. *Drug Discovery Today* **11**(13/14): 595–600.

Yang, Y., S. Takeuchi, W. K. Hofmann, T. Ikezoe, J. J. van Dongen, T. Szczepanski (2006). Aberrant methylation in promoter-associated CpG islands of multiple genes in acute lymphoblastic leukemia. *Leukemia Research* **30**(1): 98–102.

Yasuda, T., M. Kanamori, S. Nogami, T. Hori, T. Oya, K. Suzuki (2009). Establishment of a new human osteosarcoma cell line, UTOS-1: cytogenetic characterization by array comparative genomic hybridization. *Journal of Experimental and Clinical Cancer Research* **28**: 26.

Yeoh, E. J., M. E. Ross, S. A. Shurtleff, W. K. Williams, D. Patel, R. Mahfouz (2002). Classification, subtype discovery, and prediction of outcome in pediatric acute lymphoblastic leukemia by gene expression profiling. *Cancer Cell* **1**(2): 133–143.

Yeung, K. Y. and R. E. Bumgarner (2003). Multiclass classification of microarray data with repeated measurements: application to cancer. *Genome Biology* **4**(12): R83.

Yu, K., B. Kwon, J. Ni, Y. Zhai, R. Ebner and B. S. Kwon (1999). A newly identified member of tumor necrosis factor receptor superfamily (TR6) suppresses LIGHT-mediated apoptosis. *Journal of Biological Chemistry* **274**(20): 13733–13736.

Yu, L. and H. Liu (2004). Efficient feature selection via analysis of relevance and redundancy. *Journal of Machine Learning Research* **5**: 1205–1224.

Zhang, A. (2006). *Advanced analysis of gene expression microarray data*. Singapore: World Scientific.

Zhang, W. and B. T. Chait (2000). ProFound: An expert system for protein identification using mass spectrometric peptide mapping information. *Analytical Chemistry* **72**(11): 2482–2489.

Zhang, X., X. Lu, Q. Shi, X.-Q. Xu, H.-C. E. Leung, L. N. Harris (2006). Recursive SVM feature selection and sample classification for mass-spectrometry and microarray data. *BMC Bioinformatics* **7**(197).

Zheng, D., A. Frankish, R. Baertsch, P. Kapranov, A. Reymond, S. W. Choo (2007). Pseudogenes in the ENCODE regions: Consensus annotation, analysis of transcription, and evolution. *Genome Research* **17**(6): 839–851.

Zhou, X. and D. P. Tuck (2007). MSVM-RFE: extensions of SVM-RFE for multiclass gene selection on DNA microarray data. *Bioinformatics* **23**(9): 1106–1114.

Zhu, H., M. Bilgin, R. Bangham, D. Hall, A. Casamayor, P. Bertone (2001). Global analysis of protein activities using proteome chips. *Science* **293**(5537): 2101–2105.

Zou, H. and T. Hastie (2007). Model building and feature selection with genomic data. In: Liu, H. and H. Motoda, editors. *Computational methods of feature selection*. Boca Raton, FL: Taylor & Francis, pp. 393–411.

INDEX

Data Mining for Genomics and Proteomics. By Darius M. Dziuda